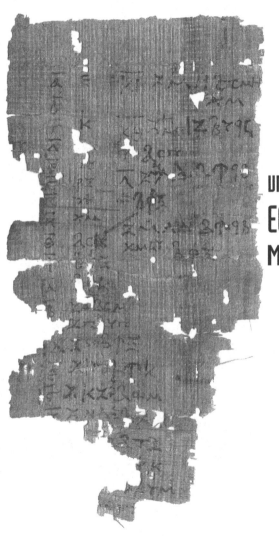

unexpected LINKS between
EGYPTIAN and BABYLONIAN
Mathematics

unexpected LINKS between EGYPTIAN and BABYLONIAN Mathematics

Jöran Friberg
Chalmers University of Technology, Gothenburg, Sweden

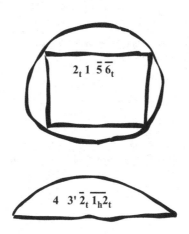

NEW JERSEY • LONDON • SINGAPORE • BEIJING • SHANGHAI • HONG KONG • TAIPEI • CHENNAI

Published by

World Scientific Publishing Co. Pte. Ltd.
5 Toh Tuck Link, Singapore 596224
USA office: 27 Warren Street, Suite 401-402, Hackensack, NJ 07601
UK office: 57 Shelton Street, Covent Garden, London WC2H 9HE

British Library Cataloguing-in-Publication Data
A catalogue record for this book is available from the British Library.

Photo on title page by Mohamed Ibrahim, © IFAO.
Photo on back cover by Wong Mei Fong.

UNEXPECTED LINKS BETWEEN EGYPTIAN AND BABYLONIAN MATHEMATICS

Copyright © 2005 by World Scientific Publishing Co. Pte. Ltd.

All rights reserved. This book, or parts thereof, may not be reproduced in any form or by any means, electronic or mechanical, including photocopying, recording or any information storage and retrieval system now known or to be invented, without written permission from the Publisher.

For photocopying of material in this volume, please pay a copying fee through the Copyright Clearance Center, Inc., 222 Rosewood Drive, Danvers, MA 01923, USA. In this case permission to photocopy is not required from the publisher.

ISBN-13 978-981-256-328-6
ISBN-10 981-256-328-8

Printed in Singapore

Preface

Ancient Mesopotamian mathematics is known from hundreds of texts recorded on clay tablets in the *cuneiform* script. Some of the mathematical cuneiform texts are quite large and contain many exercises or long tables of numbers or measures. The great majority of these texts are *Old Babylonian*, from the first half of the second millennium BCE.[1] A few are *Kassite*, from the latter half of the second millennium BCE, some are *Late Babylonian/Seleucid*, from the latter half of the first millennium BCE, and others are *pre-Babylonian*, from various periods within the third millennium, or the last part of the fourth millennium. New clay tablets with mathematical cuneiform texts keep appearing from time to time, excavated in the field, extracted from the archives of large museums in Europe, America, and the Near East, or offered for sale in the antiquities market. Therefore, the writing of the history of Mesopotamian mathematics is a dynamic, never-ending process.[2]

Egyptian mathematics, on the other hand, is known from a comparatively much smaller number of original documents, belonging to three distinct groups. The first group consists of texts from the earlier part of the second millennium BCE, written in the *hieratic* script. It contains two mathematical papyrus rolls, *P.Rhind = P.BM 10057/8* (Peet, *RMP* (1923), Chace, Bull, and Manning, *RMP* (1927-29), Robins & Shute, *RMP*

---

1. Among the oldest known OB mathematical texts are those from the southern cities Ur, Uruk, and Larsa, before their destruction by Samsuiluna in 1739 BCE, and texts from Nippur before its destruction in 1721 BCE. Other early texts are those from Eshnunna, before 1763, and those from Mari, destroyed by Hammurabi in 1757. The mathematical cuneiform texts from northern sites, like Sippar, are later, and so are the mathematical texts from Susa. The OB period ended in 1595 BCE. (All these dates are given in the Middle Chronology.)

(1987)), and *P.Moscow E 4676* (Struve, *QSA 1* (1930)), the *Mathematical Leather Roll P.BM 10250* (Glanville, *MLR* (1927)), the papyrus fragments *P.Berlin 6619* (Schack-Schackenburg, *ZÄS* 38 (1900)), and the *Lahun mathematical fragments*, formerly known as the *Kahun fragments* (Griffith, *HPKG* (1898), Imhausen and Ritter, *UCLLP* (2004)). There are also two wooden tablets *WT.Cairo 23567/8* and two ostraca.[3] Texts belonging to this first group, in the following referred to as "hieratic mathematical papyri", will be discussed in Chapter 2.

The second group of known Egyptian mathematical texts consists of documents from the Hellenistic and Roman periods, mostly from the last part of the first millennium BCE, written in the *demotic* script. The group consists of one large papyrus, *P.Cairo*, and six smaller texts or fragments, all published by Parker in *JNES* 18 (1959), *Cent.* 14 (1969), *DMP* (1972), and *JEA* 61 (1975), plus several ostraca. A number of exercises from Parker's "demotic mathematical papyri", will be discussed in Chapter 3.

The third group of known Egyptian mathematical texts consists of documents from the Hellenistic and Roman periods, that is from the last part of the first millennium BCE and the first half of the first millennium CE, written in *Greek*. A small subgroup including 6 ostraca, a papyrus roll, and three papyrus fragments, all related in one way or another to Euclid's *Elements*, will not be considered here. However, the third group also includes

---

2. See the fascinating story of the development of the history of Mesopotamian mathematics, as described by Høyrup in *HSci* 34 (1996). See also the annotated bibliography Friberg, *HMAP* (1985), with an updated edition on CD-ROM (2000). Recently published works on Mesopotamian mathematics not mentioned in those bibliographies are Chambon, *FlM* 6 (2002), Damerow, *ChV* (2001), Englund, *ChV* (2001), Foster and Robson, *ZA* 94 (2004), Fowler and Robson, *HM* 25 (1998), Friberg, *BaM* 30 (1999), *AfO* 46/47 (1999/2000), *BaM* 31 (2000), *ChV* (2001), *MCTSC* (2005), *CDLJ* (2005/2), Høyrup, *HM* 29 (2002), *UOS* (2002), Jursa and Radner, *AfO* 42/43 (1995/96), Melville, *UOS* (2002), Muroi, *SCIAMVS* 1 (2000), *HSci* 10 (2001), *SCIAMVS* 2 (2001), *HSci* 12 (2002), *SCIAMVS* 4 (2003), *HSci* 13 (2003), Nemet-Nejat, *UOS* (2002), Oelsner, *ChV* (2001), Proust, *RHM* 6 (2000), *FlM* 6 (2002), *TMN* (2004), Quillien, *RHM* 9 (2003), Robson, *SCIAMVS* 1 (2000), *UOS* (2002), *HMT* (2003), *SCIAMVS* 5 (2004).

3. Cf. the timeline in Imhausen, *ÄA* (2003), Table 1: The majority of the Egyptian hieratic mathematical texts are from the time of the Middle Kingdom, Dyn. 11-12 (2119-1794/93 BCE). Only *P.Rhind* is from the Second Intermediary Period, Dyn. 13-17 (1794/93-1550 BCE), although the preface of *P.Rhind* states that the papyrus is a copy of a text from the time of a king of the twelfth dynasty.

texts that show almost no signs of having been influenced by high level Greek mathematics. The most interesting examples of such "non-Euclidean" Greek mathematical texts include a codex of six papyrus leaves, *P.Akhmîm* (Baillet, *BMA* (1892)), a large papyrus roll, *P.Vindob. G. 19996* (Gerstinger and Vogel, *GLP 1* (1932)), six smaller papyri or papyrus fragments, an ostracon, and a wooden tablet. These "Greek-Egyptian mathematical documents" will be discussed in Chapter 4.

All the mentioned *hieratic* mathematical texts had already been published by 1930, the *demotic* mathematical texts by 1975, and the *Greek-Egyptian* mathematical texts by 1981. Since then not much has happened in the study of Egyptian mathematics. The few books and papers that have been written about "Egyptian mathematics" have been concerned exclusively with the hieratic mathematical texts[4] and have mostly reiterated the interpretations and presentations of those texts that were offered already in the original publications.[5] Very little[6] seems to have been written about the demotic mathematical texts since they were published by Parker, and not much[7] about the Greek-Egyptian mathematical texts.

My original impetus to search for links between Egyptian and Babylonian mathematics came from an observation that two small but particularly interesting mathematical texts from the Old Babylonian city Mari have clear Egyptian parallels, one in an exercise in the well known hieratic *Papyrus Rhind*, the other in a relatively unknown Greek-Egyptian papyrus fragment. The details will be presented below in Chapter 1.

My observation that there seems to exist clear links between Egyptian and Babylonian mathematics is in conflict with the prevailing opinion in formerly published works on Egyptian mathematics, namely that practically no such links exist. However, in view of the mentioned dynamic

---

4. Recently published works on the subject of hieratic mathematical texts are Cavéing, *Essai* (1994), Clagett, *AES 3* (1999), Couchoud, *ME* (1993), Imhausen *UOS* (2002), *HM* 30 (2003), *ÄA* (2003), Ritter, *EHS* (1989), *AHST* (1995), *HNWM* (2000), *UOS* (2002), Robins and Shute, *RMP* (1987).

5. Cf. Høyrup's poignant statement in his review of Couchoud, *ME* (1993) in *MR* (1997), that the book "presents the state of the art as it has looked without fundamental change since the early 1930s".

6. Known to me are only Zauzich, *BiOr* 32 (1972), Kaplony-Heckel, *OLZ* 76 (1981), Knorr, *HM* 9 (1982), Fowler, *MPA* (1987 (1999)), Sec. 7.3(e), and Melville, *HM* 31 (2004).

7. See, in particular, Fowler, *MPA* (1987 (1999)), Secs. 7.1(d), 7.2, 7.3(c)-(e).

character of the history of Mesopotamian mathematics, not least in the last couple of decades, it appeared to me to be *high time to take a renewed look at Egyptian mathematics against an up-to-date background in the history of Mesopotamian mathematics!* That is the primary objective of this book.

My search for links between Egyptian and Babylonian mathematics has been unexpectedly successful, in more ways than one. Not only has the search turned up numerous possible candidates for such links, but the comparison of Egyptian and Babylonian mathematics has in many cases led to a much better understanding of the nature of important Egyptian mathematical texts and of particularly interesting exercises that they contain. In addition, my careful examination of a great number of individual Egyptian hieratic, demotic, and Greek mathematical exercises has made this book into a useful survey of a substantial part of the whole corpus of Egyptian mathematics.

Several of the techniques and concepts that I have developed in the course of my intensive study of mathematical cuneiform texts during the last 25 years have proven themselves to be eminently suitable also for a study of Egyptian mathematical texts. An obvious example of a helpful technique is the use of "conform" transliterations for detailed outlines of mathematical texts.[8] A particularly useful concept is that of a "mathematical recombination text", which is an appropriate name for a large mathematical text with a somewhat chaotic collection of individual exercises.

The detailed comparison in this book of a large number of known Egyptian and Mesopotamian mathematical texts from all periods has led me to the conclusion that the level and extent of mathematical knowledge must have been comparable in Egypt and in Mesopotamia in the earlier part of the second millennium BCE, and that there are also unexpectedly close connections between demotic and "non-Euclidean" Greek-Egyptian mathematical texts from the Ptolemaic and Roman periods on one hand and Old or Late Babylonian mathematical texts on the other.

---

8. Compare the conform transliterations of Babylonian mathematical cuneiform texts in Figs. 1.1.2-3, 1.1.5-6, 1.2.1-2, 2.1.9-11, 2.1.17, 2.2.1, 2.2.3, 2.3.2, 3.1.8, 3.3.1, and 3.3.6 below with the similar conform transliterations (in mirror images because the Egyptian direction of writing was from right to left) of hieratic mathematical texts in Figs. 1.1.7, 2.1.4, 2.1.7, 2.2.2, 2.2.5-.6, 2.3.3-5, and 3.1.3, and of demotic mathematical texts in Figs. 3.1.5, 3.1.9, 3.1.12, 3.2.1, 3.3.2-4, 3.5.1, 3.7.1, and 3.7.6.

Contents

Preface   v

Contents   ix

1. Two Curious Mathematical Cuneiform Texts from Old Babylonian Mari   1
   1.1. M. 7857. A Fanciful Interpretation of a Geometric Progression. 2
      1.1 a. M. 7857. A text with three kinds of counting numbers. 2
      1.1 b. OB texts with ascending or descending geometric progressions. 5
      1.1 c. A Late Babylonian text using the trailing part algorithm. 8
      1.1 d. The sum of a geometric progression in an Old Babylonian text 8
      1.1 e. The sum of a geometric progression in a Seleucid text 10
      1.1 f. *P.Rhind* # 79: a parallel to M. 7857 in a hieratic papyrus. 11
      1.1 g. Summary. The Mesopotamian roots of a *Mother Goose* riddle 12
   1.2. M. 8631. Another Curious Mathematical Text from OB Mari 14
      1.2 a. M. 8631. A fanciful interpretation of 30 doublings 14
      1.2 b. The Old Babylonian doubling and halving algorithm 18
      1.2 c. *P.IFAO 88*: A parallel to M. 8631 in a Greek-Egyptian papyrus 19
      1.2 d. Summary. The Mesopotamian roots of a well known legend 22

2. Hieratic Mathematical Papyri and Cuneiform Mathematical Texts   25
   2.1. Themes in *P.Rhind*, a Large Hieratic Mathematical Papyrus Roll 26
      2.1 a. *P.Rhind*, a hieratic mathematical recombination text 26
      P.Rhind: Contents 27
      2.1 b. Theme E: division problems (*P.Rhind* ## 24-38). 29
      2.1 c. Theme F: sharing problems (*P.Rhind* ## 39-40, 63-65, 68) 36
      2.1 d. Theme G: geometry problems (*P.Rhind* ## 41-46, 48-60) 40
      2.1 e. Theme H: baking or brewing numbers (*P.Rhind* ## 69-78) 59
      2.1 f. Theme I: a combined price problem (*P.Rhind* # 62). 67
   2.2. Themes in *P.Moscow*, a Smaller Hieratic Mathematical Papyrus Roll. 69
      P.Moscow: Contents 69
      2.2 a. *P.Moscow*: Another Egyptian mathematical recombination text 69
      2.2 b. *P.Moscow* # 23: A combined work norm 69
      2.2 c. *P.Moscow* ## 6-7, 17: Metric algebra 71
      2.2 d. *P.Moscow* # 14: The volume of a truncated pyramid 74
      2.2 e. *P.Moscow* # 10: The area of a semicircle(?) 78

2.3. Hieratic Mathematical Papyrus Fragments 81
  2.3 a. *P.Berlin 6619* # 1: Metric algebra................................. 81
  2.3 b. *P.UC 32160* (= *Kahun IV. 3*) # 1: A cylindrical granary 85
  2.3 c. *P.UC 32160* # 2: An arithmetic progression with 10 terms 89
  2.3 d. *P.UC 32161* (= *Kahun XLV. 1*): A list of large numbers 90
2.4. Conclusion
  New thoughts about the nature of *P.Rhind* 92
  Summary: Comparison of hieratic Egyptian and OB mathematics........ 102

3. Demotic Mathematical Papyri and Cuneiform Mathematical Texts   105
  3.1. Themes in *P.Cairo* (Ptolemaic, 3rd C. BCE.) 105
    3.1 a. *P.Cairo*, a hieratic mathematical recombination text 105
    3.1 b. *P.Cairo* § 8 (*DMP* ## 24-31). A pole against a wall 107
    3.1 c. *P.Cairo* § 1 (*DMP* ## 2-3). Two closely related division problems..... 109
    3.1 d. *P.Cairo* § 2 (*DMP* ## 4-6). Completion problems.................. 112
    3.1 e. *P.Cairo* § 3 (*DMP* ## 7, 15-16, 18). A rectangular sail 115
    3.1 f. *P.Cairo* § 4 (*DMP* ## 8-12, 14, 17). Reshaping a rectangular cloth..... 115
    3.1 g. *P.Cairo* § 7 (*DMP* # 23). Shares in a geometric progression.......... 120
    3.1 h. *P.Cairo* § 9 (*DMP* ## 32-33). Diameter of a circle with given area 122
    3.1 i. *P.Cairo* § 10 (*DMP* ## 34-35). Metric algebra..................... 124
    3.1 j. *P.Cairo* § 11 (*DMP* ## 36, 38). An equilateral triangle in a circle 126
    3.1 k. *P.Cairo* § 12 (*DMP* # 37). A square inscribed in a circle............. 130
    3.1 l. *P.Cairo* §§ 13-14 (*DMP* ## 39-40). Pyramids with a square base 136
    3.1 m. *P.Cairo* §§ 15-16 (## "32-33"). Metric algebra 136
  3.2. *P.British Museum 10399* (Ptolemaic, Later than *P.Cairo*) 137
    3.2 a. *P.BM 10399* § 1 (*DMP* # 41). A circle-and -chord problem 137
    3.2 b. *P.BM 10399* § 2 (*DMP* ## 42-45). Masts (truncated cones) 141
    3.2 c. *P.BM 10399* § 3 (*DMP* ## 46-51). The reciprocal of $1 + 1/n$, *etc.* 143
  3.3. *P.British Museum 10520* (Early(?) Roman) 145
    3.3 a. *P.BM 10520* § 1 (*DMP* # 53). The iterated sum of 1 through 10 145
    3.3 b. *P.BM 10520* § 2 (*DMP* # 54). A multiplication table for 64 148
    3.3 c. *P.BM 10520* § 3 (*DMP* # 55). A new multiplication rule............. 149
    3.3 d. *P.BM 10520* § 4 (*DMP* # 56). 2/35 expressed as a sum of parts 149
    3.3 e. *P.BM 10520* § 5 (*DMP* ## 57-61). Operations with fractions 150
    3.3 f. *P.BM 10520* § 6 (*DMP* ## 62-63). The square side rule.............. 154
    3.3 g. *P.BM 10520* § 7 (*DMP* ## 64-65). The quadrilateral area rule 156
  3.4. *P.British Museum 10794* (Date Uncertain) 165
    3.4 a. Multiplication tables for 90 and 150 (*DMP* ## 66-67)................ 165
  3.5. *P.Carlsberg 30* (Probably 2nd C. BCE) 166
    3.5 a. *P.Carlsberg 30* # 1 (*DMP* # 69). The diagonal of a square, *etc.*........ 167
    3.5 b. *P.Carlsberg 30* # 2 (*DMP* ## 71-72). A system of linear equations 168
  3.6. *P.Griffith Inst. I. E. 7* (Late Ptolemaic or Early Roman).................. 173
    3.6 a. A theme text with linear equations 173
  3.7. *P.Heidelberg 663* (Ptolemaic, 2nd or 1st C. BCE) 174
    3.7 a. *P.Heidelberg 663* # 1. A vertically striped trapezoid 174

3.7 b. *P.Heidelberg 663* # 2. A horizontally striped trapezoid............ 177
   The lengths of the transversals..................................... 177
   The lengths of the partial heights.................................. 179
   The lengths of the sloping sides.................................... 180
3.7 c. *P.Heidelberg 663*. Parallel texts............................... 180
3.7 d. Conclusion 189
   The importance of *P.Cairo* for the history of mathematics.......... 189
   The other demotic mathematical papyri 191

4. Greek-Egyptian Mathematical Documents and Cuneiform Mathematical Texts   193
   4.1. *O.Bodl. ii 1847* (30 BCE). An Ostracon with Schematic Field Plans....... 194
   4.2. *P.Oxyrhynchus iii 470* (3rd C. CE). A Water Clock 195
   4.3. *P.Vindobonensis G. 26740*. Five Illustrated Geometric Exercises 196
      4.3 a. *P.Vindob. G. 26740* # 1. A segment of a circular band 196
      4.3 b. *P.Vindob. G. 26740* ## 2-4. A circle area found in three ways 198
      4.3 c. *P.Vindob. G. 26740* # 2....................................... 198
      4.3 d. *P.Vindob. G. 26740* # 3....................................... 198
      4.3 e. *P.Vindob. G. 26740* # 4....................................... 198
      4.3 f. *P.Vindob. G. 26740* # 5. The area of a semicircle................. 199
   4.4. *P.Mich. 620* (2nd C. CE). Systems of Linear Equations. Tabular Arrays.... 200
      4.4 a. *P.Mich. 620* # 1. A system of linear equations: four unknowns........ 200
      4.4 b. *P.Mich. 620* # 2. A system of linear equations: two unknowns 202
      4.4 c. *P.Mich. 620* # 3. A system of linear equations: three unknowns 202
   4.5. *P.Akhmîm* (7th C. CE). Calculations with Fractions................... 208
      4.5 a. *P.Akhmîm*: Contents 208
      4.5 b. *P.Akhmîm*. Ten tables of fractions 210
      4.5 c. *P.Akhmîm* § 1. The capacity measure(?) of a truncated cone......... 211
      4.5 d. *P.Akhmîm* § 2. The capacity measure of a rectangular granary 211
      4.5 e. *P.Akhmîm* §§ 3, 5. Unequal sharing, and division exercises 212
      4.5 f. *P.Akhmîm* §§ 4-10. Examples of counting with fractions............. 212
      4.5 g. *P.Akhmîm* § 9. Prices and market rates 214
   4.6. *WT.Michael. 62* (7th? C. CE). Prices and Market Rates 215
   4.7. Problems for Right Triangles and Quadrilaterals 220
      4.7 a. *P.Genève 259*. Problems for the sides of a right triangle 220
      4.7 b. *P.Chicago litt. 3 (1st C. CE?)*. Three types of non-symmetric trapezoids 221
      4.7 c. *P.Cornell 69*. Non-symmetric trapezoids, and a birectangle 226
   4.8. *P.Vindobonensis G. 19996* (1st C. CE?). Stereometric Exercises.......... 233
      4.8 a. *P.Vindob. G. 19996*: Contents 234
      4.8 b. *P.Vindob. G. 19996* # 10. A pyramid with a triangular base 235
      4.8 c. *P.Vindob. G. 19996* # 13. A truncated triangular pyramid............ 241
      4.8 d. *P.Vindob. G. 19996* # 18. A square pyramid 241
      4.8 e. *P.Vindob. G. 19996* # 19. A truncated square pyramid 241
      4.8 f. *P.Vindob. G. 19996* # 24. A truncated circular cone................ 242
      4.8 g. Pyramids and cones in Old Babylonian mathematical texts.......... 243
         The volume of a truncated square pyramid........................... 243

　　　　　The volume and grain measure of a ridge pyramid................... 245
　　　　　The grain measure of a ridge pyramid truncated at mid-height.......... 251
　　　　　Systems of linear equations for the length u and the ridge r............ 254
　　　　　Rectangular-linear systems of equations........................... 256
　　　　　Problems for a circular cone and its 'feed' 257
　　　　　Problems for truncated circular cones............................ 262
　　4.9. Conclusion... 268
　　　　　Relations between Greek-Egyptian and Babylonian mathematics 268

New Thoughts About the Early History of Mathematics　　　　　　　　　　269

Index of Texts　　　　　　　　　　　　　　　　　　　　　　　　　　　271

Index of Subjects　　　　　　　　　　　　　　　　　　　　　　　　　277

Bibliography　　　　　　　　　　　　　　　　　　　　　　　　　　　283

Chapter 1

Two Curious Mathematical Cuneiform Texts from Old Babylonian Mari

Mari was the center of a small kingdom on the middle Euphrates, independent until it was conquered by Hammurabi in 1757 BCE. Important for the discussion below is that Mari's location in the north-western corner of Mesopotamia may have allowed it to promote an exchange of ideas between Mesopotamia and its neighbors to the west, maybe even between Mesopotamia and cities along the coast of the Mediterranean, and ultimately Egypt. Old Babylonian cuneiform texts from a royal archive at Mari are in the process of being published by a team of French scholars. Among already published cuneiform texts from Mari are several texts of mathematical interest, in particular

    a) some mathematical table texts published by D. Soubeyran in *RA* 78 (1984), among them a text with 30 terms of a geometric progression (Sec. 1.2 a below),

    b) a round hand tablet published by D. Charpin in *MARI* 7 (1993), inscribed with an outline of a city wall and with numbers indicating the volumes of the four sides of the city wall and the sizes of the four teams of workers needed to erect them,

    c) a rectangular hand tablet published by M. Guichard in *MARI* 8 (1997), with 30 terms of a geometric progression expressed in three kinds of numbers (Sec. 1.1 b),

    d) three metrological texts published by G. Chambon in *FlMar* 6 (2002), among them a clay cylinder of a rare type with a metrological list of weight measures.

The map in Fig. 1.1.1 below shows the location of Mari, as well as of Ebla, another ancient city in Syria, and of Ugarit on the coast of the Mediterranean. Cuneiform texts with metrological tables of Old Babylonian style have been found at Ugarit (Nougayrol, *Ugaritica* 5 (1968)), and interesting mathematical cuneiform texts from the late third millennium BCE have been found at Ebla (Friberg, *VOr* 6 (1986)).

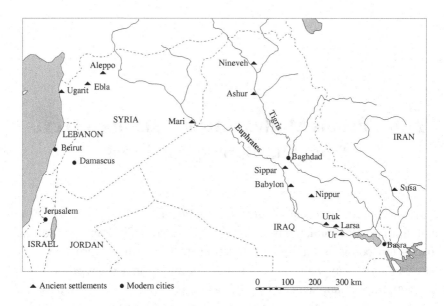

Fig. 1.1.1. A map of Mesopotamia and its neighbors.

1.1. M. 7857. A Fanciful Interpretation of a Geometric Progression

1.1 a. M. 7857. A text with three kinds of counting numbers

M. 7857, the mathematical text from Mari published by Guichard, was poorly understood by him and described as "an account of ants". Actually, the obverse of the clay tablet contains the computation of five terms of a geometric progression, with the first term 99 and the common ratio 9. The computation is carried out twice, first in sexagesimal place value numbers, then in "mixed decimal-sexagesimal" numbers. These were the OB learned and lay ways, respectively, of expressing numbers. The other side of the clay tablet (the reverse) contains a fanciful reformulation of the computation, expressed (not quite successfully) in the "centesimal" numbers used locally at Mari. (This interpretation of the text was discovered independently by the present author and by C. Proust, *FlM* 6 (2002).)

In the Babylonian *sexagesimal place value system*, there are special cuneiform number signs for the ones, from 1 to 9, and for the tens, from 10 to 50. In the OB (non-positional) *mixed decimal-sexagesimal system*, in

1.1. M. 7857. A Fanciful Interpretation of a Geometric Progression

the local variant used in texts from Mari, there are signs for the number words 'a hundred' (*me,* abbreviation for *mêtum*), 'a thousand' (*līm,* pl. *līmī*), and 'a great' (Sum. gal), meaning either 'ten thousand' or, equivalently, 'a hundred hundred'. Numbers below 100 are written as sexagesimal numbers, with or without the word *šu-ši* 'sixty'. The Mari *centesimal place value system*, on the other hand, operates in the same way as the Babylonian sexagesimal place value system, but with the base 100 instead of 60, and with cuneiform signs not only for the tens from 10 to 50, but also for 60, 70, 80, and 90 (written with from six to nine oblique wedges).

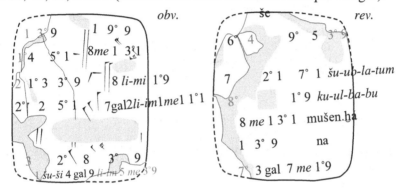

Fig. 1.1.2. M. 7857. An OB mathematical text from Mari, in conform transliteration.

A "conform transliteration" of the text of M. 7857 is presented in Fig. 1.1.2, within an outline of the clay tablet. In such a conform transliteration, the cuneiform signs are replaced by their numerical or phonetic values, placed in the same positions as the original cuneiform signs. The notations 1°, 2°, ···, 9° are used as conform transliterations of the cuneiform signs for the tens, from 10 to 90.

As the figure shows, the surface of the clay tablet is damaged, pieces of it are missing near the corners, and there are traces of a previous, incompletely erased inscription. Anyway, most of the text is preserved, and the lost parts of the text can be reconstructed (grey numbers and letters).

Below is given a standard transliteration of the text in the two columns on the *obverse* of M. 7857 (within the frame), together with a direct translation (underneath the frame). In the standard transliteration, zeros are inserted where needed. Reconstructed parts of the text are placed within straight brackets in the transliteration, but are written with italics in the translation. The exclamation marks indicate corrections: 28 in line 6 should be 25, and 1 99 in line 1 should be 1 39 (= 99), an interesting error.

M. 7857

obv. 1	[1 3]9	‖	1 99
2	[1]4 51	‖	8 *me* 1 31
3	[2] 13 39	‖	8 *li-mi* 19
4	20 02 51	‖	7 gal 2 *li-im* 1 *me* 1 11
5		erasure	
6	[3] 28 39		
7	[1] *šu-ši* 4 gal 9 [*li-mi* 5 *me* 39]		

         1 39   ‖   1 3¹9
         14 51   ‖   8 hundred 1 31
         2 13 39 ‖   8 thousand 19
         20 02 51 ‖   7 great 2 thousand 1 hundred 1 11
               erasure
    3 00 25⁖ 39
    1 sixty 4 great 9 *thousand 5 hundred* 39

The computations producing the numbers in the two columns on the obverse can be explained as follows:

    1 39   ‖   99
   14 51   (= 9 · 1 39)   ‖   891   (= 9 · 99)
  2 13 39 (= 9 · 14 51) ‖   8,019 (= 9 · 891)
20 02 51 (= 9 · 2 13 39) ‖ 72,171 (= 9 · 8,019)
3 00 25 39 (= 9 · 20 02 51) ‖ 649,539 (= 9 · 72,171)

There is only a single column of text on the *reverse* of M. 7857:

edge	še	
rev. 1	64 95 [39] /	
2	7 21 71	*šu-ub-la-tum* /
3	[80] 19	*ku-ul-ba-bu* /
4	8 *me* 1 31	mušen.ḫa /
5	1 39	na /
6	73 gal 7 *me* 19	

    64 95 *39* barley-corns
    7 21 71   ears of barley
    80 19   ants
    8 hundred 1 31 birds
    1 39   people(?)
    73 great 7 hundred 19

The one who wrote the text, probably a school boy, was clearly confused by the bewildering variety of number systems. In line 1 on the obverse he hesitated between writing 99 as 1 39 in the mixed decimal-sexagesimal system or as 99 in the centesimal system, and ended up writing 1 99. In lines 4-6 on the reverse, he forgot that he was supposed to use the centesimal place value system and reverted to the mixed decimal-sexagesimal system used in the right column on the obverse.

The switching to the centesimal number system on the reverse is not the only difference between the reverse and the obverse on M. 7857. On the obverse the terms of the geometric progression increase from 99 to $9^4 \cdot 99$ = 649,539, while on the reverse the recorded numbers decrease from 649,539 in line 1 to 99 in line 5. In addition, the numbers of the geometric progression have been given a fanciful interpretation, with a series of appended Sumerian or Akkadian (Babylonian) words. Finally, the number recorded in line 6 on the reverse is the sum of the five terms of the geometric progression, while there is no sum recorded on the obverse. Indeed, in the centesimal system (with the mistakes on the reverse corrected), the sum can be computed as follows:

64 95 39	barley-corns	(Sum. še)
7 21 71	ears of barley	(Akk. *šublātum*)
80 19	ants	(Akk. *kulbābū*)
8 91	birds	(Sum. mušen.ḫá)
+ 99	people	(Sum. na[?]; the translation is problematic)
73 07 19	diverse items	

It is tempting to try to reconstruct a whimsical story that can have accompanied the text on the reverse. It may have gone like this:

There were 645, 539 barley corns, 9 barley-corns on each ear of barley, 9 ears of barley eaten by each ant, 9 ants swallowed by each bird, and 9 birds caught by each man. How many were there altogether?

1.1 b. OB texts with ascending or descending geometric progressions

There is no known OB cuneiform text that is a direct parallel to the Mari text M. 7857. There are, however, quite a few known OB clay tablets on which are recorded *ascending* or *descending* geometric progressions of various kinds. The simplest examples of texts with ascending geometric progressions are inscribed with a small number of terms, usually 10, of a "table of powers" (a geometric progression in which the common ratio is equal to the first term). One such text is **Ist. O 3826** (Neugebauer *MKT 1* (1935), 77) inscribed with the first 10 powers of 9, followed by 5 powers of 1 40, from the 6th to the 10th power. Another example is **BM 22706**, in Nissen, Damerow, and Englund, *ABk* (1993), 150 (10 powers of 1 40, followed by 10 powers of 5).

The most recently published text of this kind is **IM 73355**, in Arnaud, *TL* (1994) (10 powers of 3 45, followed by 10 powers of 16, the sexagesimal reciprocal of 3 45). See Fig. 1.1.3 below.

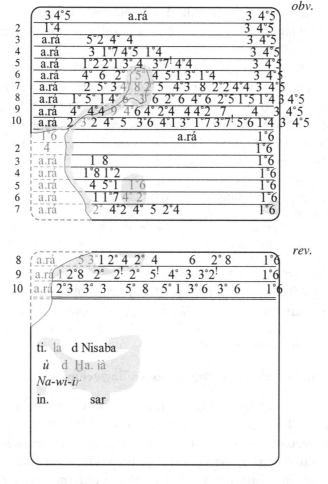

Fig. 1.1.3. IM 73355. Two OB tables of powers, for 3 45 and for its reciprocal 16.

1.1. M. 7857. A Fanciful Interpretation of a Geometric Progression

The text of IM 73355 can be translated as:

3 45 times 3 45 /	14	3 45	3 45 · 3 45	=	14 03 45
times	52 44	3 45	3 45 · 14 03 45 =		52 44 03 45
times	3 17 45 14	3 45	3 45 · 52 44 03 45 = 3 17 14 03 45		
etc.			etc.		

In the text, a.rá is Sumerian for 'times', literal meaning possibly 'steps'. After the two tables of powers on IM 73355 there follows a brief subscript:

'By the life of (the god) Nisaba and (the god) Haia. Nawir wrote it.'

Note that all the powers of 3 45 are sexagesimal numbers with the "trailing part" 3 45, while all the powers of 16 are numbers with the trailing part 16. For emphasis, all the powers of 3 45 are written with the trailing part 3 45 in each line of the table written close to the right edge of the tablet. All the powers of 16 are written in a similar way, with the trailing part 16 in each line of the table written close to the right edge of the tablet.

There are two known examples of OB clay tablets on which are recorded a finite number of terms of a *descending* geometric progression, both in Friberg, *MCTSC* (2005). One of them is **MS 3037** (*op. cit.*, Fig. 1.4.1), on which is recorded, in descending order, the first 12 powers of 12. The other is **MS 2242** (Fig. 1.1.4 below), with the first 6 powers of 3 45, in descending order. It is likely that these two texts were answers to assignments, namely, to find the factorization of a given many-place regular sexagesimal number through successive elimination, one at a time, of factors visible as the trailing parts of the successively computed sexagesimal numbers. (See the discussion of "regular sexagesimal numbers" and of "the trailing part algorithm" in Friberg, *RlA 7* (1990), Secs. 5.2 b and 5.3.)

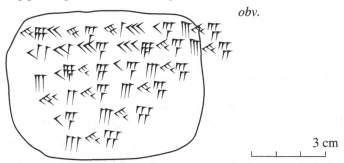

Fig. 1.1.4. MS 2242, *obv.* (*rev.* blank). An OB descending table of six powers of 3 45.

1.1 c. A Late Babylonian text using the trailing part algorithm

In the discussion above of the descending table of powers on the OB clay tablets MS 3037 and MS 2242 it was suggested that the computations in those texts were applications of a certain *trailing part algorithm*. Repeated applications of the same trailing part algorithm can be found also in a *Late Babylonian* (LB) text, the round clay tablet **W 23021** (Friberg, *BaM* 30 (1999), *BaM* 31 (2000)). In W 23021, the trailing part algorithm is used to compute the reciprocals of eight "many-place" regular sexagesimal numbers, all between 52 40 29 37 46 40 and 49 00 07 12.

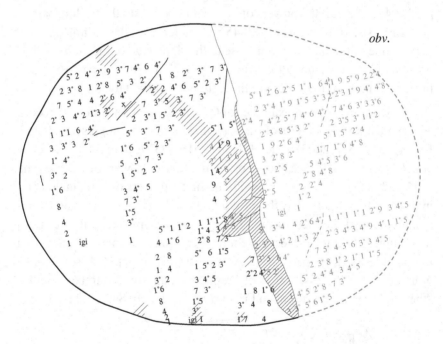

Fig. 1.1.5. W 23021. A LB text with eight applications of the trailing part algorithm.

1.1 d. The sum of a geometric progression in an Old Babylonian text

MS 1844 is a massive OB round hand tablet with what appears to be the numerical solution algorithm for an inheritance problem (Friberg, *MCTSC* (2005), Fig. 7.4.2). The tablet is inscribed with 6 terms of a *decreasing* geometric progression with the *last term* 2, and with the *constant ratio* be-

1.1. M. 7857. A Fanciful Interpretation of a Geometric Progression

tween the terms equal to 1 − 1/7. The constant ratio is described with the following words in a somewhat cryptic subscript in the last line of the text:

igi 7.gál.bi tur.šè   for a 7th-part less.

The sum of the geometric progression is inscribed in the first line of the text. It is, for a certain reason, given in the following curious form:

23 15 20 36
12 08 53 20.

There is a numerical error in the number recorded in line 3. It is easy to check that this error is *propagated upwards*, to the numbers recorded in lines 1 and 2. This means that the numbers in the algorithm table were computed *in reverse order*, beginning with the number '2' in line 8.

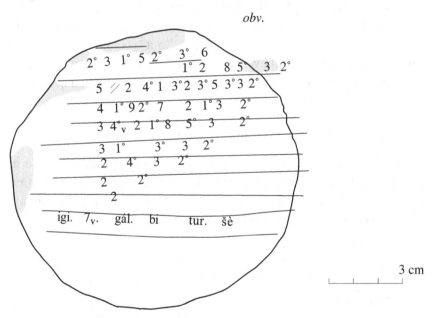

Fig. 1.1.6. MS 1844. An OB computation of the sum of a geometric progression.

The reason for the reverse order of computation is obvious. Since 7 is not a regular sexagesimal number, Babylonian mathematicians would have found it difficult to *count* with a number like 1 − 1/7. On the other hand, it is known through explicit examples (see Sec. 3.1 e below) that they were familiar with a counting rule of the type

if $b = a - a \cdot 1/7$, then $a = b + b \cdot 1/6$.

The correctness of this counting rule is obvious, at least in the case when *a* is a multiple of 7. Indeed,

if $a = n \cdot 7$ and $b = a - a \cdot 1/7$, then
$b = n \cdot 7 - n = n \cdot 6$, and $b + b \cdot 1/6 = n \cdot 6 + n = n \cdot 7 = a$.

In view of this simple counting rule, the requirement that each number in the algorithm table shall be equal to the number in the preceding line, diminished by 1/7 of its value, can be replaced by the equivalent requirement that each number in the table shall be equal to number below it, increased by 1/6 of its value. This reformulation of the requirement is a great simplification, since 6 is a regular sexagesimal number with the reciprocal 10. (In modern notations, this means that 1/6 = 10/60.) Therefore, increasing a given number by 1/6 of its value is equivalent to multiplying the given number by 1 10 in Babylonian relative (floating) place value notation, or by 1;10 in absolute place value notation. So it would be easier for the author of the text to count upwards, multiplying with the factor '1 10' in each step of the algorithm, than to count downwards, subtracting seventh parts. That is also precisely what he did.

1.1 e. The sum of a geometric progression in a Seleucid text

AO 6484 (Neugebauer, *MKT 1* (1935), 96) is a mathematical cuneiform text of mixed content from the Seleucid period, the last third of the first millennium BCE. The first exercise in that text shows through an example how to compute the sum of a geometric progression:

AO 6484 # 1

1	ta 1 en 10 gar a.na 2-*ma* bal-*it* gar.gar-*ma* 8 [32 gar]
2	[1 ta 8 32 lá-*ma*] / *re-ḫe* 8 31
	8 31 *a-na* 8 32 tab-*ma* 17 03 [⋯]

From 1 to 10 set, always by 2 surpass, heap (add together), then *8 32 set.*
1 from 8 32 subtract, then 8 31 remains.
The 8 31 to 8 32 repeat (add on), then 17 03 ⋯.

The phrasing is quite obscure, and the translation above is only tentative. Anyway, what is going on here seems to be that 10 numbers are given. Each number is twice the one before it, and the first number is 1. What is then the sum of the 10 numbers, from 1 to ($2^9 = 512 =$) 8 32?

1.1. M. 7857. A Fanciful Interpretation of a Geometric Progression

This means that the 10 numbers form a geometric progression with the first term 1 and the common ratio 2. How the sum S of the 10 terms is computed in the text can possibly be explained (in modern terms) as follows:

$S = 1 + 2 + \cdots + 8\ 32.$
$2 \cdot S = 2 + 4 + \cdots + 8\ 32 + 17\ 04 = S + 17\ 04 - 1.$
$S = 17\ 04 - 1 = 8\ 32 + (8\ 32 - 1).$

1.1 f. *P.Rhind* # 79: a parallel to M. 7857 in a hieratic papyrus

The largest and best known Egyptian hieratic mathematical papyrus is ***P.Rhind***. The only exercise in that text concerned with the sum of a geometric progression is *P.Rhind* # 79. In Fig. 1.1.7 below is shown a copy of the original hieratic text of that exercise, borrowed from Chase, *et al. RMP* (1929), together with a conform transliteration. The conform translation is in the form of a *mirror image* of the hieratic text, which is written from right to left. Moreover, since the decimal numbers in the hieratic text are written with non-positional number notations, it is appropriate to let this be apparent also in the conform translation. The way this is done in Fig. 1.1.7, and in the following, is by use of *subscripts*, 't' for tens, 'h' for hundreds, 'th' for thousands, and 'tth' for ten thousands.

The text in column *ii* of *P.Rhind* # 79 can be explained as follows:

houses	7		7	$(= 1 \cdot 7)$
cats	$4_t\ 9$		49	$(= 7 \cdot 7)$
mice	$3_h\ 4_t\ 3$		343	$(= 49 \cdot 7)$
emmer	$2_{th}\ 3_h\ 1$		$2,401^1$	$(= 343 \cdot 7)$
heqats	$1_{tth}\ 6_{th}\ 8_h\ 7$		+ 16,807	$(= \underline{2,401 \cdot 7})$
total	$1_{tth}\ 9_{th}\ 6_h\ 7$		19,607	$(= 2,801 \cdot 7)$

Column *i* contains the computation $7 \cdot 2,801 = 19,607$. In the usual *binary arithmetic* of *P.Rhind*, the product is computed as follows:

$7 \cdot 2,801 = (1 + 2 + 4) \cdot 2,801 =$
  1   2,801
  2   5,602
  4   + 11,204
  total   19,607

The computation can be explained as follows:

Presumably, it was known beforehand that $1 + 7 + \cdots + 2,401 = 2,801$.
Consequently, $7 + \cdots + 16,807 = 7 \cdot (1 + 7 + \cdots + 2,401) = 7 \cdot 2,801 = 19,607$.

Essentially, *P.Rhind* # 79 is the computation of the sum of a geometric progression of five terms with the first term 7 and the common ratio 7. However, this computation has been given a fanciful interpretation, reminding very much of the fanciful interpretation in M. 7857 of the sum of another geometric progression. One can try to reconstruct the whimsical story associated with the text of *P.Rhind* # 79. It may have gone like this:

> There were 7 houses, in each house 7 cats, each cat caught 7 mice, each mouse ate 7 bags of emmer, and each bag contained 7 *heqat*. How many were there altogether?

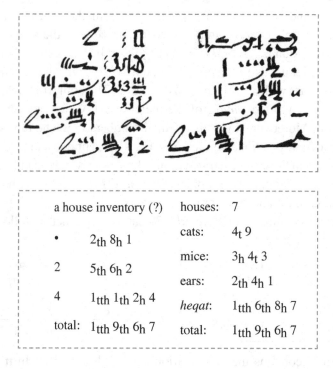

a house inventory (?)		houses:	7
•	2_{th} 8_h 1	cats:	4_t 9
		mice:	3_h 4_t 3
2	5_{th} 6_h 2	ears:	2_{th} 4_h 1
4	1_{tth} 1_{th} 2_h 4	*heqat*:	1_{tth} 6_{th} 8_h 7
total:	1_{tth} 9_{th} 6_h 7	total:	1_{tth} 9_{th} 6_h 7

Fig. 1.1.7. *P.Rhind* # 79. Col. *i*: The sum, computed as 7 times 2,801.
Col. *ii*: The five terms and their sum.

1.1 g. Summary. The Mesopotamian roots of a Mother Goose riddle

The mathematical cuneiform text M. 7857 from Mari (Fig. 1.1.2) seems to be a crucial link in a chain of related texts that begins with some OB algorithm texts and ends with a nursery rhyme still familiar today. The OB algorithm texts in question are several known examples of tables of

1.1. M. 7857. A Fanciful Interpretation of a Geometric Progression

powers, either ascending like the one on IM 73355 (Fig. 1.1.3) or descending like the one on MS 2242 (Fig. 1.1.4). In Mesopotamia proper, the last manifestation of a text of this kind is W 23021, a Late Babylonian round clay tablet from Uruk (Fig. 1.1.5) with eight applications of the trailing part algorithm.

In addition to such OB and LB purely *abstract and numerical* tables of powers there is also MS 1844 (Fig. 1.1.6 above), an OB example of "applied" or "practical" mathematics, where a descending geometric progression and its sum are interpreted as the progressively smaller shares of a given sum of silver divided between seven partners or brothers.

In OB Mari, in the north-western periphery of Mesopotamia, the text genre seems to have been transformed into something else, according to the testimony of M. 7857. In that text, which features both ascending and descending geometric progressions, counting with Babylonian sexagesimal numbers is replaced by counting with decimal-sexagesimal or even centesimal numbers. Moreover, what starts out on the obverse of the clay tablet as a no-nonsense abstract computation of a geometric progression with five terms is turned on the reverse into a whimsical computation with *barley-corns, ears, ants, birds, and people*, and a totally nonsensical summation of those five disparate categories.

A similar whimsical interpretation of a geometric progression with five terms can be found in the hieratic *P.Rhind* # 79 (Fig. 1.1.7 above), which even, just like M. 7857, ends with a meaningless summation, this time of *houses, cats, mice, ears, and grains*. Although both the objects counted and the numbers are different in the two exercises, the texts are so similar in spirit that it is inconceivable that they were devised independently.

As is well known, the next reappearance of the text genre is in Leonardo Pisano's 13th century treatise **Liber abaci**, (fol. 138 *recto*), in a problem which starts with the words *Septem uetule uadunt romam* '7 old women go to Rome', and ends asking for *summa omnium predictorum* 'the sum of all those mentioned above' (old women, mules, sacks, breads, knives, and sheaths).

The final reappearance of the same topic, with an added twist, is the **Mother Goose** riddle "As I was going to St. Ives, I met a man with seven wives. Each wife had seven sacs, each sack had seven cats, each cat had seven kits. Kits, cats, sacks, and wives, how many were going to St. Ives?"

14   Unexpected Links Between Egyptian and Babylonian Mathematics

Note that, just as in *P.Rhind* # 79, both the first term and the common ratio are 7 in the *Liber abaci* problem and in the *Mother Goose* riddle, although the number of terms is not the same in the three texts.

1.2. M. 8631. Another Curious Mathematical Text from OB Mari

1.2 a. M. 8631. A fanciful interpretation of 30 doublings

M. 8631 (Fig. 2.1.1) is a fragment of an OB mathematical text from Mari, published by Soubeyran in *RA* 78 (1984), together with some multiplication tables and a table of squares.

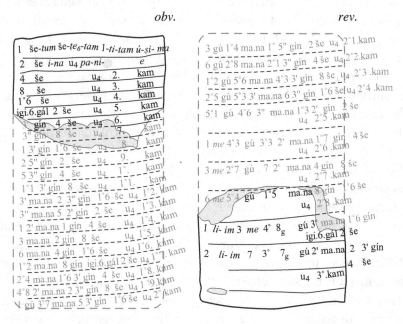

Fig. 1.2.1. M. 8631. An initial capital of 2 barley-corns, doubled 29 times.

The text on M. 8631 begins with the following introductory phrase in line 1 of the obverse:

1 še-*tum* <*a-na*> še-*te₆-tam* 1-*ti-tam ú-ṣi-ma*
1 barley-corn (to) a single barley-corn I added, then ⋯

Here barleycorn (Sum. še, Akk. *uṭṭetum*) is written as a Sumerian logogram še followed by an Akkadian phonetic complement -*tum*. The corre-

1.2. M 8631. Another Curious Mathematical Text from OB Mari.

sponding inflected form is *uṭṭetam*, written as š e followed by the Akkadian phonetic complement -*ṭe₆-tam*. Similarly, 1-*ti-tam* is an inflected form (feminine, accusative) of the Akkadian word *ištēnum* 'one', written as the digit 1 (also a Sumerian logogram!) followed by an Akkadian phonetic complement. The Akkadian verb hiding behind the inflected form *ú-ṣi-(ma)* is *(w)aṣābum* 'to add (something) to (something)'.

The meaning of the whole phrase is, as shown by the ensuing text, that an initial capital of 1 barley-corn is doubled each day for a month of 30 days. The barley-corn (appr. 0.05 g) was the smallest unit in the Sumerian/Old Babylonian system of weight measures. The relation of the barley-corn to the higher units of the system can be described by the following chain of equations:

1 gú (talent, man's-load) = 60 ma.na,
1 ma.na (mina, c. 500 g) = 60 gín,
1/3 gín (shekel) = 60 š e (barley-corns).

The curious form of the third equation, which defines the size of the barley-corn, is due to the fact that originally (in the Early Dynastic IIIa period in Mesopotamia in the third millennium BCE) the barley-corns and a unit equal to 60 barley-corns belonged to a system of weights suitable for the weighing of small quantities of some precious metal (silver or gold), while the higher units belonged to another system of weights, suitable for measuring a less valuable metal like copper, or other heavy objects. (See Friberg *JCS* 51 (1999), 133, Friberg, *CDLJ* (2005, to appear)). Since at that time silver was 180 times more valuable than copper, 1 barley-corn of silver was worth as much as 1 shekel of copper.

It is likely that the author of M. 8631 started his work by constructing a preliminary table in abstract numbers for a geometric progression of sexagesimal place value numbers with the first term 20 (because 1 barley-corn = 1/3 of 1/60 shekel = ;00 20 shekel) and the common ratio 2.

Here follows first a standard transliteration of the 21 lines of text on the *obverse* of M.8631, with an explanation of how the successive weight numbers may have been computed. Reconstructed parts of the text are as usual within brackets in the transliteration, but in italics in the translation.

M. 8631, obv.

1	1 še-*tum* še-*ṭe₆-tam* 1-*ti-tam ú-ṣi-ma* /				
	2 še	*i-na* u₄ *pa-ni-e* /			
	4 še			u₄	2.kam /
	8 še			u₄	3.kam /
	16 še			u₄	4.kam /
5	igi.6.gál	2 še		u₄	5.kam /
	[3'] gín	4 še		u₄	6.kam /
	[3" gín	8 še		u₄]	7.[kam] /
	[1 3' gín	16 še		u₄	8.kam] /
	[2 6" gín	2 še		u₄	9.kam] /
10	[5 3" gín	4 še		u₄	10.kam] /
	[11 3' gín	8 še		u₄	11.kam] /
	[3' ma.na	2 3" gín	16 še	u₄	12.kam] /
	[3" ma.na	5 2' gín	2 še	u₄	13.kam] /
	[1 2' ma.na	1 gín	4 še	u₄	14.kam] /
15	[3 ma.na	2 gín	8 še	u₄	15.kam] /
	[6 ma.na	4 gín	16 še	u₄	16.kam] /
	[12 ma.na 8 gín igi.6.gál		2 še	u₄	17.kam] /
	[24 ma.na	16 3' gín	4 še	u₄	18.kam] /
	[48 2' ma.na	2 3" gín	8 še	u₄	19.kam] /
20	[1 gú 37 5/6 ma.na	3' gín	16 še	u₄	20.kam] /

1 barley-corn to a single barley-corn I added, then		
2 b.c.	in the first day	(= ;00 40 shekel)
4 b.c.	the 2nd day	(= ;01 20 sh.)
8 b.c.	the 3rd day	(= ;02 40 sh.)
16 b.c.	the 4th day	(= ;05 20 sh.)
1/6 shekel 2 b.c.	the 5th day	(= ;10 40 sh.)
1/3 shekel 4 b.c.	the 6th day	(= ;21 20 sh.)
2/3 shekel 8 b.c.	the 7*th day*	(= ;42 40 sh.)
1 1/3 shekels 16 b.c.	*etc.*	(= 1;25 20 sh.)
2 5/6 shekels 2 b.c.		(= 2;50 40 sh.)
5 2/3 shekels 4 b.c.		(= 5;41 20 sh.)
11 1/3 shekels 8 b.c.		(= 11;22 40 sh.)
1/3 mina 2 2/3 shekels 16 b.c.		(= 22;45 20 sh.)
2/3 minas 5 1/2 shekels 2 b. c		(= 45;30 40 sh.)
1 1/2 minas 1 shekels 4 b.c.		(= 1 31;01 20 sh.)
3 minas 2 shekels 8 b.c.		(= 3 02;02 40 sh.)
6 minas 4 shekels 16 b.c.		(= 6 04;05 20 sh.)
12 minas 8 1/6 shekels 2 b.c.		(= 12 08;10 40 sh.)
24 minas 16 1/3 shekels 4 b.c.		(= 24 16;21 20 sh.)
48 1/2 minas 2 2/3 shekels 8 b.c.		(= 48 32;42 40 sh.)
1 talent 37 minas 5 1/3 shekels 16 b.c.		(= 1 37 05;25 20 sh.)

1.2. M 8631. Another Curious Mathematical Text from OB Mari.

In the transliteration above, the following notations are used for the four Sumerian/Old Babylonian "basic fractions": 3' (= 1/3), 2' (= 1/2), 3" (= 2/3), and 6" (= 5/6). These notations are intended to turn the readers' attention to the fact that special Sumerian/Old Babylonian cuneiform signs existed only for these fractions. All other fractions were of the form igi n 'the opposite of n' (= $1/n$), n being a regular sexagesimal number. (A sexagesimal number n is called "regular" if there exists another sexagesimal number n', such that n and n' is a "pair of reciprocals" in the sense that $n \cdot n'$ = '1' in floating sexagesimal place value notation, that is if $n \cdot n'$ = a power of 60.) Fractions of the form n/m do not appear in OB cuneiform texts.

On the reverse of M. 8631 there were, originally, ten more lines for days 21 through 30. In the same way as in the case of the text on the obverse, the computation of the weight numbers in those ten lines on the reverse can be explained as follows, by use of sexagesimal arithmetic:

3 talents	14 minas	10 5/6 sh.	2 b.c.	day 21	(= 3 14 10;50 40 sh.)
6 talents	28 1/3 minas	1 2/3 sh.	4 b.c.	day 22	(= 6 28 21;41 20 sh.)
12 talents	56 2/3 minas	3 1/3 sh.	8 b.c.	day 23	(= 12 56 43;22 40 sh.)
25 talents	53 1/3 minas	6 2/3 sh.	16 b.c.	day 24	(= 25 53 26;45 20 sh.)
51 talents	46 5/6 minas	3 1/2 sh.	2 b.c.	day 25	(= 51 46 53;30 40 sh.)
1 43 talents	33 2/3 minas	7 sh.	4 b.c.	day 26	(= 1 43 33 47;01 20 sh.)
3 27 talents	7 1/2 minas	4 sh.	8 b.c.	day 27	(= 3 27 07 34;02 40 sh.)
6 54 talents	15 minas	8 sh.	16 b.c.	day 28	(= 6 54 15 08;05 20 sh.)
13 48 talents	30 minas	16 1/6 sh.	2 b.c.	day 29	(= 13 48 30 16;10 40 sh.)
27 37 talents	1/2 mina	2 1/3 sh.	4 b.c.	day 30	(= 27 37 00 32;21 20 sh.)

As noticed already in Soubeyran's original publication of M. 8631, there are some surprising notations in the last two lines of the cuneiform text. Thus, the weight numbers recorded for days 29-30 are

1 *li-im* 3 *me* 48$_g$ gú 30 [ma.na 16 gín] / igi.6.gál 2 še
2 *li-im* 7 37$_g$ gú 2' ma.na 2 3' gín / 4 še

(The subscripts in the numbers 48$_g$ and 7 37$_g$ before gú 'talent' are meant to be reminders of the fact that in the OB cuneiform script the digits 1 through 9 are written with *horizontal* wedges when preceding the cuneiform sign for the weight unit gú (or the capacity unit gur), but with *vertical* wedges in most other circumstances.)

What is remarkable here is that in the line for day 29 the weight number '13 48 talents' is written as '1 thousand 3 hundred 48 talents', that is with the OB notation for 'a thousand' used to denote 'ten sixties', and with the

OB sign for 'a hundred' used to represent 'sixty'. Similarly in the line for day 30 the weight number '23 37 talents' is written as '2 thousand 7 37 talents', again with the normal notation for 'a thousand' used to represent 'ten sixties'. It is not clear if these curious deviations from the OB norm were mistakes or if they were intentional. Anyway, just like the use of mixed decimal-sexagesimal and centesimal numbers alongside with sexagesimal numbers in M. 7857, so the use of decimal notation for sexagesimal numbers in M. 8631 testifies that the scribes in Mari were not comfortable with the use of sexagesimal numbers, and that they, possibly for that reason, were experimenting with new ways of writing numbers in cuneiform.

1.2 b. The Old Babylonian doubling and halving algorithm

The 30 weight numbers recorded in the Mari text M. 8631 form a geometric progression of a special kind, 30 successive *doublings* of an initial weight number. Indirect parallels are several well known OB algorithm tables in which an initial regular sexagesimal number is doubled a number of times. The best known example of such a text is **UM 29.13.21** (Neugebauer and Sachs, *MCT* (1945), 13; Friberg, *MCTSC* (2005), App. 3), a small fragment of a large table text from Nippur with several applications of the OB "doubling and halving algorithm" (Sachs, *JCS* 1 (1947), Friberg, *RlA 7* (1990) Sec. 5.3 b). A reconstruction of most of that text is possible. A conform transliteration of the reconstructed text within an outline of the clay tablet is shown in Fig 1.2.2 below.

The *doubling and halving algorithm* is based on the observation that if n and n' are a pair of reciprocals, then $2 \cdot n$ and $1/2 \cdot n'$ are another pair of reciprocals. Successive doublings of n and halvings of n' will produce a never ending sequence of pairs of reciprocals. In the first application of this algorithm in UM 29.13.21, a table with 30 pairs, the first pair is chosen as n = 2 05 (the 3rd power of 5) and n' = 28 48 (the 3rd power of 12). The other 29 pairs can be characterized as *29 doublings of the initial number 2 05*, together with the corresponding reciprocals, *29 successive halvings of the initial reciprocal number 28 48*. Thus, the table proceeds from

    2 05 / igi.bi 28 48
    (2 05, its reciprocal is 28 48)

in line 1 to

1.2. M 8631. Another Curious Mathematical Text from OB Mari. 19

1 25 20 X 58 09 11 06 40 / igi.bi 41 42 49 22 21 12 39 22 30
(1 25 20 X 58 09 11 06 40, its reciprocal is 41 42 49 22 21 12 39 22 30)

in line 30. The cuneiform sign represented here by an X is a separation sign, probably introduced in line 20 of the algorithm table as a kind of sign for zero. The scribe's inability to handle this zero correctly in lines 21-30 led to an error in line 30, where the number written 1 25 20 X 58 09 11 06 40 should rightly be 1 25 20+58 09 11 06 40 = 1 26 18 09 11 06 40.

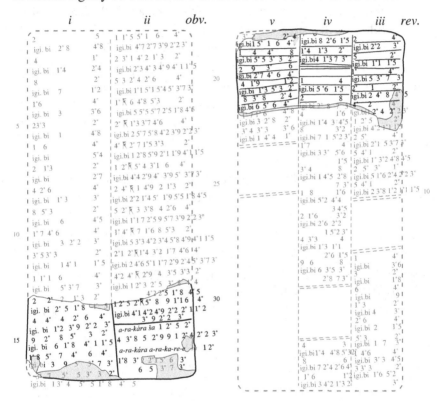

Fig. 1.2.2. UM 29.13.21. Five applications of the OB doubling and halving algorithm.

1.2 c. P.IFAO 88: A parallel to M. 8631 in a Greek-Egyptian papyrus

P.IFAO 88 is a Greek-Egyptian papyrus fragment of unknown date and origin (IFAO = *Institut Français d'Archéologie Orientale du Caire*), published by Boyaval in *ZPE* 7 (1971), in the form of a murky photograph accompanied by an only partly successful interpretation. Improved interpretations were presented later, by Rea in *ZPE* 8 (1971), and by Boyaval himself in *ZPE* 14 (1974).

The inscription on *P.IFAO 88* consists of Greek alphabetic number signs for decimal numbers, in the usual Greek non-positional decimal number notation, with α, β, γ, ⋯ for 1, 2, 3, ⋯, ι, κ, λ, ⋯ for 10, 20, 30, ⋯, ρ, σ, τ, ⋯ for 100, 200, 300, ⋯, and α´, β´ γ´, ⋯ for 1,000, 2,000, 3,000, ⋯. From line 14 on, numbers smaller than 6,000 are preceded by a special sign, presumably standing for *drachma*, a small Greek unit of weight. Referring to a suggestion by Youtie, Boyaval (*op. cit.*, 1974) reads the sign as χ^α, meaning χα(λκου) 'copper (drachma)'. Multiples of 6,000 are counted separately and are preceded by another special sign, known to represent the *talent*. Customarily, 1 *talent* = 6,000 *drachmas*.

The use of the *talent* and the copper *drachma* as monetary units is a clue to the date of the text. Those units were used in Egypt in the Ptolemaic and early Roman periods. Boyaval (*op. cit.*, 1974) dates the text to not later than the 1st century CE.

In this connection it may be worth mentioning that Greek multiplication tables for fractions, such as, for instance[9], those on the first three pages of the mathematical papyrus codex **P.Akhmim** (Baillet, *PMA* (1892); 7th c. CE; Sec. 4.5 below), or on the obverse of the mathematical wooden tablet **Michael. 62** (Crawford, *Aeg.* 33 (1953); 6th? c. CE; Sec. 4.6 below), begin by listing the fraction in question applied to 6,000, 'the number' (ἀριθμóc), as in the following examples:

3" to 'the number'	4,000	3' to 'the number'	2,000
of 1 the 3"	3"	of 1 the 3'	3'
of 2	1 3'	of 2	3"
of 3	2	of 3	1
⋯ ⋯	⋯ ⋯	⋯ ⋯	⋯ ⋯
of 1 (myriad)	6,666 3"	of 1 (myriad)	3,333 3'.

About this, Crawford (*op. cit.*, 227) has the following to say:

> "The number 6,000, the fraction of which figures in the first or second place in all versions of the tables, can only be, as has been recognized, the number of *drachmae* in the *talent*. The implication is that the tradition of starting the tables in this way dates from a time when money was always counted in *drachmae* and *talents*; in fact it must have been firmly established before the introduction of the gold-standard

---

9. Cf. the catalogue of published tables in Fowler, *MPA* (1987; 1999), Sec. 7.5.

1.2. M 8631. Another Curious Mathematical Text from OB Mari. 21

coinage. Now when this tablet (*Michael. 62*) and P.Akhmim were written and probably even when P.Mich. 621 was written (if it is the 4th Century), monetary values were normally reckoned in *nomismatia* and *keratia* ⋯ ⋯. We may therefore be confident that tables in this particular form go back at least to the 3rd Century A. D. They may of course be much earlier still."

Thus, in *Michael. 62*, which contains multiplication tables on the obverse beginning with fractions of 6,000 (*drachmas*), the problem texts on the reverse nevertheless count in terms of *nomismatia* and *keratia*. (See the text of *Michael. 62* # 2 in Sec. 4.6 below.)

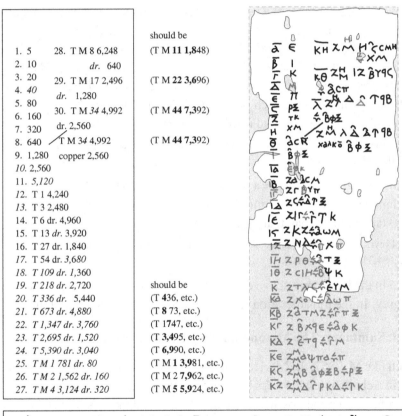

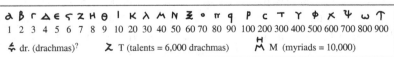

Fig. 1.2.3. *P.IFAO 88*. An initial capital of 5 copper drachmas, doubled 29 times.

The lines of *P.IFAO 88* are numbered, making it easy to determine the original extent of the text. Thus, the line numbering begins with $\overline{\alpha}, \overline{\beta}, \overline{\gamma}$ (= 1, 2, 3) in the left column and continues with $\overline{\kappa\eta}, \overline{\kappa\theta}, \overline{\lambda}$ (= 28, 29, 30) in the right column. The end result in line 30 is repeated in a final line without line number.

It is easy to see that the recorded numbers form a geometric progression of 30 terms, with the first term equal to 5 (*copper drachmas*) and with the common ratio 2. In other words, the geometric progression has been given an interpretation as an initial weight of 5 *copper drachmas*, doubled 29 times. The regular form of a geometric progression makes it easy to reconstruct the missing parts of the text, as shown in Fig. 1.2.3.

What makes the text particularly interesting is that there seems to be a trivial error in the non-preserved line 20, namely 336 talents instead of 436 talents, possibly due to the fact that the two Greek letters T = 300 and Y = 400 can be mistaken for each other. The effect of the error is avalanching through lines 21-30. The end result, in line 30, is given as

34 myriad 4,992 talents 2,560 drachmas (=344,992 talents 2,560 drachmas = 2,069,954,560 drachmas)

(Remember that 1 myriad = a hundred hundred = 10,000, 1 talent = 6,000 drachmas.) The correct result should be $2^{10} \cdot 100$ talents = 10 myriads 2,400 talents more than that, or altogether

44 myriad 7,392 talents 2,560 drachmas (= 447,392 talents 2,560 drachmas = 2,684,354,560 drachmas).

Note: In the hand copy above of *P.IFAO 88*, reconstructed parts of the text are grey. In the transliteration reconstructed parts are written in italics.

1.2 d. Summary. The Mesopotamian roots of a well known legend

Much like M. 7857, the Mari text M. 8631 (Fig. 1.2.1) seems to be a crucial link in a chain of related texts beginning with an important category of OB algorithm tables and possibly ending with a legend that is still well known today. The OB texts in question are various applications of doubling or doubling-and-halving algorithms, notably the fragment UM. 29.13.21 (Fig. 1.2.2) from Nippur (see the map in Fig.1.1.1) which begins with 30 doublings and halvings of the pair of reciprocals (2 05, 28 48).

In M. 8631, a table of 30 doublings was imaginatively reinterpreted as

1.2. M 8631. Another Curious Mathematical Text from OB Mari. 23

the growth of an initially given capital (1 barley-corn) over a month of 30 days, with a daily doubling of the capital.[10]

Then follows a large gap in the supposed chain of related texts, because there is no known hieratic or even demotic Egyptian text mentioning 30 doublings. Instead, there is the Greek-Egyptian *P.IFAO 88* (Fig. 1.2.3) with its 30 lines of doublings of an initial amount of 5 drachmas, where the final amount is expressed in terms of myriads of talents. There is no explicit mention of days in that text, only line numbers. Note, by the way, that the Sumerian/Old Babylonian talent of $60 \cdot 60$ shekels (60 shekels = 1 mina) clearly is the ancestor of the Greek talent of $60 \cdot 100$ drachmas (100 drachmas = 1 mina). Hence, the assumption of a connection between a Greek-Egyptian papyrus and a clay tablet from OB Mari is not as far-fetched as it may seem at first sight. (The circumstance that the Greek 'myriad' = $100 \cdot 100$ is a decimal number unit formed in the same way as the Mari 'great' is interesting but may have no historical significance.)

The final reappearance of the theme is in the form of the legend about the reward granted by some Indian king to the inventor of the game of chess, who demanded one grain of rice on the first square of a chess board and then twice as much on each consecutive square.[11] It may be noted, in this connection, that the ancient Egyptian game of Senet had a game board with $10 \cdot 3 = 30$ squares. So, maybe, the missing link in the chain is a hieratic mathematical text that simply has not happened to be preserved.

---

10. Note, for comparison, that the standard interest rate in OB mathematical texts was 1/5 or 1/3 of the capital, not per day but per year or per transaction. See, for instance, the mixed theme text **YBC 4698** (Fig. 2.1.17) § 1 (## 1-2), where the interest on 1 gur of barley (1 gur = 5 00 sìla = appr. 300 liters) is given as 1 barig (= 1 00 sìla) in the first example, and as 1 barig 4 bán (= 1 40 sìla) in the second example. However, there are no known OB "interest tables" for the growth of a capital, based on standard interest rates. On the other hand, there is a known Neo-Sumerian table for the regular growth of a herd of cows and bulls during a period of 10 years (**AO 5499**: see Nissen/Damerow/Englund, *ABk* (1993), Figs. 76-79).

11. A variant of this legend, mentioned by Rea in *ZPE* 8 (1971), is "the tale of the crafty blacksmith who offered to shoe the king's horse at one penny for the first nail, twopence for the second, fourpence for the third, and so on till all the nails were in. Since a horse needs twenty-eight or thirty (!) nails to keep its shoes in place, the result was a formidable accumulation".

24 Unexpected Links Between Egyptian and Babylonian Mathematics

Fig. 1.2.4. *P. IFAO 88.*

Chapter 2

Hieratic Mathematical Papyri and Cuneiform Mathematical Texts

The fact that there seems to be a definite connection between an OB mathematical text from Mari and *P.Rhind* # 79 (Fig. 1.1.7 above) suggests that it may be worthwhile to start searching for more connections between OB and Egyptian mathematics, in particular further OB parallels to exercises in *P.Rhind*. In this connection, it is important to point out that the original version of *P.Rhind* may have been somewhat older than the OB mathematical texts from Mari, which were all written before the fall of Mari in 1757 BCE. Indeed, although *P.Rhind* itself is dated in a preface to the time of a king of the Fifteenth Dynasty, during the Hyksos period, the preface also says that the papyrus is a copy of an older text from the time of what appears to be the sixth king of the Twelfth Dynasty, who reigned in the second half of the nineteenth century BCE. On the other hand, as will be shown below, *P.Rhind* # 79 is an interpolated exercise in *P.Rhind*. Therefore, it may well be of a later date than the main part of the papyrus.

No serious attempt to make a comparison between Egyptian hieratic mathematical papyri and OB mathematical cuneiform texts has been published before, although in Høyrup, *LWS* (2002), 321 the practical division problem *P.Rhind* # 37 is compared with the OB text **IM 53957** (Baqir, *Sumer* 7 (1951), 37). Høyrup's conclusion in that particular case is that "The coincidences are too numerous to be accidental.". More about this example below, in the discussion of *P.Rhind*, "Theme E". However, later in the same book (*op. cit.*, 405), Høyrup returns to the question and states emphatically that

"Apart from the family likeness between the filling problems in IM 53957 and Rhind Mathematical Papyrus # 37, no evidence suggests the slightest connection between OB mathematics and 'classical' (Pharaonic) Egyptian mathematics as found in Middle and New Kingdom papyri – nor between the surveyors' tradition and classical Egyptian mathematics." [12]

In spite of Høyrup's pessimistic attitude, it is not difficult to find further parallels between Egyptian and Babylonian mathematics. As a matter of fact, one important first result of an endeavor to look for parallels between Egyptian and Babylonian mathematical texts is the realization that the well known hieratic mathematical papyri *P.Rhind* and *P.Moscow* must both be "recombination texts" of a type represented by several large and well known Old or Late Babylonian mathematical texts. (*Recombination text* is a suitable name for some enterprising teacher's more or less systematic compilation in one text of a set of whole or partial copies of older texts that happened to be available to him.)

The fact that *P.Rhind* is a recombination text will be demonstrated in Sec. 2.1 a below, where the complicated and only partly well organized structure of the text is highlighted in a detailed table of contents. It is possible to discern eleven different themes to which most of the individual exercises in *P.Rhind* belong. Of these, at least five have OB parallels. These five themes will be discussed in Sec. 2.1 b-f.

The table of contents for *P.Moscow* in Sec. 2.2 a closely follows the division of the text into nine paragraphs suggested in Struve's initial edition of the papyrus. Babylonian parallels to exercises belonging to four of these paragraphs are discussed in Sec. 2.2 b-e.

Babylonian parallels to mathematical exercises in three small hieratic papyrus fragments (*P.Berlin 6619*, *P.UC 32160* (= *Kahun IV.3*), and *P.UC 32161* (= *Kahun XLV.1*) are discussed in Sec. 2.3.

In the conclusion in Sec. 2.4 there is, in addition to a summary of the results obtained in Chapter 2, also an attempt to explain why there are so many explicit computations in *P.Rhind*, but none in *P.Moscow*.[13]

---

12. A similar statement can be found in Ritter, *MPIWG 103* (1998), p. 13.
13. New interpretations are offered in Chapter 2 for the following texts: **BM 34800** (Sec. 2.1 b), ***P.Rhind* # 53**, **IM 43996**, **NCBT 1913** (Sec. 2.1 d), **YBC 4698 ## 3. 4** (Sec. 2.1 e-f), **IM 121613 # 1** (Sec. 2.2 b), ***P. Mosc.* # 10** (Sec. 2.2 e), ***BM 86194* ## 5, 22** (Sec. 2.3 a), ***P.UC 32161*** (Sec. 2.3 b).

2.1. Themes in *P.Rhind*, a Hieratic Mathematical Papyrus Roll

2.1 a. *P.Rhind*, a hieratic mathematical recombination text

In the search for parallels between exercises in *P.Rhind* and Babylonian mathematical exercises, the following table of contents for *P.Rhind* will prove to be useful. In the table, an effort has been made to bring related problems together into paragraphs, even in the cases when the problems are *not* consecutive in the papyrus. Note that in the established numbering, the exercises in *P.Rhind* ## 33-34, for instance, have been numbered incorrectly precisely for the reason that they are intimately connected with ## 31-32. Actually, they appear in the papyrus after ## 37-38.

P.Rhind (BM 10057 + 10058): Contents.

obv.	Title page	
§ 1	$2/n$ table, for n odd, between 3 and 101	
§ 2	$n/10$ table, for n between 1 and 9	
§ 3	Applications of the $n/10$ table: divide n loaves between 10 men,	
	for $n = 1, 2, 6, 7, 8, 9$	## 1-6
§ 4 a	Multiplication problem: $a \cdot b = ?$,	
	for $a = 1\ 2'\ 4'\ (7/4)$ and $b = 4'\ \overline{28}\ (2/7)$ [14]	# 7
§ 5 a	Multiplication problem: $a \cdot b = ?$, for $a = 1\ 3"\ 3'\ (= 2)$ and $b = 4'\ \#\ 8$	
§ 4 b-h	Multiplication problems: $a \cdot b = ?$,	
	b) $2'\ \overline{14}\ (= 4/7)$, c) $4'\ \overline{28}\ (= 2/7)$, d) $\overline{7}\ (= 2/7)$, e) $\overline{14}\ (= 1/14)$	## 9-12
	f) $\overline{16}\ \overline{112}\ (= 1/14)$, g) $\overline{28}\ (= 1/28)$, h) $\overline{32}\ \overline{224}\ (= 1/28)$	## 13-15
§ 5 b-f	Multiplication problems: $a \cdot b = ?$,	
	for $a = 1\ 3"\ 3'\ (= 2)$ and $b =$ b) $2'$, c) $3'$, d) $6'$, e) $\overline{12}$, f) $\overline{24}$	## 16-20
blank space		
§ 6 a-b	Completion problems: Complete a to 1,	
	for a) $a = 3"\ \overline{15}\ (= 11/15)$, b) $a = 3"\ \overline{30}\ (= 21/30)$	## 21-22
§ 7	Completion problem: Complete a to $3"\ (= 30/45)$,	

---

14. As mentioned above, in OB cuneiform texts special cuneiform signs are used for the Babylonian basic fractions $3'\ (= 1/3)$, $2'\ (= 1/2)$, $3"\ (= 2/3)$, and $6"\ (= 5/6)$. In a similar way, special notations are used in hieratic mathematical papyri for the hieratic basic fractions $6'\ (= 1/6)$, $4'\ (= 1/4)$, $3'\ (= 1/3)$, $2'\ (= 1/2)$, and $3"\ (= 2/3)$. All other fractions, not counting fractions of measures, are written as "parts" (also called "unit fractions") with dots over the numbers. In order to distinguish basic fractions from parts, the parts will be transliterated in this paper as $\overline{5}\ (=$ the 5th part), $\overline{7}\ (=$ the 7th part), $\overline{8}\ (=$ the 8th part), *etc.*

	for $a = \overline{4}\,\overline{8}\,\overline{10}\,\overline{30}\,\overline{45}$ (= 23 2' 4' $\overline{8}$ / 45)	# 23

blank space

§ 8 a-c	Division problems: a) 1 $\overline{7}$ · a = 19, b) 1 2' · a = 16	## 24-25
	c) 1 4' · a = 15, 1 3' · a = 21	## 26-27
§ 9 a-b	Iterated division problems: a) (1–3') · 1 3" · a = 10,	
	b) 3' · 1 3' · 1 3" · a = 10 (incomplete)	## 28-29
§ 8 d-f	Division problems:	
	d) 3" $\overline{10}$ · a = 10, e) 1 3" 2' $\overline{7}$ · a = 33, f) 1 3' $\overline{4}$ · a = 2	## 30-32
§ 10 a-d	Applied division problems:	
	a) 3 3' · a = 1 *heqat*, b) 3 3' $\overline{5}$ · a = 1 (*heqat*)	## 35-36
	c) 3 3' (3' · 3') $\overline{9}$ · a = 1 *heqat*, d) 3 $\overline{7}$ · a = 1 *heqat*	## 37-38
§ 8 g-h	Division problems: g) 1 3" 3' $\overline{7}$ · a = 37, h) 1 2' 4' · a = 10	## "33-34"
§ 11 a	Unequal sharing: 100 loaves for 10 men;	
	6 small shares equal to 4 large shares	# 39
§ 12 a	Arithmetic progression: 100 loaves for 5 men	# 40

large blank space

§ 13 a-c	Round granary: content in khar and quadruple *heqat*	## 41-43|
§ 14 a-c	Square granary: content in *khar* and quadruple *heqat*	## 44-46|
§ 15	Table of fractions of a large capacity measure, 100 quadruple *heqat*	# 47|
§ 16	Areas of a square and its inscribed circle	# 48|
§ 17 a-d	Areas of four fields, a) a rectangle, b) a circle, c) a triangle	## 49-51|
	d) a cut-off triangle (trapezoid)	# 52|
§ 18 a	Applied division problem: a) one tenth of 7 *setat*	
	removed from 10 fields	# "54"|
§ 19	Three-striped triangle with an almost round area number	
	(incomplete exercise)	# 53 a|
§ 20	Rectangle with an almost round area number (incomplete exercise)	# 53 b|
§ 18 b	Applied division problem: one fifth of 3 *setat* removed from 5 fields	# 55|
§ 21 a-e	Inclination of a pyramid	## 56-59b|
§ 22	Inclination of a cone(?)	# 60

rev.

§ 23	Table and rule for computing 2/3 of a fraction	# 61
§ 24	Combined price for equal amounts of gold, silver, and lead;	
	relative prices: 12, 6, 3	# 62
§ 11 b	Unequal sharing: 700 loaves for 4 men in given proportions;	
	relative rations 3", 2', 3', 4'	# 63
§ 12 b	Arithmetic progression: 10 *heqat* of barley for 10 men;	
	common difference 8'$_h$ (1/8 *heqat*)	# 64
§ 11 c	Unequal sharing: 100 loaves for 10 men, 3 with double rations	# 65|
§ 25	Applied division problem: 10 *heqat* of fat for a year, share per day	# 66|
§ 26	Fraction of a herd of cattle (3" · 3' of 70 animals)	# 67|
§ 27	Equal sharing: 100 great quadruple *heqat* of grain,	

2.1. Themes in P.Rhind, a Hieratic Mathematical Papyrus Roll

	4 teams of 12, 8, 6, and 4 men	# 68|
§ 28 a-b	Baking number: a) 3 $2'_h$ *heqat* of flour, 80 loaves,	
	b) 7 $2'_h$ $4'_h$ $8'_h$ *heqat*, 100 loaves	## 69-70|
§ 29	Brewing number of beer, diluted with water	# 71|
§ 30 a-b	Two baking numbers, a) 100 loaves, b) 100 loaves	## 72-73|
§ 31	House inventory: the sum of a geometric progression	
	(houses, cats, mice, ···)	# "79"|
§ 30 c-d	Two baking numbers, c) 1000 loaves, d) 155 loaves	## 74-75|
§ 32	Three baking numbers, 1000 loaves	# 76|
§ 33 a-b	Baking and brewing numbers: loaves in exchange for beer	## 77-78|
§ 34 a-b	Tables for converting *heqat* fractions into *hinu*#	# 80-81|
§ 35 a-b	Feed for geese, daily and monthly rates	## 82-83
§ 36	Feed for oxen, monthly rate	# 84
large blank space; end of the mathematical part of the manuscript		
	"Enigmatic writing"	# 85
	Account (upside down, placed at the end of the papyrus)	# 86
	Diary entries	# 87

(The marks to the right of ## 41-55 and ## 65-81 are there as a reminder that the layout of the parts of the papyrus where these exercises occur is shown in Figs. 1.6.3-4 below.)

A coarser division of the problems into a variety of themes:

A	The $2/n$ table, the $n/10$ table, and applications of the $n/10$ table	§§ 1-3
B	Multiplication problems	§§ 4-5
C	Subtraction problems (completions)	§§ 6-7
D	Tables of fractions of capacity measures	§§ 15, 34
E	Division problems	§§ 8-10
F	Equal or unequal sharing problems	§§ 11-12, 27
G	Plane and solid geometry problems	§§ 13-14, 16-22
H	Baking and brewing numbers	§§ 28-30, 32-33
I	Combined price problem	§ 24
J	The sum of a geometric progression	§ 31
K	Feed for geese or oxen	§§ 35-36
L	Various other problems	§§ 23, 25-26

The first four of these themes (A-D) have no direct Babylonian parallels because they are exclusively about counting with either typically Egyptian sums of parts or fractions of Egyptian capacity measures. There are also no Babylonian parallels to the odd themes K-L. Theme J (§ 31) consists of the single problem # 79, discussed in Sec. 1.1 g above. OB parallels to exercises belonging to the remaining six themes (E-I) will be discussed below.

2.1 b. Theme E: division problems (*P.Rhind* ## 24-38)

It is not difficult to find Babylonian parallels to the division problems in § 8, but that does not mean much, in view of how simple most of those division problems are. Somewhat more interesting are the "iterated" division problems in *P.Rhind* § 9, the two problems ## 28-29. The text of *P.Rhind* is clearly corrupt here. The beginning of # 29 is missing, but the incomplete exercise is brought together with # 28 so that ## 28-29 together look like a complete exercise (Peet, *RMP* (1923), 63). (In the reproduction below of the hand copy in Chase, *et al.*, *RMP* (1929), red signs in the original are shown as grey. In the transliterations, red signs are shown as bold.)

The preserved part of **P.Rhind # 29** contains only the last part of the solution, and a verification. Here is a (mirror image) conform translation of the text of # 29:

```
1       10
4'      2 2'
10      1
sum 13 2'
3"      9 sum    22 2'   3"    20
3'      7 2' sum   30    3'    10
```

In the first three lines, the following product is computed:

1 4' 10 · 10 = 10 + 2 2' + 1 = 13 2'.

This is last part of the *solution* to the lost question. *Verification*:

13 2' + 3" of 13 2' = 13 2' + 9 = 22 2', 22 2' + 3' of 22 2' = 22 2' + 7 2' = 30, 3' of 30 = 2' of 3" of 30 = 2' of 20 = 10.

(For some reason 1/3 of a number was normally computed in hieratic mathematical texts as 1/2 of 2/3 of that number.)

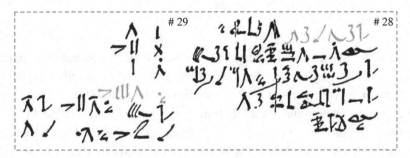

Fig. 2.1.1. *P.Rhind* ## 28-29. Two incomplete exercises written closely together.

The verification in # 29 allows the following reconstruction of the lost question:

> 3" of a quantity is added to the quantity, 3' of the sum is added to the sum, 3' of that is 10.
> What is the quantity?

In modern terms:

$a \cdot 1\,3" = b,\ \ b \cdot 1\,3' = c,\ \ c \cdot 3' = 10,\ \ a = ?$

The way in which the solution was obtained can be explained as follows:

> [Suppose the quantity is 27 (= 3 · 3 · 3).
> Then 27 + 3" · 27 = 27 + 18 = 45, 45 + 3' · 45 = 45 + 15 0 60, 60 · 3' = 20.
> Now, 20 goes 1 4' $\overline{10}$ times into 27.
> Therefore, if the quantity is 1 4' $\overline{10}$ the result will be 1.]
> Hence, if the quantity is 1 4' $\overline{10}$ times 10 = 10 + 2 2' + 1 = 13 2',
> then the result will be 10, as required.

This is, of course, an application of the rule of false value, often used in OB mathematical texts. (See. Friberg, *RlA 7* (1990) Sec. 5.7 d.)

The verification in # 29 can be explained like this:

> According to the computation, the quantity is 13 2' = a.
> Therefore, a + 3" · a = 13 2' + 3"· 13 2' = 13 2' + 9 = 22 2' = b,
> so that b + 3' · b = 22 2' + 3' · 22 2' = 22 2' + 7 2' = 30 = c,
> and finally c · 3' = 30 · 3' = 10, as required.

Here is a translation of **# 28**:

> **3" is what goes in, 3' is what goes out**, 10 remains. /
> Make $\overline{10}$ of this 10, it becomes 1. The remainder is 9. /
> 3" of it, 6, is going into it, sum 15.
> 3' of it is 5. /
> See, 5 is going out, the remainder is 10. /
> The doing as it becomes.

Line 1 in # 28 contains the question, which is a concise way of stating that if 2/3 of a quantity is added to the quantity, and if 1/3 of the sum is subtracted from the sum, then 10 remains. In modern terms:

$a \cdot 1\,3" = b,\ \ b \cdot (1 - 3') = 10,\ \ a = ?$

Line 2 contains a verbal description of the solution, namely that the unknown quantity is (1 − 1/10) times the given remainder 10, and lines 3-4 contain the verification. The solution method can be explained (as in the case of # 29) as an application of the rule of false:

[Suppose the quantity is 9 (= 3 · 3).
Then   9 + 3" · 9 = 9 + 6 = 15,   and   15 − 3' · 15 = 15 − 5 = 10.
Now, 10 goes 1 − $\overline{10}$ times into 9.
Therefore, if the quantity is 1 − $\overline{10}$, the result will be 1.]
Hence, if the quantity is 1 − $\overline{10}$ times 10 = 10 − 1 = 9,
then the result will be 10, as required.

(Note that, apparently, in this exercise 9 divided by 10 is not represented by 3" $\overline{5}$ $\overline{30}$, a sum of parts, as in the $\overline{10}$ · *n* table in *P.Rhind* and in *P.Rhind* # 6, but by 1 − $\overline{10}$, a difference!)

The verification in # 28 can be explained like the verification of # 29:

According to the computation, the quantity is 9 = *a*.
Therefore,   *a* + 3" · *a* = 9 + 3" · 9 = 9 + 6 = 15 = *b*,
so that   *b* − 3' · *b* = 15 − 3' · 15 = 15 − 5 = 10, as required.

A parallel to *P.Rhind* ## 28-29 is the OB text **YBC 4652** (Neugebauer and Sachs, *MCT* (1945),100; Melville, *HM* 29 (2002)), a cuneiform text belonging to the Goetze/Høyrup/Friberg Group 2 a, and therefore possibly from Ur and contemporaneous with the mathematical texts from Mari. (See Friberg, *RA* 94 (2000), 162, 173.) YBC 4652 is a well organized "theme text", the theme being 21 iterated division exercises. Here is an example:

YBC 4652 # 9

1	na₄ ì.pà ki.lá nu.na.tag igi.7.gál ba.zi igi.11.gál bí.daḫ /
2	[igi 1]3.gál ba.zi ì.lá 1 ma.na sag na₄ en.nam /
3	1 ma.na 9 2' gín 2 2' še

A (weight) stone I found, it was not marked.
A 7th-part I tore off, an 11th-part I joined,
a 13th-*part* I tore off. I weighed it: 1 mina.
The original stone was what?
1 mina 9 1/2 shekels 2 1/2 barley-corns.

The question here can be reformulated, in modern terms, as:

$a \cdot (1 - 1/13) = b$, $b \cdot (1 + 1/11) = c$, $c \cdot (1 - 1/7) = 1$ mina, $a = ?$

The answer is given, but not the solution procedure. It would almost certainly have started, as in other similar situations in OB mathematical texts, by assuming a suitable false value:

2.1. Themes in P.Rhind, a Hieratic Mathematical Papyrus Roll   33

Suppose that $a = 7 \cdot 11 \cdot 13 = 16\ 41\ (1{,}001)$.
Then $a - 1/13) \cdot a = 16\ 41 - 1\ 17 = 15\ 24\ (= 7 \cdot 11 \cdot 12) = b$,
so that $b + 1/11 \cdot b = 15\ 24 + 1\ 24 = 16\ 48 = (7 \cdot 12 \cdot 12) = c$,
and finally $c - 1/7 \cdot c = 16\ 48 - 2\ 24 = 14\ 24 = (6 \cdot 12 \cdot 12)$.
The reciprocal of 14 24 is ;00 04 10 (= ;05 · ;05 · ;10).
Therefore, if the wanted result is 1, then the corrected value of a is
$a = ;00\ 04\ 10 \cdot 16\ 41 = 1;09\ 30\ 50$.
Since the wanted result is, actually, 1 mina,
the true value of a is $a = 1;09\ 30\ 50$ mina = 1 mina 9 2' shekels 2 2' barley-corns.
(Remember that 1 barley-corn = ;00 20 shekel.)

The problem type was still around a millennium and a half later, as shown by the Late Babylonian/Seleucid fragment **BM 34800** (Sachs, *LBAT* (1955) # 1647).

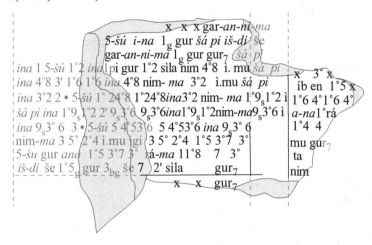

Fig. 2.1.2. BM 34800. A Late Babylonian/Seleucid repeated division problem.

Most of the question is lost in this text, as well as a large part of the solution procedure. Therefore, no interpretation has been suggested before. However, there seems to be references to various parts of the question in the solution procedure, all preceded by the phrase *šá pi* 'as instructed'. These repeated references make it possibly to reconstruct the question. Apparently, it was formulated more or less as follows:

A granary. Its barley I don't know.
I removed a 5th of it, then I removed a 3rd of it, then I removed two 5ths of it,
then I removed half of it, then I removed three 5ths of it, and 1 gur remained.
What was the original barley?

34 *Unexpected Links Between Egyptian and Babylonian Mathematics*

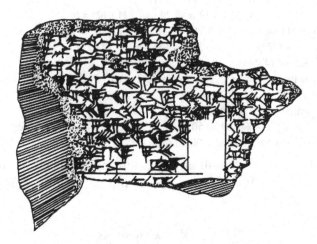

Fig. 2.1.3. BM 34800. Hand copy of the fragment..

BM 34800

1'	[⋯ ⋯ ⋯ ⋯] x x x gar-*an-ni-ma* /
2'	[⋯ ⋯] 5-*šú ina* 1$_g$ gur *šá pi* *iš-di* [*še* ⋯] /
3'	[⋯ ⋯] gar-*an-ni-ma* 1$_g$ gur gur$_7$
4'	[*šá pi*] / [*ina* 1 5-*šú* 12] [*ina* 1 *p*]*i* gur 12 sìla nim 48 ì.mu
5'	[*šá pi*] / [*ina* 48 3' 16] [16 *ina*] 48 nim-*ma* 32 ì.mu
6'	*šá pi* / [*ina* 32 2 • 5-*šú* 12 48] 12 48 *ina* 32 nim-*ma* 19 12 ì<.mu> /
7'	[*šá pi ina* 19 12 2' 9 36] 9 36 *ina* 19 12 nim-*ma* 9 36 ì<.mu> /
8'	[<*šá pi*> *ina* 9 36 3 • 5-*šú* 5 45 36]
9'	5 45 36 *ina* 9 36 / [nim-*ma* 3 50 24 ì.mu] [igi] 3 50 24 15 37 30 /
10'	[5 *šá* 1$_g$ gur *ana* 15 37 30 r]á-*ma* 1 18 07 30 / [*iš-di* še 15$_g$ gur 3$_{bg}$] 7 2' sìla gur$_7$

⋯ ⋯ x x x you set it for me, then
⋯ ⋯ its 5th from 1 gur, as instructed.
The base of the *barley* ⋯
⋯ you set it for me, then 1 gur, the granary.
As instructed, from 1 its 5th, 12.

2.1. Themes in P.Rhind, a Hieratic Mathematical Papyrus Roll

*From 1 ba*rig 12 sìla lift, 48 it gives.
As instructed, from 48 the third, 16.
16 from 48 lift, 32 it gives.
As instructed, *from 32, 2 · its 5th, 12 48.*
12 48 from 32 lift, then 19 12 it <gives>.
As instructed, from 19 12 the half, 9 36.
9 36 from 19 12 lift, then 9 36 it <gives>.
<As instructed,> from 9 36, 3 · its 5th, 5 45 36.
5 45 36 from 9 36 *lift, then 3 50 24 it gives.*
The opposite of 3 50 24 (is) 15 37 30.
5 of 1 gur to 15 37 30 go, then 1 18 07 30.
The base of the barley, 15 gur 3(ba rig) 7 1/2 sìla,
the granary.

In modern notations, the question can be rephrased as follows:

$a \cdot (1 - 1/5) = b$, $b \cdot (1 - 3') = c$, $c \cdot (1 - 2/5) = d$, $d \cdot (1 - 1/2) = e$,
$e \cdot (1 - 3/5) = 1\ 00$ gur $= 5\ 00$ sìla. $a = ?$

(Unfortunately, it is not clear how the fractions 1/3, 1/5, 2/5, and 3/5 were expressed in this text.)

Just as in *P.Rhind* ## 28-29, the solution procedure in BM 34800 makes use of the rule of false value, in (essentially) the following way:

Suppose the original barley was 1 barig (= 1 00 sìla).
Then a 5th less is 1 00 s. − 12 s. = 48 <s.> (= 4/5 · 1 00).
A 3rd less is 48 − 16 = 32 (= 4/5 · 2/3 · 1 00).
Two 5ths less is 32 − 12;48 = 19;12 (= 4/5 · 2/3 · 3/5 · 1 00).
A half less is 19;12 − 9;36 = 9;36 (= 4/5 · 2/3 · 3/5 · 1/2 · 1 00).
Three 5ths less is 9;36 − 5;45 36 = 3;50 24 (= 4/5 · 2/3 · 3/5 · 1/2 · 2/5 · 1 00).
The reciprocal of 3;50 24 is ;15 37 30 (= 5/4 · 3/2 · 5/3 · 2 · 5/2 · ;01).
Hence, if the original barley were 15;37 30 sìla, the final barley would be 1 sìla.
However, the final barley should be 1 gur = 5 00 sìla.
Therefore, the original barley in the granary was
5 00 · 15;37 30 sìla = 1 18 07;30 sìla = 15 gur 3 barig 7 1/2 sìla.

The applied division problems in **P.Rhind § 10** (## 35-38) are also interesting. The first one, # 35, for instance, begins with the question

I went down into the *heqat* 3 times, 3' of me to me, and I was full. Who says it?

This is an eccentric way of asking for the size of an unknown measuring vessel if it is known that 1 *heqat* is 3 1/3 times larger. The answer is computed in # 35 in three different ways, first in terms of ordinary sums of parts (abstract numbers), then in terms of multiples of the *ro* = 1/320 of 1 *heqat*, and finally in terms of *binary fractions of 1 heqat*, for which there existed special hieratic signs, and for which it may be useful to employ the following special notations: $2'_h$, $4'_h$, $8'_h$, $16'_h$, $32'_h$, $64'_h$ (= 5 ro). [15]

The computation, employing the rule of false value, can be explained as follows:

> Suppose the unknown size of the vessel is 1. Then 3 3' times that is 3 3'.
> Now, 3 3' goes $\overline{5}\ \overline{10}$ times in 1.
> (In modern terms, the reciprocal of 3 3' = 10/3 is 3/10 = 1/5 1/10.)
> Therefore, the vessel contains $\overline{5}\ \overline{10}$ (of a *heqat*).
> This is the same as $\overline{5}\ \overline{10} \cdot 320\ ro = 96\ ro$.
> Finally, since $64'_h = 5\ ro$, $32'_h = 10\ ro$, etc.,
> 96 ro = (80 + 10 + 5 + 1) ro = $4'_h$ $32'_h$ $64'_h$ 1 ro.

In the third of the applied division problems, *P.Rhind* # 37, the *heqat* is 3 3' (3' · 3') 1/9 times larger than the unknown vessel. According to Høyrup, *LWS* (2002), 321, this is the only occurrence of an "ascending continued fraction" in the rich Egyptian record. That is one of several reasons why he concludes that there must be a factual relation between *P.Rhind* # 37 and the OB mathematical exercise **IM 53957** (Baqir, *Sumer* 7 (1951), 37), where the question is phrased as follows:

> If [someone] asks (you) thus:
> To 3" of my 3" I appended 100 sìla and my 3".
> 1 gur was completed.
> The *tallum*-vessel of my grain is corresponding to what?

2.1 c. Theme F: sharing problems (*P.Rhind* ## 39-40, 63-65, 68)

It is not difficult to find OB parallels to the problems in § 11 and § 27 of *P.Rhind*, but that does not mean much in view of the simple nature of the problems occurring there. The problems about sharing in arithmetic progression in *P.Rhind* § 12 (## 40 and 64) are more interesting.

The question in ***P.Rhind* # 64** is

> Method of dividing the excesses. If it is said to you: 10_h *heqat* of barley for 10 men, the excess of each man to his second, in barley, is $8'_h$.

---

15. These hieratic signs for the binary *heqat* fractions have traditionally been interpreted as the hieratic versions of the originally hieroglyphic "Horus eye fractions". That this interpretation is false is convincingly demonstrated in the fascinating essay Ritter, *UOS* (2002).

2.1. Themes in P.Rhind, a Hieratic Mathematical Papyrus Roll

The usual sign for '1' is used here as a sign for 10 *heqat* (10_h)! Later on in this exercise, a dot (1_h) is used as a sign for 1 *heqat*, and the mentioned special number signs denote binary fractions of the *heqat*. (See Gardiner, *EG* (1927) § 266, for a survey of Egyptian notations for capacity measures.) What the obscurely phrased question means is that 10 men share 10 *heqat* of barley in an arithmetic progression, with the constant difference between the shares equal to 1/8 *heqat* ($8'_h$).

The solution proceeds (essentially) as follows:

> The average share is $10_h/10 = 1_h$. The largest share is equal to the average share plus 9 times half the difference, that is $1_h + 9 \cdot 16'_h = 1_h \, 2'_h \, 16'_h$. Hence, the 10 shares are, in decreasing order, $1_h \, 2'_h 16'_h$, $1_h \, 4'_h \, 8'_h \, 16'_h$, *etc.*

The shares are ordered with the biggest share first. This is also the way in which shares in an arithmetic progression normally are ordered in OB mathematical texts. (See, for instance, **MS 1844** in Fig. 1.1.6 above.)

An OB (imperfect) parallel to the arithmetic progression exercise in *P.Rhind* # 64 is exercise # 1 in **Str. 362** (Neugebauer, *MKT 1* (1935), 239) a small cuneiform text of mixed content, (probably) from Uruk. In Str. 362 # 1 the question is formulated as follows:

> 10 brothers, 1 3" mina of silver. Brother over brother is always going above, how much he is always going above I do not know. The share of the 8th brother is 6 shekels. Brother over brother, how much is he always going above?

The solution procedure is (essentially) the following:

> The average share is 1 40 / 10 = 10 (shekels). Double the average share minus double the 8th share is 20 – 12 = 8. Hence, the common difference is 8/5 = 1;36 (shekel).

The solution procedure is obviously based on the observation that the difference between the average share and the 8th share is 5 times half the common difference.

The question in *P.Rhind* § 12 a (# 40) is phrased as follows:

> 100 loaves for 5 men. 7̄ of the 3 above to the 2 men below.
> What is the difference (between the shares)?

The solution procedure is clearly corrupt, with the initial part missing. It begins by saying that the difference is 5 2', and then immediately assumes that the five shares are 23, 17 2', 12, 6 2', and 1. The sum of these assumed shares is found to be 60, instead of the required 100. The necessary correction factor is obviously equal to 100/60 = 1 3". Consequently, the true value of the first share is 23 + 3" · 23 = 38 3', the second share is 17 2' + 3" · 17 2' = 29 6', and so on.

The preserved part of the solution is a straightforward application of the rule of false value. The missing argument can probably be reconstructed in (essentially) the following form. (The alternative reconstructions suggested in Chase, *et al.*, *RMP 1* (1927), 102, Gillings, *MTP* (1982), 170, and Couchoud, *ME* (1993), 157, are not convincing.)

> Assume that the smallest share is 1 (loaf).
> Then the sum of the three largest shares is 3 plus 9 differences, while 7 times the sum of the two smallest shares is 14 plus 7 differences. If the two sums are equal, then 11 is equal to 2 differences.
> Therefore, the (common) difference is 5 2'. The largest share is then 1 plus 4 · 5 2' = 23, the next one is 23 − 5 2' = 17 2', etc.

Two OB parallels to the arithmetic progression exercise in *P.Rhind* # 40 are **YBC 9856** (Neugebauer and Sachs, *MCT* (1945), 99) and **VAT 8522 # 2** (Neugebauer, *MKT 1* (1935), 368). These difficult texts were first adequately explained in Muroi, *HSci* 34 (1988).

YBC 9856 consists of a brief and obscurely phrased question followed by an answer but no solution procedure. Here is a tentative translation:

> 1 mina of silver, his 5 brothers. Two-thirds the small brothers.
> May brother constantly be above brother.
> 1 4, 2 8, 3 12, 4 16, 5 20.

Apparently what this means is that five shares form an arithmetic progression with the sum 1 mina. In addition, the shares of the four younger brothers is 2/3 of all five shares. The missing solution procedure can be reconstructed in (essentially) the following form:

> Assume that the smallest share is 1. Then the sum of all the five shares is 5 plus 10 differences, while 1 1/2 times the sum of the four smallest shares is 6 plus 9 differences. If the two sums are equal, then 1 is equal to 1 difference.
> Therefore, if smallest share is 1, then the five shares are 1, 2, 3, 4, 5, and, the sum is 15, instead of the prescribed 1 mina. Hence, the correction factor is 4 shekels, and the five shares are 4, 8, 12, 16, and 20 shekels.

VAT 8522 # 2 consists of a partly preserved text, followed by a badly organized scribbled solution procedure. Here is a tentative translation of the text with the missing parts reconstructed:

> 5 brothers, 6" [mina 9 shekels of silver]. Half of [the big brother the small brother

took], and 4 30 (a strange error for 45) the 5 brothers divided. A third-part of what the big brother over the next brother was above may brother over brother be constantly above. How much silver did they take?

The scribbled solution suggests that the problem was solved in (essentially) the following way:

> Of the equally divided 45 (shekels), each brother receives 9 shekels. The remaining parts of the shares have to be computed so that the following conditions are satisfied:
> 1) the constant difference between the four younger brothers is 1/3 of the difference between the two largest shares,
> 2) the smallest share is one half of the largest share,
> 3) the sum of the five shares is 59 − 45 = 14 shekels of silver.
> These three conditions are taken care of, one at a time, in the following three steps of the computation:
> 1) Assume that the smallest share is 1 and that the constant difference is 1. Then the shares are 5, 4, 3, 2, 1. However, since the first difference should be 3 times bigger, the shares must be, instead, 7, 4, 3, 2, 1.
> 2) Since the largest share should be only twice the smallest share, let the smallest share be $1 + a$ instead of 1. Then the largest share is $7 + a$ and must be equal to $2 \cdot (1 + a)$. Hence, $a = 5$, and the five shares are 12, 9, 8, 7, and 6.
> 3) The sum of 12, 9, 8, 7, and 6 is 42 instead of the prescribed 14 shekels. Hence, the correction factor is ;20 shekel, and the five shares are 4, 3, 2;40, 2;20, and 2 shekels. Add to this the equal shares of 9 shekels, and the total shares of the 5 brothers are 13, 12, 11;40, 11;20, and 11 shekels.

The problem type continued to be popular. The next time it surfaces is in the Chinese mathematical classic *Jiu Zhang Suan Shu* (Nine Chapters on Mathematics), which was written at some time towards the end of the first millennium BCE. The text of ***JZSS* 6:18** (Shen, Crossley, and Lun, *NCMA* (1999), 333-336) can be translated as follows:

> Now given 5 persons are to share 5 coins.
> Let the sum of the two greater (shares) be equal that of the three lesser.
> Tell: How much does each get?
> Answer: *A* gets 1 2/6 coins; *B*, 1 1/6 coins; *C*, 1 coin; *D*, 5/6 coin; and *E*, 4/6 coin.

The solution procedure is similar to the one used in *P.Rhind* # 40 and in VAT 8522 # 2:

> Method: Lay down the cone-shaped rates (an arithmetic progression) for the distribution. The sum of the two greater is 9, while that of the three lesser is 6. 6 is less than 9 (by) 3. Add 3 to each of the rates for the shares. Take the sum as divisor. Multiply the coins to be shared by each of the shares for each dividend. Divide, giving the coins required.

2.1 d. Theme G: geometry problems (*P.Rhind* ## 41-46, 48-60)

The areas of four basic plane geometric figures are computed in *P.Rhind* § 17 (## 49-52). The linear dimensions of the figures are given in multiples of 1 *khet* = 100 cubits. This means that the geometric figures were thought of as actual *fields*, not as tiny figures drawn on papyrus. In accordance with this interpretation of geometric figures, the word used for the size of the surface of a plane figure was ꜣḥ.t 'field'.

The situation is similar in OB mathematical texts, where the basic plane geometric figures normally have linear dimensions in tens or sixties of 1 ninda = 12 cubits (= appr. 6 meters), and where the word used for the size of the surface of a figure is aša$_5$ (or gán) 'field'

Areas in exercises in *P.Rhind* are normally expressed as multiples of 1 *setat* = 1 square *khet*. Other units of area measure that are used in computations of areas are the 'cubit strip' = 1 cubit · 1 *khet*, and the 'thousand-cubit-strip' = 1,000 cubits · 1 *khet* = 10 *setat*.

Take, for instance, *P.Rhind* § 17 c (# 51), where the area of a triangle of length (or long side) 10 *khet* and width (or short side) 4 *khet* is computed (essentially) as follows. (The literal meanings of the words used for 'triangle', 'length', and 'width' are 'sharp', 'quay', and 'mouth')

1/2 · 400 = 200	(half the width	in cubits)
1,000	(the length	in cubits)
2,000	(the length times half the width	in cubit-strips)
2	(the area	in thousand-cubit-strips).

Instead of interpreting the result '2' as 2 thousand-cubit-strips, one may interpret it as '20 *setat*', written with a special sign 20$_s$. There are, as a matter of fact, special signs for 1 through 9 *setat*. It is convenient to use special notations also in transliterations of those signs, say 1$_s$, 2$_s$, ⋯ , 9$_s$.

A major difference between hieratic and OB mathematics is that they use different methods to compute the area of a circle. The *hieratic* "circle area rule" is explicitly exhibited in ***P.Rhind*** § 16 (# 48). The drawing accompanying this text shows a crudely drawn circle (looks like an octagon), with its circumscribed square. (The copy of the hieratic text in Fig. 2.1.4 below is borrowed from Chase, *et al.*, *RMP 2* (1929).) The diameter of the circle is given as 9 (*khet*), as indicated by the number 9 recorded inside the circle, along a horizontal diameter. The text consists of two computations, one showing that 8 · (8 *setat*) = 64 *setat*, the other that 9 · (9 *setat*) = 81 *setat*. Clearly, what this means is that a circle of diameter 9 *khet* has (approximately) the area 64 *setat*, while a square of side 9 *khet* has the area 91 *setat*.

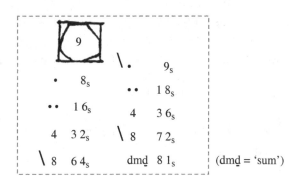

Fig. 2.1.4. *P.Rhind* # 48. The areas of a circle and its circumscribed square.

The reasoning behind this Egyptian approximate value for the area of a circle is not known. The tentative explanation illustrated in Fig. 2.1.5 below is an elaboration by Gillings, *MTP* (1982), 144, of an idea due to Neugebauer, *QSB 1* (1931), 429. The figure makes it clear that the area of a circle of diameter 9, say, is approximately equal to the area of a square of side 9 minus the area of 4 half-squares of side 3 (= the combined area of 2 rectangles with the sides 1 and 9), which again is approximately as much as the area of a square of side 8.

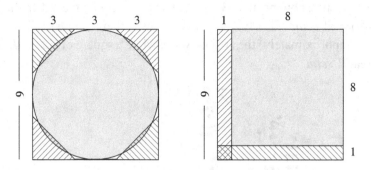

Fig. 2.1.5. Possibly the naive geometric derivation of the Egyptian circle area rule.

The possibility cannot be excluded that this is the way in which the Egyptians found their area rule. It may not be by accident that the circle in *P.Rhind* # 48 looks like an octagon and has the diameter 9!

***P.Rhind* § 17 b (# 50)** is illustrated by a drawing of a circle, inside of which are written the signs for 9 *khet*. The text of the exercise is the Egyptian circle area rule in explicit verbal form:

Example of making a round field of 9 *khet*. What is the amount of it as field?
You take away $\overline{9}$ of it, 1, the remainder is 8.
You make the multiplication, 8 times 8, it becomes 64.
The amount of it, as field, is 60 4_s.

In quasi-modern notations, this means that the area of a circle of diameter *d* is given by the equation

$A = $ sq. $(d - d/9)$.

(The abbreviation sq. stands, here and in the following, for 'the square of'. It would be too anachronistic to write $(d - d/9)^2$.)

In an OB table of constants the following constants for *circles* are mentioned:

5 igi.gub	šà gúr	5, the *constant*	of a circle	BR 2
20 dal	šà gúr	20, the transversal	of a circle	BR 3
10 [*pi*]-*ir-ku*	šà gúr	10, the cross-line	of a circle	BR 4

(The following notations are used here and in the following for cited OB tables of constants: **BR** = ***TMS*** **3**, Bruins and Rutten, *TMS* (1961), text 3, **NSd** = **YBC 5022**, Neugebauer and Sachs, *MCT* (1945), text Ud, **NSe** = **YBC 7243**, Neugebauer and Sachs, *MCT* (1945), text Ue.)

2.1. Themes in P.Rhind, a Hieratic Mathematical Papyrus Roll

What this means is, as shown by various applications of the constants, that if A is the area of the circle, a the arc (circumference), d the diameter, and p the radius orthogonal to the diameter, then

$A = ;05 \cdot$ sq. a, $d = ;20 \cdot a$, $p = ;10 \cdot a$.

In other words, $A = 1/12 \cdot$ sq. a, $d = 1/3 \cdot a$, and $p = 1/6 \cdot a$. Thus, the constants ;05, ;20, and ;10 are somewhat crude, but very convenient, approximations to what we would call $1/4\pi$, $1/\pi$, and $1/2\pi$.

There are similar constants for *semicircles* in OB tables of constants. (See, for instance, Muroi, *SK* 143 (1994).)

15	gán [u₄.sakar]	15	crescent-field	NSe 2
10	gán u₄.[sakar ki.2]	10	crescent-field #2	NSe 3
45	ša gán u₄.sakar ki.3	45	of a crescent-field #3	NSd 54
40 dal	šà ús-ka₄-ri	40	the transversal of a crescent	BR 8
20 pi-ir-ku	šà ús-ka₄-ri	20	the cross-line of a crescent	BR 9

What this means is that if a is the arc (the semi-circumference) of a semicircle, then there are three ways of computing the area, namely as

$A = 1/4 \cdot a \cdot d$, or $A = 1/6 \cdot$ sq. a, or $A = 3/4 \cdot d \cdot p$.

Moreover, the diameter and the radius orthogonal to the diameter are

$d = 2/3 \cdot a$ and $p = 1/3 \cdot a$.

It is not known, of course, how OB "mathematicians" found these constants and how they demonstrated their validity to their students. It is likely, however, that they used naive geometric demonstrations of the kind shown in Fig. 1.3.5 below (in the case of the semicircle):

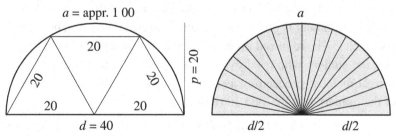

The circumference of a semicircle approximated by half the circumference of a (regular) '6-side'.

The area of a semicircle approximated by the area of half a (regular) polygon with many equal sides.

Fig. 2.1.6. Possible naive derivations of the OB constants for semicircles.

The Egyptian circle area rule is used in three problems in of **P.Rhind** § 13 (## 41-43) dealing with the contents of cylindrical granaries. In *P.Rhind* # 41, the diameter of the granary is 9, and the height is 10 (cubits). The volume is computed as $V = \{\text{sq. }(9 - 1/9 \cdot 9)\} \cdot 10 = 640$ (cubic cubits). Then this result is multiplied by 1 1/2, so that the volume is expressed instead in the volume unit *khar*:

$V = 640$ cubic cubits $\cdot$ 1 1/2 *khar* / cubic cubit = 960 *khar*.

Finally, the volume is divided by 20, so that the content of the granary becomes expressed in capacity measure:

$C = 960$ *khar* $\cdot$ 1/20 of a 100-quadruple-*heqat* / *khar* = 48 100-quadruple-*heqats*.

(It is clear from this computation that 1 *heqat* is equal to 1/30 of a cubic cubit. With 1 cubit = appr. 52 cm., it follows that 1 *heqat* = appr. 4.7 liters.)

In *P.Rhind* # 42, both the diameter and the height are 10 (cubits), which makes the computation messier. In *P.Rhind* # 43, the diameter is 6 and the height 9 (cubits). The solution method in this example differs from the straightforward method in the two preceding examples. Unfortunately, the text is corrupt, but a parallel text discussed below (*P.UC 32160 = Kahun IV.3 # 1*; Fig. 2.3.3) shows what should have been written in # 43.

Three problems in **P.Rhind** § 14 (## 44-46) are concerned with rectangular granaries. In # 44, first the volume of a granary with length, width, and height all equal to 10 (cubits) is computed, and then the content is computed in 100-quadruple-*heqats*. A naive drawing of the granary shows only a rectangle with three numbers 10, one at one side, one on top, and one inside. In # 45, the problem is reversed (# 46 is similar, with different numbers). The content is given, and the question asked is "Of it, how much by how much?" The reversed problem should rightly lead to a cubic equation, but the student who tried to solve this problem preferred a simpler, less correct method, proceeding as if he knew already that the square bottom of the granary had the side 10 (cubits).

The four exercises in *P.Rhind* § 17 (## 49-52), concerned with the areas of plane geometric figures, are followed by **P.Rhind** § 18 (## 54-55), two exercises about subtracting equal pieces of land with a given total area from a given number of fields. This requires the conversion of ordinary fractions into the special *setat* fractions. Thus, in # 54 one tenth of 7 *setat*

2.1. Themes in P.Rhind, a Hieratic Mathematical Papyrus Roll

is converted to *setat* fractions and multiples of the cubit-strip = 1/100 *setat* as follows:

$\overline{10} \cdot 7 = 2' \overline{5}$,
$2' \overline{5} \cdot 1$ setat $= 2'_s \, 8'_s \, mh \, 7 \, 2'(1/2 + 1/5 = 70/100 = 1/2 + 1/8 + 7 \, 1/2 \,/100)$

P.Rhind # 54 is very short and leaves place for a second exercise immediately to the left of it on the papyrus. This is the puzzling exercise *P.Rhind* # 53, or, more correctly, **P.Rhind # 53 a-b**, which is illustrated by a drawing of a triangle divided into three stripes by two transversals parallel to the short side of the triangle. (See the layout of a section of *P.Rhind* in Fig. 2.4.3 below.)

Three different stumbling blocks have thwarted all earlier attempts to understand what is going on in this exercise. (Only Robins and Shute explicitly admit (*RMP* (1987), 477) that they do not understand what is going on, with the words "No. 53 is also concerned with triangles but unfortunately, through faulty copying, is now incomprehensible.")

First, it has not been observed that the two columns of computations directly under the drawing of the striped triangle belong to a different problem (here called **# 53 b**, while the problem associated with the drawing will be called **# 53 a**).

Secondly, it has not been clear that six dashes along two of the parallel lines in the triangle are not number notations. (See the drawing in Neugebauer, *QSB 1* (1931), 449, and Neugebauer's commentary (*ibid.*), 421, that "Daß die Figur alle Teilgebiete durch ein Dreieck umschließt, paßt offenbar nicht zu den angegebenen Maßen (z. B. Basis = erste Querlinie = 6)". Chase, *et al.* write (*RMP* (1927), 94): "In his figure the author puts down 6 as the length of both of these lines, but in his calculations he seems to take 4 1/2 for the base." Similarly, Couchoud writes (*ME* (1993), 60), with regard to the middle part of the triangle, that "il s'agit d'un trapèze ayant une base de 6 et un sommet de 2 1/4" and makes a copy of the drawing where two of the parallel sides are indicated as having the common length 6. Most recently, Clagett, *AES 3* (1999), 382-383, compounded the mistake, misinterpreting 3 1/4 and 5 as the heights of two sections of the striped triangle, and changing the drawing of # 53 a into a strange drawing of a figure composed of a triangle, a trapezoid, and a rectangle.) Instead, *the dashes along the two parallel lines probably indicate that the lengths of these lines are unknown.*

Finally, insufficient attention has been paid to the fact that in Egyptian as well as in Babylonian mathematical texts, the values of number signs are *context dependent*, and that numbers recorded in the interior of some part of a geometric figure usually denote the area of that part of the figure. Thus, the numbers written inside the two trapezoidal parts of the triangle are to be understood as the area numbers 50_s (50 *setat* or 5 thousand-cubit strips) and $30_s\,4'_s$ (30 1/4 *setat*), not as the ordinary numbers 5 and 3 1/4! (Cf. *P.Rhind* # 48 (Fig. 1.3.3 above), where, for instance, 64 *setat* is written as 6 followed by a special sign 4_s for 4 *setat*.)

With these clarifications, it is possible to show that the exercise is a cleverly composed geometric-arithmetic problem, and that the data were chosen by the author of the problem, with some difficulty, so that the three partial areas, the area numbers 50 *setat*, 30 1/4 *setat*, and 7 1/2 1/4 1/8 *setat*, would be close to *round* area numbers, namely 50, 30, and 8 *setat*.

(In the reproduction below of the hand copy in Chase, *et al.*, *RMP* (1929), red signs in the original are shown as grey.)

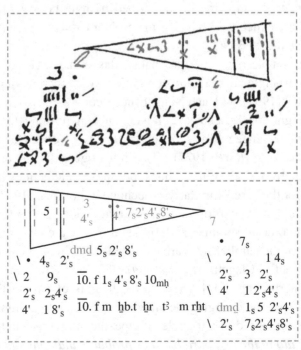

Fig. 2.1.7. *P.Rhind* # 53 a-b. A combination of two unrelated and incomplete exercises.

2.1. Themes in P.Rhind, a Hieratic Mathematical Papyrus Roll 47

The only preserved part of the text of the exercise illustrated by the drawing of the striped triangle is the following computation recorded near the tip of the triangle:

	1	7_s
\	2	$1\ 4_s$
	$2'_s$	$3\ 2'_s$
\	$4'_s$	$1\ 2'_s\ 4'_s$
	sum	$1_s\ 5\ 2'_s 4'_s$ setat
	$2'_s$	$7_s\ 2'_s\ 4'_s\ 8'_s$ setat

There are several errors of notation in this brief text, in particular that 15 *setat* is written with a curved line (meaning *setat*) over the digit '1'instead of over the digit '5'. Nevertheless, it is clear that what is computed is 1/2 · 2 4' (*khet*) · 7 (*khet*) =1/2 · 15 2' 4' *setat* = 7 2' 4' 8' *setat*. This is the area of the small triangle near the tip of the triangle. The computed area, the length 7, and the width 2 4' of the small triangle are all written as red digits in the drawing (reproduced as grey in Fig. 2.1.7), the sign for 7 a bit off-center. Note that, as could be expected, the whole triangle must be sized as a fairly large field, with sides measured in hundreds of cubits. Note also that, precisely as in OB mathematical texts, what looks in the drawing like the length 7 of the *side* of the small triangle must be understood instead as the length of the *height* of that triangle.

The interpretation suggested here is that *P.Rhind* # 53 a is an assignment, where the student was given the drawing of the triangle together with four given numbers: the two sub-areas $A_1 = 50$ *setat* and $A_2 = 30\ 4'$ *setat*, the second transversal $d_2 = 2\ 4'$ (*khet*), and the length of the small triangle, $l_3 = 7$ (*khet*). His task was to compute the lengths of the two parallel lines marked with dashes, namely the front s of the triangle and the first transversal d_1. (See again Fig. 2.1.7 below.) He started his work with the computation of the area A_3 of the small triangle. Then he quit, or else the remainder of his solution procedure has been lost. However, it is not difficult to complete his work for him. This will be done here.The solution can be based on the idea of using the "growth factor" f of the triangle and the naive-geometric observation that in a striped triangle like the one in Fig. 2.1.7 below *the whole triangle and its sub-triangles are similar triangles*. The value of f can be computed as the ratio of the given sides of the small triangle (see Fig. 2.18 for the notations used here):

$f = d_2/l_3 = 2\ 4'\ /\ 7 = 4'\ \overline{14}\ (= 9/28)$, $1/f = 3\ \overline{9}\ (= 28/9)$.

For the next step of the computation, the following equation is needed:

sq. $d_1 = d_1 \cdot (l_2 + l_3) \cdot f = (A_2 + A_3) \cdot 2 \cdot f$,

where

$A_2 + A_3 = 30\ 4'_s$ (given) $+ 7\ 2'_s\ 4'_s\ 8'_s$ (computed) $= 38\ 8'_s$.

Therefore, the length d_1 of the first transversal is given by the equation

sq. $d_1 = (A_2 + A_3) \cdot 2 \cdot f = 38\ \overline{8} \cdot 2 \cdot 4'\ \overline{14} = 24\ 2'\ \overline{112}$ (setat) = appr. 25 (setat).

It follows that

d_1 = appr. 5 (khet).

Similarly, the short side s of the whole triangle is given by the equation

sq. $s = (A_1 + A_2 + A_3) \cdot 2 \cdot f = 88\ \overline{8} \cdot 2 \cdot 4'\ \overline{14} = 56\ 2'\ \overline{7}\ \overline{112}$ (setat).

It follows that

s = appr. 7 2' (khet).

Therefore, the length of the whole triangle must be

$l = s \cdot 1/f = 7\ 2' \cdot 3\ \overline{9} = 23\ 3'$ (khet).

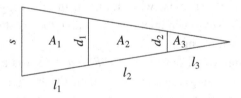

Given: Computed:
$A_1 = 50_s$ $A_3 = 7_s\ 2'_s\ 4'_s\ 8'_s$
$A_2 = 30_s\ 4'_s$ $f = d_2/l_3 = 4'\ \overline{14}$
$d_2 = 2\ 4'$ (khet) d_1 = appr. 5 (khet)
$l_3 = 7$ (khet) s = appr. 7 2' (khet)
                        $l_1 + l_2 + l_3$ = appr. 23 3' (khet)

Fig. 2.1.8. Given and computed numbers in P.Rhind # 53 a.

It is easy to find OB parallels to P.Rhind # 53 a, in particular **Str. 364** (Neugebauer, *MKT 1* (1935), 248), which is a well organized theme text from Uruk, where the theme is problems for striped triangles leading to quadratic equations. (See Friberg, *RlA 7* (1990) Sec. 5.4 i, and Friberg,

2.1. Themes in P.Rhind, a Hieratic Mathematical Papyrus Roll

MCTSC (2005), Fig. 10.2.12.) A more specifically parallel text is **IM 43996**, of which a photo was published in Bruins, *CCPV 1, Part 3* (1964), pl. 2. This is an OB square hand tablet with a geometric assignment on the obverse in the form of a drawing, showing a striped triangle with two transversals and associated numbers (see Fig. 2.1.9 below). The given numbers are the measures of the three sub-areas, the two transversals, and the first segment of the length (or height):

$A_1 = 9\ 22;30$ (sq. ninda), $A_2 = 20\ 37;30$ (sq. ninda), $A_3 = 10\ 00$ (sq. ninda),
$d_1 = 17;30$ (ninda), $d_2 = 10$ (ninda), $l_1 = 30$ (ninda).

The two remaining segments of the length were originally given, too, but were then erased by the tip of a finger, indicating that the unknown numbers should be computed by the student, in the same way as the unknown numbers to be computed by the student were marked by dashes in *P.Rhind* # 53 a. (Faint traces remain of the number 1 30, the size of the second segment of the length.)

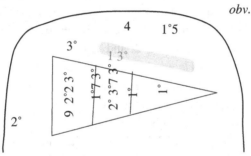

Fig. 2.1.9. IM 43996. An OB hand tablet with a geometric assignment.

The solution to the assignment presented on the obverse of IM 43996 is not explicitly given but is easy to reconstruct. Probably the first step was to compute the size of the short side s of the whole triangle as follows (the notations are the same as in the drawing in Fig. 2.1.8 above):

sq. s = (sq. d_2) · $(A_1 + A_2 + A_3)/A_3$
= sq. 10 (ninda) · 40 00 / 10 00 = sq. 10 (ninda) · 4 = sq. 20 (ninda),
hence $s = 20$ (ninda).

(Note the number 20 near the left edge of the tablet, which can be a left-over from the construction of the problem.) The next step probably was to compute the growth rate f, as follows:

50   *Unexpected Links Between Egyptian and Babylonian Mathematics*

$f = (s - d_1)/u_1 = (20 - 17;30)/30 = 2;30/30 = ;05 = 1/12$.

After that, it was easy to compute the lengths of the unknown segments:

$l_3 = d_2 \cdot 1/f = 10$ (ninda) $\cdot 12 = 2\ 00$ (ninda),
and $l_2 + l_3 = d_1 \cdot 1/f = 17;30$ (ninda) $\cdot 12 = 3\ 30$ (ninda),
so that $l_2 = 1\ 30$ (ninda).

The number 4 and its reciprocal 15, near the upper edge of the tablet, are probably left-overs from the construction of the problem, too, when the growth rate f was computed as $s/l = 20 \cdot 1/4(00) = 20 \cdot (;00)\ 15 = ;05$.

On the reverse of the tablet, the answer to the assignment is given in the form of a drawing of the striped triangle, with all the numerical data recorded, except the length of the front. The inscription on the reverse seems to have been done when the clay of the tablet was nearly dry.

As so often in the case of other OB mathematical texts, there is more to say also in this case about the teacher's actual *construction* of the problem. The given areas of the two sub-trapezoids, 9 22;30 and 20 37;30 are relatively close to two round area numbers, 10 00 and 20 00. Suppose that it was the teacher's intention to construct a striped triangle where the three sub-areas were proportional to 1, 2, and 1, for instance exactly 10 00, 20 00, and 10 00 sq. ninda. He would then soon realize that for this to happen it was necessary to let the front (the short side) and the two transversals be proportional to 2, sqr. 3 (the square root of 3), and 1, say, for instance, 20, 10 · sqr. 3, and 10. Sure enough, the standard OB approximation to sqr. 3 is 7/4 = 1;45, and 10 · 1;45 = 17;30, the value he chose for the length of the first transversal. The three segments of the length of the triangle would then have to be proportional to 20 − 17;30 = 2;30, 17;30 − 10 = 7;30, and 10, so he chose to let them be 30, 1 30, and 2 00. Then, the three sub-areas would be

$A_1 = 30 \cdot (20 + 17;30)/2 = 9\ 22;30$,
$A_2 = 1\ 30 \cdot (17;30 + 10)/2 = 20\ 37;30$,
$A_3 = 10 \cdot 2\ 00/2 = 10\ 00$.

In other words, the whole construction of the data in IM 43996 can be explained as a consequence of the teacher's decision to *try* to let the three sub-areas be proportional to 1, 2, 1.

All this suggests that it may be a good idea to turn back to *P.Rhind* # 53 a, to see how the data for *that* problem were constructed! What is immediately clear is that the first sub-area, 50 *setat*, is a round area number, and that the second sub-area, 30 1/4 *setat*, is very close to a round area number. As for the length of the whole triangle, it is 23 1/3 (*khet*), a length number that can be analyzed as $7 \cdot 3\ 1/3 = (1 + 1/6) \cdot 20$ (*khet*). This is interesting, because there are several examples of drawings of geometric figures on OB mathematical hand tablets where one of the sides of the figure can be expressed as $(1 + 1/6)$ times a round length number. Two such examples will be discussed below. The first example is **NCBT 1913** (Neugebauer and Sachs, *MCT* (1945), 10):

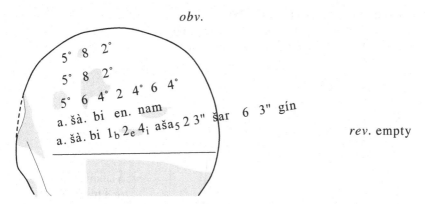

Fig. 2.1.10. NCBT 1913. Computation of the area of a square, an almost round number.

This is a round hand tablet (a lentil) where the text begins with the computation of the square of the number 58 20, which can be assumed to mean 58;20 ninda. The computed number is 56 42 46 40, obviously meaning 56 42;46 40 sq. ninda, because the text continues with the two lines

a.šà.bi en.nam / a.šà.bi $1_{bùr}$ $2_{èše}$ 4_{iku} 2 3" šar 6 3" gín,

that is, in English translation,

Its field (= area) is what? / Its field is 1(bùr) 2(èše) 4(iku) 2 3" šar 6 3" gín.

The OB area units figuring here are, with their Sumerian names,

1 bùr = 3 èše, 1 èše = 6 iku, 1 iku = 1 40 (100) šar, 1 šar = 1 sq. ninda, and 1 gín (shekel) = 1/60 šar.

The given value $s = 58;20$ ninda for the side of the square can be analyzed as follows:

$s = 58;20$ n. $= 7 \cdot 8;20$ n. $= (1 + 1/6) \cdot 50$ n.

Another interesting feature of NCBT 1913 is that the computed area number 56 42;46 40 sq. ninda is relatively close to the *round* area number 1 00 00 sq. ninda = 12 èše = 2 bùr.

A second example of a similar kind is given by the square hand tablet **YBC 7290** (Neugebauer and Sachs, *MCT* (1945), 44. It has on the obverse a drawing of a trapezoid, around which are recorded the lengths $s_1 = 2\ 20$ and $s_2 = 2\ (00)$ of the short sides, and the length $l = 2\ 20$ of the long side, and inside which is recorded the area number $A = 5\ 03\ 20$. These data can be analyzed as follows:

$l = 2\ 20$ n. $= 7 \cdot 20$ n. $= (1 + 1/6) \cdot 2\ 00$ n.,
$(s_1 + s_2)/2 = 2\ 10$ n. $= 13 \cdot 10$ n. $= (1 + 1/12) \cdot 2\ 00$ n.,
$A = 5\ 03\ 20$ sq. n. $=$ appr. $5\ 00\ 00$ sq. n. $= 10$ bùr.

obv.

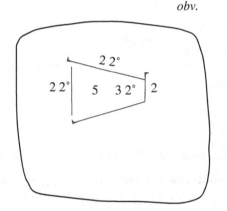

rev.: a small drawing of a trapezoid

Fig. 2.1.11. YBC 7290. A trapezoid with an almost round area number.

A sexagesimal area number that is close to a round area number and that can be factorized as the product of a regular sexagesimal area number and one or two small non-regular numbers, such as 7 or 11 or 13, can be called an "almost round area number". (In YBC 7290, $A = 5\ 03\ 20$ sq. n. $= 7 \cdot 13 \cdot 3\ 20$ sq. n.) The history of the use of almost round numbers in mathematical texts goes all the way back to the end of the fourth millennium BCE, the time of the earliest known written records. The "proto-cuneiform" texts of the proto-literate period in Mesopotamia are inscribed clay tablets from the strata Uruk IV-Uruk III. Two examples are shown in Fig. 2.1.12 below, borrowed from Friberg, *AfO* 44/45 (1997/98).

2.1. Themes in P.Rhind, a Hieratic Mathematical Papyrus Roll

It is easy to read these early texts, due to the amazing circumstance that the systems of length and area measures known from OB mathematical texts are (essentially) identical with the corresponding systems of measures used in the proto-cuneiform texts. A trivial difference is that the proto-cuneiform number signs were inscribed with a round stylus, while ordinary cuneiform number signs are written with a sharpened stylus. The direction of writing had also changed, so that, for instance, the vertical number signs in **W 14148** (Fig. 2.1.12 left) had been replaced in the OB period by rotated, horizontal number signs. (See Fig. A4.10 in Friberg, *MCTSC* (2005), App. 6, which shows the historical development of the "factor diagram" for Mesopotamian area numbers.)

Fig. 2.1.12. W 14148, and W 20044, 28, two proto-cuneiform field texts from Uruk.

Thus, the inscription on W 14148 can be immediately understood as an area number with the interesting factorization:

$A = 6$ bùr 1 èše 3 iku $= 6\ 1/2$ bùr $= (1 + 1/12) \cdot 6$ bùr.

At the same time, this area number is close to a round area number:

$A = 6$ bùr 1 èše 3 iku $= 19\ 1/2$ èše $=$ appr. 20 èše.

Hence, the number recorded on W 14148 is an almost round area number.

In a similar way, it does not take much imagination to see what is recorded on **W 20044, 28** (Fig. 2.1.12 right), namely four length numbers for a field, two for sides "along" and two for sides "across". Apparently, the field in question is in the form of a trapezoid, with the sides

$l_1 = l_2 = 1(\text{géš})\ 50\ \text{n.},\quad s_1 = 1(\text{géš})\ 10\ \text{n.},\ s_2 = 1(\text{géš})\ \text{n.},$

and the area

$A = 1(\text{géš})\ 50\ \text{n.} \cdot (1(\text{géš})\ 10 + 1(\text{géš}))/2\ \text{n.} = 1(\text{géš})\ 50 \cdot 1(\text{géš})\ 5\ \text{sq. n.}$
$= 1\ \text{šár}\ 59(\text{géš})\ 10\ \text{sq. n.} = 4\ \text{bùr} - 1/2\ \text{iku}.$

(Note that the proto-cuneiform length numbers are written with non-positional sexagesimal numbers, as multiples of an un-named length unit, clearly the ninda. In the transcription above, the signs for 60 and 60 · 60 are written, for convenience, by use of the Sumerian words for 'sixty' and 'sixty times sixty', which are géš and šár.)

It is now easy to see that the area A of the trapezoid with the sides given in W 20044, 28can be analyzed in the following two ways:

a) $A = (1 + 1/10) \cdot (1 + 1/12) \cdot 10\ \text{èše},$ and b) $A = \text{appr. } 12\ \text{èše}.$

In other words, the area of the trapezoid is an almost round area number.

The occurrence of almost round area numbers in several proto-cuneiform texts was explained in Friberg, *AfO* 44/45 (1997/98) (see now also Friberg, *MCTSC* (2005), Sec. 1.1 c) as a consequence of the existence of a "proto-literate field expansion procedure" used (for some unknown reason) to construct fields of approximately a given area and with sides in approximately a given ratio. The simplest way to explain how the procedure operates is to work through the details of its application in an explicit example.As it happens, *P.Rhind* # 53 a can serve as just such an example!

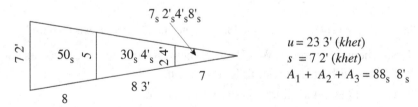

Fig. 2.1.13. The full set of parameters for the striped triangle in *P.Rhind* # 53 a.

Suppose that a teacher wanted to construct a triangle with the area 88 (*setat*) and with its growth rate (that is, the ratio between its sides) equal to $\overline{5}\ \overline{10}$ (= 3/10). (His ulterior motive was, apparently, to divide the resulting triangle into three stripes with the area of the second stripe equal to 2' $\overline{10}$ (= 3/5)of the area of the first stripe, and with the area of the third stripe equal to 1/10 of the combined area of the two first stripes.) If the teacher

2.1. Themes in P.Rhind, a Hieratic Mathematical Papyrus Roll

knew how to work with "metric algebra" of the kind which is richly represented in OB mathematical texts (cf. also *P.Moscow* ## 6-7, 17; Sec. 2.2 c below), he could have solved his construction problem as follows:

If $1/2 \cdot 1 \cdot s = A = 88$ (*setat*) and $s/l = f = \overline{5}\,\overline{10}$,
then sq. $s = 2 \cdot A \cdot f = 53 - \overline{5}$ (*setat*),
so that s = appr. 7 4', and l = appr. 24 6'.

However, if the teacher was not familiar with metric algebra, or if he chose not work that way, he could instead use the following procedure:

Start with a smaller initial triangle with the sides in the right ratio,
say one with the length $l_1 = 20$ (*khet*) and the short side $s_1 = 6$ (*khet*).
The area $A_1 = 60$ (*setat*), which is 28 (*setat*) less than the target area $A = 88$ (*setat*).
The area deficit 28 (*setat*) is nearly 1/2 of A_1.
In order to eliminate nearly half this deficit, increase s_1 by 1/4 of its length.
The front then increases from $s_1 = 6$ (*khet*) to $s = 6 + 1/4 \cdot 6$ (*khet*) = 7 2' (*khet*),
and the area increases from $A_1 = 60$ (*setat*) to $A_2 = 60 + 1/4 \cdot 60$ (*setat*) = 75 (*setat*).
The new area deficit is 13 (*setat*), which is about 1/6 of A_2.
In order to eliminate most of this deficit, increase l_1 by 1/6 of its length.
The length then increases from $l_1 = 20$ (*khet*) to $l = 20 + 1/6 \cdot 20$ (*khet*) = 23 3' (*khet*),
and the area increases from $A_2 = 75$ (*setat*) to $A = 75 + 1/6 \cdot 75$ (*setat*) = 87 2' (*setat*).

(The purpose of the procedure seems to have been not exclusively to solve a metric algebra problem relatively accurately, but rather to find a geometric figure with interesting, and not too simple data, to be used in the construction of a series of mathematical assignments.)

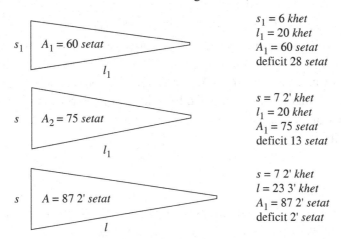

Fig. 2.1.14. The field expansion procedure in the case of *P.Rhind* # 53 a.

After having determined the sides of the triangle in this way, the teacher continued his construction by dividing the length 23 3' of the triangle into three segment of lengths 8, 8 3', and 7, obviously with the intention of dividing the triangle into three parts with sub-areas (approximately) equal to 50, 30, and 8. (The construction was not completely successful, due to the necessary approximations. This is why the sum of the three sub-areas differs slightly from the area of the whole triangle.)

After this lengthy discussion of # 53 a, it remains to explain the meaning of ***P.Rhind* # 53 b**, consisting of the following computations in the two columns of text immediately under the drawing of the triangle (Fig. 2.1.7):

\	1	$4_s 2'_s$	sum	$5_s 2'_s 8'_s$	
\	2	9_s	$\overline{10}$ of it	$1_s 4'_s 8'_s$	10 cubit(-strips)
	$2'_s$	$2_s 4'_s$	$\overline{10}$ of it taken away, then this is the amount.		
(\)	4'	$1\ 8'_s$			

There are two simple notational errors here: in the first column, $2'_s$ should be 2'. In the second column, $1\ 8'_s$ should be $1_s 8'_s$.

Apparently, what is going on here is that first the following area number is computed:

3 4' (*khet*) · 4 2' (*khet*) = $(4_s 2'_s + 9_s + 1_s 8'_s)$ = $14_s 2'_s 8'_s$

In the text, the sum $14_s 2'_s 8'_s$ is incorrectly written more like $5_s 2'_s 8'_s$. This is either a simple copying error, or an error caused by the following kind of incorrect addition in two steps:

$4_s + 9_s + 1_s = 1\ 3_s + 1_s = 5_s$ (with $1 = 10_s$ misread as 1_s).

Next, 1/10 is computed, strangely enough not of the incorrect value $5_s 2'_s 8'_s$, but of the correct value $14_s 2'_s 8'_s$. Finally, it is stated that this tenth of the area is to be subtracted (presumably from the area).

It is clear that what is computed here, in the first part of the procedure, is the area of a rectangle. It is also clear that that area is an almost round area number, since

l = 4 2' *khet* = (1 + 1/8) · 4 *khet*,
s = 3 4' *khet* = (1 + 1/12) · 3 *khet*,
A = $14_s 2'_s 8'_s$ = appr. 15_s.

The construction of the data for the rectangle in question by use of the field expansion procedure is shown in Fig. 2.1.15 below.

2.1. Themes in P.Rhind, a Hieratic Mathematical Papyrus Roll 57

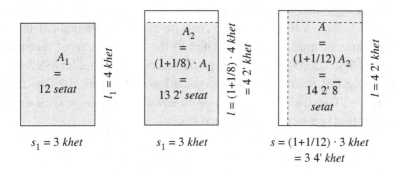

Fig. 2.1.15. *The field expansion procedure in the case of P.Rhind # 53 b.*

Suppose that a student was asked to find a rectangle with the area (close to) 15 *setat* and the short side (approximately) equal to 2' 4' (3/4) times the long side. He could then start with a rectangle with the sides 4 and 3 *khet* as a first approximation and use the field expansion procedure as in Fig. 2.1.15. Alternatively, he could use metric algebra to solve the system of equations $l \cdot s = 15$, $s = 2'\ 4' \cdot l$. In the latter case, the result would be

sq. $s = 15 \cdot 2'\ 4' = 11\ 4'$, $s = $ sqr. $11\ 4' = $ appr. $3\ 4'\ 8''$, $l = 3\ 4'\ 8'/2'\ 4' \cdot = 4\ 2'$.

Thus, the use of metric algebra would yield nearly the same result as the field expansion procedure!

P.Rhind # 53 b is only a fragment of a geometric exercise. The question is missing, so there is no way of knowing why the preserved part of the text states that 1/10 of the area should be subtracted from it. There may be some connection with the subtraction problems in § 18 (*DMP* ## 54-55).

Four problems in **P.Rhind § 21** (## 56-59) are concerned with the *seked*, the inclination of the sides of a pyramid, measured as the ratio of half the width to the height, and expressed as palms or fingers per cubit. (In OB mathematical texts, the inclination of a wall, or the side of a canal, *etc.*, is sometimes called the 'feed'. It is measured, like the *seked*, as the increase or decrease of the width for each cubit of vertical descent, and is expressed in ninda, cubits, or fingers per cubit.) In *P.Rhind* # 56, the height of a pyramid is 250 (cubits), the width of the square base 360 (cubits), and the *seked* $7 \cdot 180/250 = 5\ \overline{25}$ (fingers per cubit). In *P.Rhind* # 57, where the reverse problem is considered, the width of the base and the *seked* are given, and the height of the pyramid is computed. The drawings illustrating *P.Rhind* ## 56-59 show *upright* pictures of pyramids (see Fig. 2.1.16 below, left). This is in contrast to OB mathematical texts, where pictures of mud walls are rotated to the left. (See, most recently, Friberg, *MCTSC* (2005), Fig. 10.2.1, an outline of the clay tablet MS 3052, an interesting OB mathematical recombination text, where the first four exercises are illustrated by drawings of trapezoids and triangles, meant to represent rotated cross sections of mud walls.)

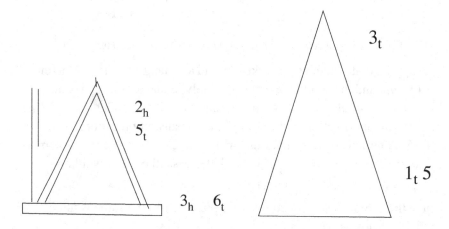

Fig. 2.1.16. Drawings accompanying *P.Rhind* ## 56 and 60. A pyramid and a cone(?).

The single problem in *P.Rhind* § 22 (# 60) is concerned with the *seked* of a 'pillar', probably meaning a circular cone, with the diameter 15 (cubits) at the base and the height 30 (cubits). Until recently, no OB mathematical texts were known that mentioned cones. However, now **BM 96954 + BM 102366 + SÉ 93**, three fragments of a large clay tablet from Sippar, have been published. Several of the exercises in that text seem to be dealing with whole or truncated cones. All three fragments are published in hand copy, transliteration, and (incomplete) translation in Robson, *MMTC* (1999) App. 3, but with no explanations or commentaries offered in the case of the exercises dealing with cones. Instead, improved transliterations and translations, and extensive commentaries to the exercises in BM 96954+ dealing with whole or truncated cones, can be found in Sec. 4.8e below. See, in particular, Fig. 4.8.5.

2.1 e. Theme H: baking or brewing numbers (*P.Rhind* ## 69-78)

In this group of exercises, the basic notion is the *pesu* of (loaves of) bread. The meaning of this term is somewhat elusive. Thus Chase, *et al.*, *RMP 1* (1927), 106, write that the *pesu* is "the number of loaves that one heqat will make". Gillings, *MTP* (1982), 129, says that the *pesu* "is a measure of the strengths of beer or bread, after either of them is made", and Robins and Shute, *RMP* (1987), claim that "The bread might be more or less aerated or otherwise expanded, and the beer more or less dilute; ⋯ . Hence the *pesu* unit, which measures lack of quality ⋯ ."

In *P.Rhind* # 69, the first of this group of exercises, for instance, the question is stated as follows:

Flour, $3_h\ 2'_h$ *heqat*, made into 80 (units of) bread.
Let me know the amount of a unit of it in flour, let me know their *pesu*.

The answer, obtained through division of 80 by 3 2', is that the *pesu* is 22 3" $\overline{7}\ \overline{21}$ (units of 'bread' per *heqat*). Other values of the *pesu* mentioned in this group of exercises are 12 3" $\overline{42}\ \overline{126}$ (# 70), 10 and 45 (#72), 10 and 15 (# 73), 5, 10, and 20 (# 74), 10, 20, and 30 (# 76), 5 (# 77), and 10 (# 78). In *P.Rhind* ## 71 and 77-78 it is also supposed to be known that two jars of beer of normal strength can be made out of 1 *heqat* of grain or flour, which gives the beer a *pesu* equal to 2. It is also assumed to be known that if the beer is diluted, then its *pesu* will be greater.

In Chase, *et al.*, *RMP 2* (1929) # 72, foot-note 4, it is mentioned that "From later texts it appears that the full form of the phrase reads 'loaves which in the baking are so many from the *heqat*', or 'a *des* (jar) of beer of a brewing: X from the *heqat*'". Hence, it may be appropriate to talk about the "baking number" of bread or the "brewing number" of beer, as an attempt to translate the term *pesu*.

Most of the *pesu* exercises are mathematically uninteresting, involving nothing more advanced than simple divisions. An exception is *P.Rhind* # 76, which begins like this:

Another (question). Bread of 10, 1000, exchanged for a number of bread of 20 and 30. Let him hear.

What this vaguely phrased question means is that flour enough to make 1,000 units of bread of *pesu* 10 is to be used to make instead *equal* numbers of units of bread of two different *pesu* values, 20 and 30. The solution procedure begins with two computations. The first computation shows that $\overline{20}$ $\overline{30}$ = 2 2' · $\overline{30}$, so that the flour needed to make 1 unit of *pesu* 20 and one unit of *pesu* 30 together is as much as that needed to make 2 1/2 units of *pesu* 30. The next computation is the division 30 / 2 2' = 12, showing that while 30 units of *pesu* 30 can be made out of 1 *heqat* of flour, only a 2 1/2 times smaller number, that is 12, units of *pesu* 20, and 12 of *pesu* 30 can be made out of 1 *heqat*. The result can be described by saying that if 20 and 30 are two given baking numbers, then 12 is the corresponding "combined baking number".

The solution procedure continues by stating that making 1,000 units of *pesu* 10 needs 100 *heqat* of flour, and that 100 *heqat* of flour is enough to make 1,200 units of *pesu* 12. Hence there is flour enough to make 1,200 units of *pesu* 20 and 1,200 of *pesu* 30. For verification, it is checked that 1,200 units of *pesu* 20 takes 60 *heqat* of flour, written as 2' 10_h, meaning 1/2 · 100 + (*heqat*), and that 1,200 units of *pesu* 30 takes 40 *heqat*, written as 4' 10_h 5_h meaning 1/4 · 100 + 15 (*heqat*).

An indirect parallel to *P.Rhind* # 76 can be found in the OB mathematical recombination text **YBC 4698** (Neugebauer, *MKT 3* (1937), pl. 5), an unprovenanced text belonging to Group 2 b (Friberg, *RA* 94 (2000), 164), hence possibly a text from Ur. See Fig. 2.1.17 below.

Here is a transliteration and translation of the brief third exercise in this text, merely a question without solution procedure or answer:

YBC 4698 § 2 a (# 3).

```
3 sìla.ta ì.sag / 1_bán 2sìla.ta ì.giš /
1 gín kù.babbar sì /
ì.giš ù ì.sag / íb.si₈-ma šàm
```

3 sìla good oil, 12 sìla plain oil (per shekel),
1 shekel of silver is given.
Plain oil and good oil make equal and buy.

This is an example of a "combined market rate problem" (see Friberg, *RlA* 7 (1990) Sec. 5.6 i; and Friberg, *MCTSC* (2005), Sec. 7.2).

2.1. Themes in P.Rhind, a Hieratic Mathematical Papyrus Roll

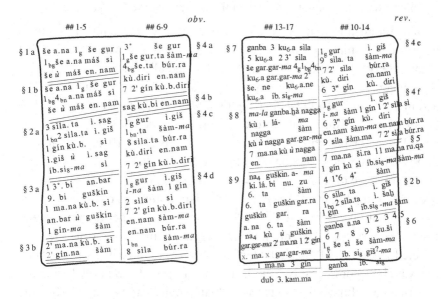

Fig. 2.1.17. YBC 4698. A recombination text dealing with prices and market rates.

OB texts with solutions to combined market rate problems are usually in the form of a small hand tablet with a tabular array, in which the *market rates* (units per shekel) for several commodities are listed in col. *i*, the corresponding *unit prices* (shekels per unit) in col. *ii*, the *prices* paid for *N* units of each commodity in col. *iii*, and the *number N* of units bought, the same for all commodities, in col. *iv*. The number *N* is computed so that the combined price paid will be a given amount of silver. Eight examples are listed in Neugebauer and Sachs, *MCT* (1945), 1, without explanations.

In the case of YBC 4698 # 3, there are two market rates, 3 and 12 sìla (liters) per shekel, and the given amount of silver is 1 shekel. It can be shown that the solution to this problem is that the price for 2;24 sìla of good oil is ;48 shekel, and the price for 2;24 sìla of plain oil is ;12 shekel, so that the combined price for 2;24 sìla of each kind is 1 shekel.

There are no known OB parallels to the *pesu* problems in *P.Rhind* ## 69-78. The reason for the absence of such problems in the corpus of OB mathematical texts may be that such OB texts with baking number problems were relatively uncommon and that no examples of clay tablets with such problems have happened to be excavated. It is unlikely that such problems were no longer relevant in the society of the OB period.

It is interesting that (inverted) baking and brewing numbers seem to have been so important in the *proto-literate* period in Mesopotamia that among the very few known "metro-mathematical" school texts from this period there is one that teaches how to operate with just such numbers. The text in question is **MSVO 4, 66** (Englund, *MSVO 4* (1996), Friberg, *JCS* 51 (1999), 112), an important proto-cuneiform so-called *bread-and-beer text* from Larsa (or possibly Uruk), dating from the Uruk III period, around 3000 BCE. See the hand copies in Figs. 2.1.18-19 below.

The obverse of this text contains two *registers* (horizontal columns) of text. In the upper part of the first register, large numbers of rations of various sizes of bread are listed, together with their costs per unit in (barley) flour. The proto-cuneiform sign for 'bread' seems to be a picture of a beveled-rim bowl, of the kind which has been excavated in great numbers, and which, according to Nissen, Damerow, Englund, *ABk* (1993), 14, can be suspected of having been used for "half a daily barley ration". (See the photo (*op. cit.*), Fig. 11.) In the lower part of the first register are recorded separately the total costs for bread rations of each size. Thus, for instance, in the first text box in the first register (text box i:1), 1 · 60 large bread rations, at a cost of 1 M each of barley flour, have together a cost of 60 · M = 2 d of barley flour. In the next text box (i:2), 1 · 120 bread rations at a cost of 1/2 M = m2 each of barley flour have together a total cost of 120 · m2 = 2 d. And so on. (The names of the units of the proto-literate system of capacity measure are not known. For convenience, they can be given names such as M, c, d, *etc.* The relative sizes of these units are fixed by the following equations:

1 D = 3 C, 1 C = 10 d, 1 d = 6 c, 1 c = 5 M.

More about that below.) In the second register, large numbers of jars of three kinds of beer are listed, together with the corresponding total costs in barley flour. (The cost per jar in each case is assumed to be known.)

By convention, the reverse of a text like this contained the summaries, or totals. In the present case, the lower register of the reverse contains the totals of the various kinds of rations, together with the corresponding totals of the costs. The upper register contains the grand total of the costs, plus an extra item.

2.1. Themes in P.Rhind, a Hieratic Mathematical Papyrus Roll

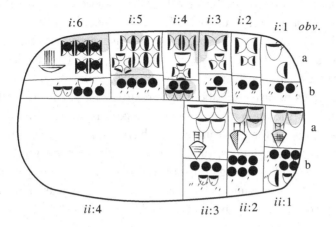

Fig. 2.1.18. *MSVO 4*, 66, *obv*. A proto-literate bread-and-beer text (Uruk III).

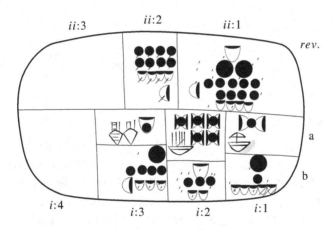

Fig. 2.1.19. *MSVO 4*, 66, *rev*. The totals of the bread-and-beer text.

That *MSVO 4*, 66 is a school text is suggested by the unusual circumstance that text box *ii*:4 on the reverse and text boxes *i*:4 and *ii*:3 on the reverse have been left empty. These are the places where in normal administrative texts non-numerical information was recorded, such as personal names, titles of officials, place names, *etc*.

The text is remarkable because it makes use of notations for numbers belonging to four different proto-cuneiform number systems, namely (non-positional) *sexagesimal numbers* (used to count jars of beer), *bisexagesimal numbers* (used to count bread rations), and two variants of *capacity numbers* (dotted numbers used for quantities of flour, and slashed numbers used for quantities of malt). In the sexagesimal number system, there are signs for 1, 10, 60, 10 · 60, *etc.* In the bisexagesimal system there are also signs for 2 · 60, 10 · 2 · 60, *etc.*, but no sign for 10 · 60. The signs used for capacity numbers are exhibited in the "factor-diagram" below, which also gives information about both relative sizes and (conjectured) absolute sizes of the main capacity units. Since it is not known what the original names were for those units, they have here been given mnemotechnically appropriate names, such as 'c' for a cup-shaped sign, 'd' for a disk-shaped sign, 'C' for a big cup-shaped sign, *etc.* (No Sumerian names are known for the proto-cuneiform capacity units, for the reason that a different system of capacity measures was used in Sumerian cuneiform texts.)

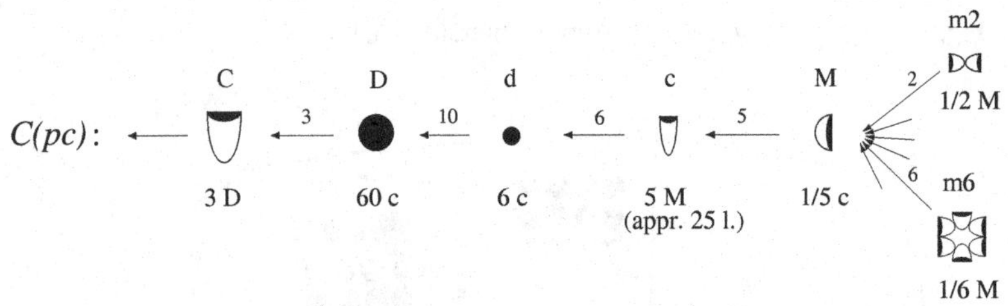

Fig. 2.1.20. Factor diagram for the proto-cuneiform system of capacity numbers.

As mentioned, there is an account of the cost in flour for large numbers of *bread rations* of various sizes in the first register on the obverse of *MSVO 4*, 66. It is, of course, much more likely that the rations were of different sizes, than that the loaves of bread were of different sizes. (In a similar way, it is possible that some of the *pesu* problems concerned with bread in theme H of *P.Rhind* ought to be reinterpreted as problems about bread rations of different sizes rather than as problems about loaves of different sizes!) The proto-cuneiform bread rations in *MSVO 4*, 66 come in six different sizes, 1 M (probably = appr. 5 liters, comparable with the Egyptian *heqat*), $m2 = 1/2$ M, $m3 = 1/3$ M, $m4 = 1/4$ M, $m5 = 1/5$ M, and $bread_1 = 1/6$ M. In *MSVO 4*, 66, there are relatively small numbers of rations of the larger sizes, but 6,000 rations of the smallest size ($bread_1$). An obvious interpretation is that the great number of small $bread_1$ rations were for ordinary workers, while the much smaller numbers of larger rations were for various categories of overseers and officials.

In the second register on the obverse, there is an account of the cost in flour for large numbers of jars of beer. The beer comes in three different strengths: the cost in flour for $3 \cdot 60$ jars of beer$_2$ is $6\,d = 6 \cdot 30$ M. Consequently, the cost in flour for 1 jar of beer$_2$ is 1 M. It can be shown in a similar way that the cost in flour for 1 jar of beer$_3$ is $1\,m3 = 1/3$ M.

The numbers appearing in *MSVO 4,* 66 were not randomly chosen. Instead, they were the result of laborious computations. (For details, see Friberg, *JCS* 51 (1999).) Thus, for instance, the total expense of flour for all bread and beer rations, recorded in text box *ii*:1 on the reverse, is

1 C 2 D 9 d 4 c 1 M = 2 C − 1 c 4 M = appr. 2 C = $2 \cdot 3 \cdot 10 \cdot 6 \cdot 5$ M = $30 \cdot 60$ M.

In other words, the total expense is very close to a *large and round* capacity number. The total cost for just the bread rations of all sizes is given in text boxes *i*:1-2 on the reverse:

1 D 1 d 5 c + 1 C 3 d 2 c = 1 C 1 D 5 d 1 c = appr. 1 C 1 D 5 d = $22\,1/2 \cdot 60$ M,
which is 3/4 of 2 C.

The total cost for all the jars of beer, given in text box *i*: 2 on the reverse, is

1 D 4 d 3 c 1 M = appr. 1 D 5 d = $7\,1/2 \cdot 60$ M,
which is 1/4 of 2 C.

And so on. As a matter of fact, the computations are so complex that there can be no doubt that learning to account for the cost of bread and beer rations was an important part of the mathematical education given to the young scribes in the scribe schools that must have existed in large cities like Uruk in the proto-literate period.

In the *pre-Sargonic and Sargonic* periods in Mesopotamia, around the middle of the third millennium BCE, bread or bread rations of different sizes were still mentioned in cuneiform texts (see the discussion in Blome, *Or* 34/35 (1928), now referred to by the somewhat obscure phrase

n ninda ba.an.né.*m*.du$_8$ *n* bread (rations), the bán divided(?) in *m* (parts).

Here, the bán is a Sumerian capacity measure, equal to 6 or 10 sìla (depending on from where and from which period the text is), hence about twice as much as an Egyptian *heqat*.

A good example is the Sargonic text fragment Thureau-Dangin, *RTC* (1903) # 125, = **Ist. O. 236**. In the last column of that text, on the reverse, the following totals of various kinds of ninda 'bread (rations)' are listed:

[... ... ...]
Total 36 ninda.gu, <the bán> divided(?) in 30. / Its flour 1 bán 2 sìla (= 12 sìla).
Total 12 sag.ninda, <the bán> divided(?) in 20. / Its flour 6 sìla.
Total 2 sag.ninda, <the bán> divided(?) in 7 <2'>, /
and total 3 sag.ninda,<the bán> divided in 15. / Their strong flour 4 2/3 <sìla>.

It is easy to check that the computations of the cost in flour are correct. It is interesting to note that in this Sargonic text, just as in the proto-cuneiform text *MSVO 4*, 66, there are many small rations and few big rations (36 of the smallest, 30.du_8, and only 2 of the biggest, 7 2'.du_8).

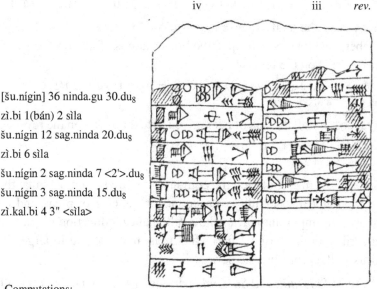

[šu.nígin] 36 ninda.gu 30.du_8
zì.bi 1(bán) 2 sìla
šu.nígin 12 sag.ninda 20.du_8
zì.bi 6 sìla
šu.nígin 2 sag.ninda 7 <2'>.du_8
šu.nígin 3 sag.ninda 15.du_8
zì.kal.bi 4 3" <sìla>

Computations:
36 · 1/30 bán = 1 bán 2 sìla
12 · 1/20 bán = 6 sìla
2 · 1/(7 1/2) bán = 2 3" sìla
3 · 1/15 bán = 2 sìla

Fig. 2.1.21. Ist. O 236. Bread (rations) of various sizes, and the cost in flour.

The notation ninda.gu 30 du_8 in this text is a direct parallel to 'bread of *pesu* 30' in *P.Rhind*. Similarly, the other kinds of bread mentioned in Ist. O 236 are parallels to bread of *pesu* 20, 15, and 7 1/2.

2.1 f. Theme I: a combined price problem (*P.Rhind* # 62)

This problem is about buying a bag filled with <equal> amounts of three precious metals, gold, silver, and lead, for a given total price of 84 *sha'ty*. The price of gold is 12 *sha'ty* per *deben*, that of silver 6 *sha'ty* per *deben*, and that of lead 3 *sha'ty* per *deben*. This problem resembles the combined *pesu* problem in # 76 but is mathematically simpler. In spite of its simplicity, it is interesting because it is so artificial. Why would anyone want to buy equal amounts of gold, silver, and lead? (Compare with the demotic mathematical text *P.carlsberg 30 # 2*, a small fragment of an exercise, which, according to the reconstruction in Sec. 3.5 b below, is a system of four linear equations for four unknowns, called 'silver', 'gold', 'copper, and 'lead'.)

The solution algorithm in # 62 proceeds, essentially, as follows:

The sum of 12, 6, and 3 *sha'ty* for 1 *deben* of gold, 1 *deben* of silver, and 1 *deben* of lead is 21 *sha'ty*.
This goes 4 times in 84 *sha'ty*, so 4 (*deben*) is what is given of each metal.
The amount of gold is 4 times 12 = 48 (*sha'ty*), that of silver is 24 (*sha'ty*), and that of lead is 12 (*sha'ty*). Sum 84 (*sha'ty*).

An OB (imperfectly) parallel text is **YBC 4698 # 4**, where iron (at that time rare and very expensive) and gold are to be exchanged for silver:

YBC 4698 § 3 a (# 4). (See Fig. 2.1.17 above.)

1 30.bi an.bar / 9.bi guškin /	Its 1 30 iron, its 9 gold.
1 ma.na kù.babbar sì /	1 mina of silver is given.
an.bar *ù* guškin /	Iron and gold,
1 gín-*ma* šàm	1 shekel, then buy.

The text is vaguely formulated and without any known OB parallel. The question seems to be that if iron and gold are 90 (sic!) and 9 times more valuable than silver, and if 1 shekel of iron and gold together is bought for 1 mina of silver, what are then the amounts of iron and gold, respectively? The question can be reformulated (in modern terms) as a system of linear equations. If a shekels is the weight of the iron and g shekels the weight of the gold, then these equations are:

$a + g = 1, \ 1\ 30 \cdot a + 9 \cdot g = 1\ 00.$

Systems of linear equations of the same type are known from the pair of

OB problem texts **VAT 8389** and **VAT 8391**. There, such systems of equations are solved by use of a variant of the rule of false value. (See Friberg, *MCTSC* (2005), Sec. 11.2 m: Fig. 11.2.14 left. See also Sec. 3.5 b below.) The first step of the solution procedure is to find a partial solution, satisfying only the first of the two given equations. Thus, in the case of YBC 4698 # 4, the first step is to give a and g the initial false values

$a^* = g^* = \;;30$.

If these values are tried in the second equation the result is that

1 30 · a^* + 9 · g^* = 49;30,

which gives a deficit of 10;30 compared to the wanted value 1 00.

To decrease the deficit, g^* is increased and a^* decreased by the small amount ;01. The result is that the deficit is decreased by a corresponding amount, namely

;01 · (1 30 – 9) = 1;21   (a *regular* sexagesimal number with the reciprocal ;44 26 40).

Hence, the whole deficit can be eliminated if g^* is increased and a^* decreased by the larger amount

;01 · 10;30 · 1/1;21 = ;10 30 · ;44 26 40 = ;07 46 40.

Therefore, the correct solution is that

a = ;37 46 40, g = ;22 13 20.

More precisely, the answer to the stated question is that the correct amounts of iron and gold are

1/2 shekel 23 1/3 barley-corns of iron   and   1/3 shekel 6 2/3 barley-corns of gold.

Note: **YBC 4698** (Fig. 2.1.17 above) is a recombination text with a mixed bag of exercises with the common topic "prices and market rates", all probably directly or indirectly borrowed from several original, well organized large theme texts. YBC 4698 # 3 has already been mentioned. It is a combined market rate exercise, where equal quantities of two kinds of oil are bought for a given amount of silver. YBC 4698 # 4 is the exercise where unequal quantities of iron and gold with a given total weight are exchanged for a given amount of silver. It is likely that the original theme text from which YBC 4698 # 4 was borrowed contained also exercises where equal quantities of iron and gold (or some other combination of metals) were bought for a given amount of silver. An exercise of that kind would be a direct parallel to the combined price problem in *P.Rhind* § 24.

2.2. Themes in *P.Moscow*, a Smaller Hieratic Mathematical Papyrus

2.2 a. *P.Moscow*: Another Egyptian mathematical recombination text

P.Moscow E 4676 is a hieratic mathematical text written on a long but narrow papyrus roll. It can be dated to the 13th Dynasty. Hence it is older than *P.Rhind*, but possibly younger than the text of which *P.Rhind* claims to be a copy. In the following table of contents for *P.Moscow*, the ordering of the exercises into nine paragraphs of related exercises is the same as in the original edition of the papyrus, in Struve, *QSA 1* (1930).

P.Moscow: Contents.

§ 1	A cedar mast (a damaged exercise; its meaning not clear)	# 3
§ 2 a-j	Baking and brewing numbers	## 5, 8-9, 12-13, 15-16, 20, 22, 24
§ 3	Mixed baking numbers	# 21
§ 4 a	Two work norms for pieces of wood	# 11
§ 4 b	Combined work norms (for making sandals)	# 23
§ 5 a-b	Division exercises: a) $1\ 2' \cdot a + 4 = 10$, b) $a + 2\ a = 9$	## 19, 25
§ 6	The area of a band of cloth?	# 18
§ 7 a-c	Metric algebra: two rectangles and a triangle	## 6-7, 17
§ 8	The volume of a truncated pyramid	# 14
§ 9 a-b	The area of a) a triangle, b) a semicircle?	## 4, 10

Exercises from four of these paragraphs are of interest in the present discussion, namely § 4 b (combined work norms), § 7 a-c (metric algebra), § 8 (the volume of a truncated pyramid), § 9 b (the area of a semicircle?).

2.2 b. *P.Moscow* # 23: A combined work norm

According to Couchoud, *ME* (1993), 171, the question in this exercise can be explained as follows:

> A shoemaker can cut 10 pairs of sandals in a day, or he can finish 5 pairs in a day.
> If he both cuts and finishes, how many pairs can he make in a day?

The solution is (essentially) the following:

> To make 10 pairs takes him 3 days (1 for the cutting, 2 for the finishing).
> Since 3 goes 3 3' times in 10, he can make 3 3' pairs in a day.

For OB parallels to this exercise, see the general discussion of "combined work norms" in Friberg, *RlA 7 (1990) Sec.* 5.6 h. Detailed discussions of specific examples can be found in Friberg, *ChV* (2001) Sec. 6: Combined work norms for work with bricks or mud, and § 8: Combined work norms for excavations or for the maintenance of canals. Most recently, combined work norms were discussed in Friberg, *MCTSC* (2005), Sec. 7.3 b, in the commentary to MS 2221. On the *obverse* of that text (see Fig. 2.2.1) are computed the "carrying numbers" for three kinds of bricks, and for mud or earth. On the *reverse*, there is a tabular array:

MS 2221 *rev.*

1	12	1 48	9
1 20	16	1 48	6 45
2 40	32	1 48	3! 22 30

The numbers recorded in this tabular array are the data for a combined work norm problem. It is helpful to note that the three numbers 9, 6 45. and 3 22 30 in the last column of the array are the carrying numbers computed on the obverse of the tablet. Here follows an expanded form of the tabular array, with detailed information about what is counted in each case:

carrying 9 · 60 bricks over 30 n.	1/5 of the time	1/5 of the bricks	carrying numbers
1 man-day	;12 man-day	1 48 bricks	9 sixties · 30 n.
1;20 man-day	;16 man-day	1 48 bricks	6;45 sixties · 30 n.
2;40 man-days	;32 man-day	1 48 bricks	3;22 30 sixties · 30 n.

The three carrying numbers 9, 6 45, and 3 22 30 have to be interpreted as 9 00, 6 45, and 3 22;30 bricks · 30 ninda. What they mean is that carrying 9 00 rectangular bricks, or 6 45 half square bricks, or 3 22 1/2 square bricks over a standard distance of 30 ninda (180 meters) is the daily work norm for one man. Inversely, it takes a man 1 day to carry 9 00 rectangular bricks the standard distance, it takes him 1;20 = 1 1/3 day to carry as many half square bricks the same distance, and it takes him 2;40 = 2 1/3 days to carry as many square bricks. The numbers 1, 1 20, and 2 40 are recorded in the first column of the tabular array.

To carry 9 00 bricks of each kind over the standard distance of 30 ninda takes a man 1 + 1;20 + 2;40 = 5 days. Therefore, in one day a man can carry 9 00/5 = 1 48 bricks *of each kind* 30 ninda, if he spends 1/5 =12/60 of the day carrying rectangular bricks,1;20/5 = 16/60 of the day carry-

2.2. Themes in P. Moscow, a Smaller Hieratic Mathematical Papyrus Roll

ing half square bricks, and 2;40/5 = 32/60 of the day carrying square bricks. This is the explanation for the numbers 1 48 in the third column of the array, and for the numbers 12, 16, and 32 in the second column.

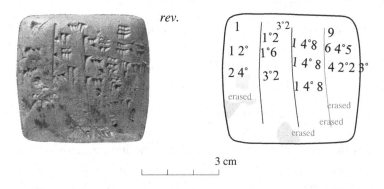

Fig. 2.2.1. MS 2221, *rev.*: A combined work norm for carrying three kinds of bricks.

2.2 c. *P.Moscow* ## 6-7, 17: Metric algebra

In these three exercises, the area and the ratio of the width to the length are given for either a rectangle or a triangle. In all three cases, the solution procedure is the same: The length is computed by use of the rule of false value, and when the length is known, it is easy to compute also the width. The most interesting of the three exercises is *P.Moscow* # 17.

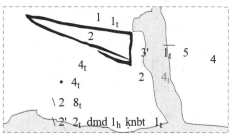

Fig. 2.2.2. *P.Moscow* # 17 (detail). A triangle with two sets of associated numbers.

Here is a rough translation of the text of # 17 (cf. Couchoud, *EM* (1993), 48, Imhausen, *ÄA* (2003), 334):

> Method of calculating a triangle. If it is said to you:
> A triangle of 20_s in field, and as for what you set as length you have 3' $\overline{15}$ as width.
> You double the 20_s, it makes 40. You count with 3' $\overline{15}$ to find 1. It makes 2 2' times.

You count 40 times 2 2', it makes 100. You count the corner (square root), it makes 10. Look, this 10 is the length.
You count 3' $\overline{15}$ of 10, it makes 4. Look, this 4 is the width.
You have found correctly.

What this means is that if the area $A = l \cdot w$ of a triangle is 20 *setat* = 2,000 cubit strips, and if the ratio of the width to the length is $w/l = f = 3'\,\overline{15}$ (= 2/5), then $2\,A = 40$ (*setat*), and $1/f = 2\,2'$ (= 5/2). Hence,

sq. $l = 1/f \cdot 2\,A = 2\,2' \cdot 40 = 100$, so that $l = 10$, and $w = 3'\,\overline{15} \cdot 10 = 4$.

The steps of the computation are repeated in the drawing, where the number '2' = 2_{thcs} (2 1,000-cubit-strips) = 20_s (20 *setat*) inside the triangle indicates the area. The numbers 1 and 3' $\overline{15}$ written near the sides of the triangle indicate the relative lengths (false values for the length and the width), while the numbers 1_t (= 10) and 4 beside them indicate the true length and the true width. There is (probably) a brief note that 2 times the area = 4_t (= 40), and, finally, the explicit computation of $2\,2' \cdot 40 = 100$ and of sqr. $100 = 10$.

An OB round hand tablet with a close parallel to the drawing in *P.Moscow* # 17 is **YBC 11126**. Its drawing of a trapezoid was published in Neugebauer and Sachs, *MCT* (1945), 44.

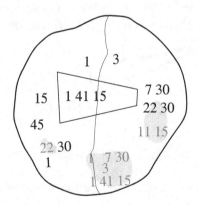

Fig. 2.2.3. YBC 11126. A trapezoid with two sets of associated numbers.

In this text, there is a drawing of a trapezoid. The area of the trapezoid is indicated by the number 1 41 15 written inside the trapezoid. The false lengths 1, 15, and 7 30, written along the sides of the trapezoid probably

2.2. Themes in P. Moscow, a Smaller Hieratic Mathematical Papyrus Roll

were meant to indicate that the "upper front" s_a and the "lower front" s_k of the trapezoid were equal to ;15 (1/4) and ;07 30 (1/8) of the length l, respectively. Consequently, the true length of l could be computed in the following way:

> If the false length is 1 00, then the false fronts are 15 and 7;30.
> Hence, he false area is 1 00 · (15 + 7;30)/2 = 22 30/2 = 11 15.
> The true area is 1 41 15, which is 9 times more, and 9 is the square of 3.
> Therefore, the true length is 3 · 100 = 3 00, and
> the true fronts are 3 · 15 = 45 and 3 · 7;30 = 22;30.

It is possible that the erasures on the tablet conceal some details of this computation. Anyway, after the computation, the true values '3', '45', and '22 30' were recorded on the tablet close to the false values '1', '7 30', and '15', just as in *P.Moscow* # 17 (Fig. 2.2.2) the true values 10 and 4 are recorded close to the false values 1 and 3' $\overline{15}$.

Another OB parallel to *P.Moscow* # 17 is exercise # 1 in the large theme text **IM 121613** (Friberg and Al-Rawi (forthcoming)).

IM 121613 # 1.

1	3" uš sag.ki
	1 èše aša₅ a.šà-*lam ab-ni*
2	uš *ù* sag.ki / *mi-nu*
	za.e
3	1 uš 40 3"-*šu* / *šu-ta-ki-il-ma*
	40 a.šà *sa-ar-ra-am ta-mar* /
4	igi 40 a.šà *sa-ar-ri-im* du₈
5	*a-na* 10 a.šà / gi.na *i-ši-ma* 15 *ta-mar*
	ba.sig₈-*e* 15 *šu-li-ma* 30 *ta-mar*
6	30 *a-na* 1 *ù* 40 / *ma-ni-a-tim i-ta-aš-ši-ma*
7	30 uš *ù* 20 sag.ki / *ta-mar*
	ki-a-am ne-pé-šum

2/3 of the length (is) the front.
1 èše (is) a field I built.
Length and front (are) what?
You:
1, the length, (and) 40, its 2/3, let eat each other, then
40, the false field, you see.
The opposite of 40, the false field, resolve,
to 10, the true field, raise (it), then 15 you see.

The equalside of 15 let come up, 30 you see.
30 to 1 and 40, your numbers, always raise, then
30, the length, and 20, the front, you see.
Such is your doing.

In this exercise, the front *s* of a rectangle is 2/3 of the length *l*, while the area of the rectangle is 1 èše 'rope' (= 10 00 sq. ninda). The solution procedure is essentially the same as the solution procedure in *P.Moscow* # 17. First, it is assumed that the length is 1 and the front ;40. Then the corresponding 'false' area is ;40. To get the true area one has to multiply this false area by 15 00 (sq. ninda), the "quadratic correction factor". The square root of this quadratic correction factor is the "linear correction factor" 30 (ninda). Therefore, the (true) length is 1 · 30 (ninda) = 30 (ninda), and the (true) front is ;40 · 30 (ninda) = 20 (ninda).

It has been shown by Høyrup (see, most recently, Høyrup, *LWS* (2002)) that OB mathematicians used geometric models in order to visualize and explain what we would call abstract or algebraic manipulations with quadratic equations or, as here, rectangular-linear systems of equations. Conceivably, the geometric model for the solution procedure in IM 121613 # 1 may have been something like this:

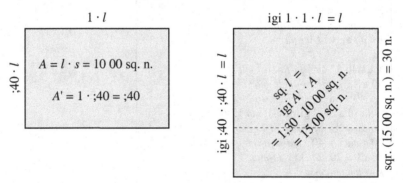

Fig. 2.2.4. A geometric model for the solution procedure in IM 121613 # 1. [16]

2.2 d. *P.Moscow* # 14: The volume of a truncated pyramid

This famous exercise contains a correct computation of the volume of a truncated pyramid. In the text, a miniature drawing of a trapezoid appears instead of the (unknown) Egyptian name for a truncated pyramid. A larger

2.2. Themes in P. Moscow, a Smaller Hieratic Mathematical Papyrus Roll

trapezoid of the same shape appears below the text, in a drawing illustrating the exercise. The pyramid has a square base of side a = '4', a square top of side b = '2', and the height '6'. In the text of the exercise, the volume is computed as follows:

V = (sq. a + 2 · a + sq. b) · 3' · h = (16 + 8 + 4) · 2 = 28 · 2 = 56.

(The mistake of doubling a instead of multiplying a with b = 2 is a kind of harmless error that can be encountered also in OB mathematical texts.)

In the drawing below the text, the computation is repeated. First, the given numbers, 2 for the side of the top, 6 for the height, and 4 for the side of the base are recorded above, inside, and below the trapezoid representing the truncated pyramid. (The inaccurate, two-dimensional picture of the three-dimensional truncated pyramid can be compared with equally inept pictures of three-dimensional objects in OB mathematical texts. See Friberg, *RlA* 7 (1990) Sec. 5.5 e.) The computation begins with the words 'square 4' after the number 2 above the drawing, with the words 'square 16' below the number 4 below the drawing, and with the multiplication 2 · 4 = 8 next to 4 and its square. The sum of 4, 16, and 8 is recorded as 'sum 28' to the right. In the last part of the computation, beside the drawing, there are two multiplications, $h/3$ = (6 ·) 3' = 2, and V = 2 · 28 = 56.

Note that it is not clear what the intended size of the pyramid actually was. If the lengths were measured in cubits, then the pyramid was very small, with a side at the base of only 4 cubits (appr. 2 meters) and a height

---

16. Cf. the following insightful commentary by Peet, *BJRL* 15(1931), 421 to the exercise *P.Moscow* # 6, long before the invention of the terms "naive geometry" and "metric algebra": "Here we are to find the sides of a rectangle of area 12, given that one side is 3/4 of the other. Stated in the form of an equation this would be $3/4\ x^2$ = 12, where the sides are x and $3/4\ x$. The Egyptian, however, uses no x and approaches the question graphically (Fig. 1). He sees that had the figure been a square whose side is the longer of the two sides of the given rectangle it would be 4/3 times as large, 4/3 being the reciprocal of 3/4. Such a square he proceeds to construct. To get the reciprocal of 3/4 he divides unity by it; result 1 1/3. Then he multiplies the given area 12 by 1 1/3, getting 16 for the area of the square. The square root of this, namely 4, will be the longer side required, and the other will be got by taking 3/4 of this. / This solution involves no algebra, nor even the use of a trial number, and the only assumption made is that if we have a rectangle, and we multiply one of its sides by k, leaving the other side constant, the area will also be multiplied by k—a theorem which follows at once from the formula for the area of a rectangle, *i.e.*, from the conception of a square measure."

of 6 cubits (appr. 3 meters). However, if the lengths were measured in *khet* (= 100 cubits), then the pyramid was very big, with a side at the base of 400 cubits (appr. 200 meters) and a height of 600 cubits (appr. 300 meters)! This can be compared with Khufu's pyramid, for which the side at the base is 230 meters and the height (originally) 145 meters.

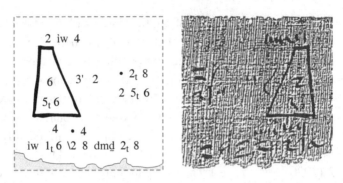

Fig. 2.2.5. *P.Moscow* # 14 (detail). A truncated pyramid, with associated numbers.

Since the publication of *P.Moscow* by Struve in 1930, a great number of authors have tried to explain how it was possible for Egyptian mathematicians to find the correct expression for the volume of a truncated pyramid, all of them assuming that this expression was not known in Babylonian mathematics. (See, for instance, Clagett, *AES 3* (1999), 83-90 and Figs. IV.9A-F.)

The situation has now been reversed, with the publication of three substantial fragments (**BM 96954 + BM 102366 + SÉ 93**) of a large OB text from Sippar with mixed problems about both whole and truncated pyramids and cones. (See Friberg, *PCHM* 6 (1996), Robson, *MMTC* (1999), App. 3, and for details Sec. 4.8 f below.) In that text, § 1 originally contained 14 exercises concerned with a roof-like "ridge pyramid" with a rectangular base, four sloping sides of equal inclinations, and a linear "ridge" instead of a single vertex. The ridge pyramid is called a 'granary' and is of a considerable size. Its rectangular base has a length of 10 ninda (60 meters) and a front of 6 ninda, the ridge is 4 ninda long, and the height measures 48 cubits (24 meters). The circular cones considered in § 4 are even more impressive, with a height of 1 00 (= 60) cubits and a circumference at the base equal to 30 ninda. The volumes of all pyramids and cones

2.2. Themes in P. Moscow, a Smaller Hieratic Mathematical Papyrus Roll

appearing in the text are correctly computed. In § 1, the grain content of the ridge pyramid is also correctly computed, in terms of a sìla measure chosen so that 1 volume-šar (=1 sq. ninda · 1 cubit) contains 1 30 (= 90) gur of 5 00 (= 300) sìla. This is a special sìla measure, about 2/3 liter, with the "storing number" 7 30 (00), apparently used only in connection with granaries.

In the first preserved exercise of BM 96954+, the ridge pyramid with the dimensions mentioned above is truncated at mid-height. The computation of the volume and grain content of the truncated ridge pyramid begins with the computation of the growth rate of the sides, which is found to be

f = 6 ninda/48 cubits = ;07 30 ninda/cubit.

Next, it is shown that the rectangular top of the truncated pyramid has a length of 7 ninda and a front of 3 ninda. Only a few words of the actual computation of the volume of the truncated ridge pyramid are preserved, yet it is clear how that volume must have been computed (see again Sec. 4.8e below). The correct expression for the volume of a truncated ridge pyramid can easily be deduced from the well known expression for the volume of a pyramid with a square base. It is as follows:

Let l and s be the length and front of the rectangular base of a truncated ridge pyramid, let l' and s' be the length and front of its rectangular top, and let h' be its height.

Then the volume is $[(l \cdot s + l' \cdot s') + (l \cdot s' + l' \cdot s)/2] \cdot h'/3$.

It is easy to check that this complicated expression is reduced to the familiar expression for the volume of a truncated pyramid in the special case when $l = s$ and $l' = s'$.

Much more can be said about this subject. The reader is referred to Friberg, *PCHM* 6 (1996) for an exhaustive survey of computations involving pyramids and cones in ancient Babylonian, Greek, Chinese, and Indian mathematical texts, and, in particular, for a discussion of proofs or derivations of the correct expression for the volume of a pyramid in an OB mathematical text (**TMS 14**), in Euclid's *Elements,* **Book XII**, and in Liu Hui's commentary to ***Jiu Zhang Suan Shu,*** **Book 5**.

2.2 e. *P.Moscow* # 10: The area of a semicircle(?)

Here is a rough translation of the text of this puzzling exercise:

Method of calculating a basket. /
If he says to you: A basket <with 9>? of mouth / and 4 2' of X.
Oh, / let me know its field.
You calculate / $\overline{9}$ of 9 because as for the basket / it is 2' of a Y. It becomes 1. /
You calculate the remainder as 8. / You calculate $\overline{9}$ of 8, / it becomes 3" 6' $\overline{18}$.
You calculate / the remainder of this 8 after / this 3" 6' $\overline{18}$. It becomes 7 $\overline{9}$. /
You calculate 7 $\overline{9}$ times 4 2'. / It becomes 32. Look, this is its field! /
You have found correctly.

There are several unfortunate circumstances that make it difficult to establish the exact meaning of this text: a) There seems to be a number missing in line 2; it is tentatively restored here as $9^?$ (following a suggestion in Peet, *JEA* 17 (1931)). b) The correct translation of the word here called X is not known. c) The word here called Y is only partly preserved. It seems to end with a determinative for a round object. d) It is not clear which object is referred to by the term *nbt* 'basket'.

All these uncertainties have led to conflicting opinions about the meaning of the text. See Clagett; *AES 3* (1999), 91-93 and 231-234, footnotes 18 and 19, for an account of various interpretation attempts, all accompanied by their respective arguments and counter-arguments. Thus, it has been suggested that the 'basket' is a *hemisphere* (half an egg), or a *semicylinder*, or *the round top of a cylinder* (a granary). In either case, it would have been an astonishing feat of an Egyptian mathematician to be able to find a correct expression for the area of such a curved surface.

The simplest interpretation, that the 'basket' in *P.Moscow* # 10 may be a *semicircle* was one of two alternative interpretations suggested in Peet (*op. cit.*). This suggestion has been uniformly rejected by other commentators. See, for instance, Clagett's objection (*op. cit.*), 234, footnote 19: "I have not taken seriously his (i. e. Peet's) reconstruction of the problem as determining the area of a semicircle, since a semicircle as a flat surface can hardly be considered as a "basket" ··· ". Still, an Egyptian picture of a basket would evidently be a semicircle (since three-dimensional objects were depicted as two-dimensional, as for instance the truncated pyramid in *P.Moscow* # 14), and the hieratic or hieroglyphic representations of the word 'basket' were pictures of semicircles followed by a phonetic complement. See lines 1-3 of the text in Fig. 2.2.6 below.[17]

2.2. Themes in P. Moscow, a Smaller Hieratic Mathematical Papyrus Roll

(A photo of the text and a hieroglyphic transliteration can be found in Struve, *QSA 1* (1930).)

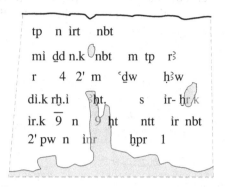

Fig. 2.2.6. *P.Moscow* # 10 (detail). Three instances of the word *nbt* 'basket'.

The computation of the area of the 'basket' in *P.Moscow* # 10 proceeds as follows:

1/9 of 9 = 1, 9 – 1 = 8.
1/9 of 8 = 3" 6' $\overline{18}$, 8 – 3" 6' $\overline{18}$ = 7 $\overline{9}$.
7 $\overline{9}$ · 4 2' = 32.

Thus, if the object of the exercise was to compute the area of a semicircle of diameter $d = 9$ (*khet*) and "thickness" (or whichever term is used) $d/2 = 4\ 1/2$ (*khet*), then it is clear that the area of the semicircle was computed as

$$A = [d \cdot (1 - 1/9) \cdot (1 - 1/9)] \cdot d/2.$$

Note that an alternative way of expressing the area of a semicircle would be as one half of the expression for the area of a circle used in *P.Rhind* ## 41-43 and 48, that is, as

---

17. Hoffmann, *ZÄS* 123 (1996) is thinking along the same lines, when he writes "In the demotic mathematical texts, however, the word denotes a circle segment, that is a two-dimensional object. It is likely that the changed meaning of the word is due to the form of the *nb*-sign in the hieroglyphic script: it shows the basket seen from the side, which makes it look like a circle segment." (My translation.) See *P.Cairo* # 36, where the word for 'circle segment' (outside an inscribed equilateral triangle) is *spdt nby*, 'triangle basket', *P.Cairo* # 37, where the word for 'circle segment' (outside an inscribed square) is *nby* 'basket', and *P.Cairo* # 38, where the chord forming the side of a circle segment is called *ḥr nby* 'over the basket'. (Cf. the glossary in Kaplony-Heckel's review, *OLZ* 76 (1981) of Parker, *DMP* (1972).)

$A = [d \cdot (1 - 1/9)] \cdot [d \cdot (1 - 1/9)] \cdot 1/2.$

The two expressions are, of course, mathematically equivalent, but the one used in *P.Moscow* # 10 has the advantage of letting the thickness (height) of the semicircle play a role in the computation.

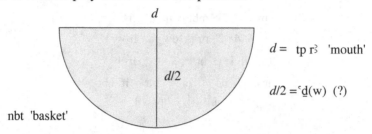

Fig. 2.2.7. Suggested explanation of three terms used in *P.Moscow* # 10.

Several items in OB tables of constants show that Babylonian mathematicians knew three different ways of computing the area of a semicircle (cf. the excerpts from the OB tables of constants **NSd** and **NSe** in Sec. 2.1 d above):

$A = ;10 \cdot$ sq. a (where a is the arc of the semicircle)
$A = ;15 \cdot a \cdot d$ (where d is the *tallu* 'transversal' of the semicircle, $d = ;40 \cdot a$)
$A = ;45 \cdot d \cdot p$ (where p is the *pirku* 'crossline' of the semicircle, $p = ;20 \cdot a$)

(In the OB table of constants BR = Bruins and Rutten, *TMS* **3** (1961), three constants are listed for each one of a series of different geometric figures, one constant for the area, one for the dal or *tallu* 'transversal', and one for the *pirku* 'crossline'. The crossline is in all cases orthogonal to the transversal.) Note that in Babylonian mathematics the 'crossline' was not thought of as a radius in general, but as the half-diameter orthogonal to the diameter.

It is not known how the arc of a circle or semicircle was related to the diameter in Egyptian "hieratic" mathematics.

Semicircles appear in two OB mathematical texts, successfully interpreted for the first time in Muroi, *SG* 143 (1994). Thus, Muroi showed that in **MLC 1354** (Neugebauer and Sachs, *MCT* (1945), 56), an exercise illustrated by a drawing of a semicircle, the arc a of the semicircle was computed as the solution to the strange quadratic equation

$;15 \cdot a \cdot (;40 \cdot a - 5) = 1\ 52;30 =$ a.šà lul, 'the false area'$(a = 30)$

The other text discussed in Muroi (*op. cit.*) is **BM 85210**, a fragment of a very long exercise, in which the following three constants for a semicircle are repeatedly mentioned:

    igi.6.gál   the reciprocal of 6 (= ;10)
    15 igi.gub.ba   15, the constant
    1 30 bal gán.u_4.sakar   1;30, the ratio of a crescent-field' (the reciprocal of ;40).

The first two of these are constants for the area of a semicircle, the third stands for the arc of a semicircle as a multiple of the diameter.

2.3. Hieratic Mathematical Papyrus Fragments

2.3 a. *P.Berlin 6619* # 1: Metric algebra

P.Berlin 6619 (Schack-Schackenburg, *ZÄS* 38 (1900)) consists of two papyrus fragments, of about the same age as *P.Moscow*. Here is a rough summary of the exercise on the obverse of the larger of the two fragments (cf. the suggested reconstructions of the text in Couchoud, *ME* (1993), 132 and Imhausen, *ÄA* (2003), 358, and see the copy of the text in Clagett, *AES* 3 (1999), 416):

> Two quantities are given, one is 2' 4' of the other.
> The sum of the squares with these quantities as side is 100.
> Which are the quantisties?
> Take a square with 1 as its side. Then the other square has 2' 4' as its side.
> The area of the first square is 1, and the area of the second square is 2' $\overline{16}$.
> The sum of the areas is 1 2' $\overline{16}$ (in the text by mistake written as 1 2' 4' $\overline{16}$).
> The corner (= the square root) of this sum is 1 4', and the corner of 100 is 10.
> You divide this 10 with this 1 4'.
> The result is 8, the first quantity.
> You multiply 8 with 2' 4', the result is 6, the other quantity.

Clearly, what is going on here is that the rule of false value is used in order to compute the solution to the following quadratic-linear system of equations:

    sq. a + sq. b = 100, $b = a \cdot$ 2' 4'.

The solution is found to be $a = 8$, $b = 6$, which is correct, since sq. 8 + sq. 6 = 64 + 36 = 100.

This exercise is interesting in several ways. It shows that Egyptian students of mathematics were supposed to be able to compute square roots (actually *squaresides*, that is sides of squares). In the present case, the square roots can have been computed by use of the OB *squareside rule*, incorrectly known under the name of "Heron's rule", although frequently used in OB mathematics long before the time of Heron:

sqr. 1 2' 16 = appr. 1 + 2' 16 / (2 · 1) = appr. 1 4'.
Verification: 1 4' · 1 4' = 1 4' + 4' 16 = 1 2' 16.

The choice of data in the exercise is interesting, because it shows that Egyptian mathematicians knew that sq. 6 + sq. 8 = sq. 10. Besides, what can be the origin of a mathematical problem of this type, other than an application of the "diagonal rule" which says that *the square on the diagonal of a rectangle is equal to the sum of the squares on the length and the width of the rectangle.* This rule was well known and thoroughly understood in OB mathematics. See the thorough discussion in Friberg, *MCTSC* (2004), App. 7 of the famous table text **Plimpton 322** and of other applications of the diagonal rule in OB mathematical texts. (Euclid's *Elements* I:47 is a Greek reformulation of this OB rule in terms of the sides of a right-angled triangle.)

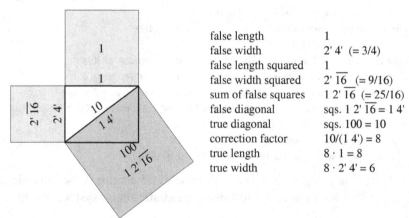

false length	1
false width	2' 4' (= 3/4)
false length squared	1
false width squared	2' 16 (= 9/16)
sum of false squares	1 2' 16 (= 25/16)
false diagonal	sqs. 1 2' 16 = 1 4'
true diagonal	sqs. 100 = 10
correction factor	10/(1 4') = 8
true length	8 · 1 = 8
true width	8 · 2' 4' = 6

Fig. 2.3.1. A geometric model for the metric algebra problem in *P.Berlin 6619* # 1.

As mentioned already, it has been shown by Høyrup (see, for instance, Høyrup, *LWS* (2002) Sec. IV) that OB mathematicians used geometric models for their visualization of quadratic equations. If Egyptian mathematicians relied on geometric models, too, then their model for the problem in *P.Berlin 6619* # 1 may have been a figure like the one in Fig. 2.3.1.

An OB parallel to *P.Berlin 6619* # 1 is exercise # 13 in **BM 13901** (Neugebauer, *MKT 3* (1937), 1]) a large OB theme text with 24 exercises:

BM 13901 # 13.

1	a.šà šita mi-it-ḫa-ra-ti-ia ak-mur-ma
	28 20 /
2	mi-it-ḫar-tum ra-bi-a-at mi-ḫa-ar-tim /
3	4 ù 1 ta-la-pa-at
4	4 ù 4 tu-uš-ta-kal 16 /
	1 ù 1 tu-uš-ta-kal <1>
	1 ù 16 ta-ka-mar-ma 17! /
5	igi 17 ú-la ip-pa-ṭa-ar
6	mi-nam a-na 17 lu-uš-ku-un / ša 28 20 i-na-di-nam
	1 40.e 10 íb.si₈ /
7	10 a-na 4 ta-na-ši-ma 40 mi-it-ḫar-tum iš-ti-a-at /
8	10 a-na 1 ta-na-ši-ma 10 mi-it-ḫar-tum ša-ni-tum

The fields of my two equalsides I added together: 28 20.
Equalside is a fourth of equalside.
4 and 1 you write down.
4 and 4 you let eat each other, 16.
1 and 1 you let eat each other, <1>.
1 and 16 you add together: 17.
The opposite of 17 does not resolve.
What to 17 shall I set, so that 28 20 it will give me?
1 40, (which) makes 10 equalsided.
10 to 4 you raise: 40, one equalside.
10 to 1 you raise: 10, the second equalside.

In this exercise, the rule of false value is used to compute the solution to the system of equations

sq. a + sq. b = 28 20, $b = a/4$.

If $a = 4$, then $b = 1$, and the sum of the squares of 4 and 1 is 17. Since 17 is not a regular sexagesimal number, its reciprocal cannot be computed. Nevertheless, 17 goes 1 40 times into 28 20, and sqr. 1 40 = 10. Hence, the true values of a and b are 10 times the false values 4 and 1.

This solution procedure is identical with the one in *P.Berlin 6619* # 1.

Another OB parallel to *P.Berlin 6619* # 1 is the reverse of the round hand tablet ***UET 6/2 274*** (Friberg, *RA* 94 (2000) § 2e), with numbers arranged in several rows and four columns.

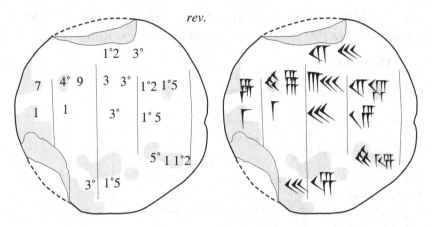

Fig. 2.3.2. *UET 6/2* 274, *rev.* A tabular array for a quadratic-linear system of equations.

This tabular array of numbers gives in a brief but translucent form the solution algorithm for the following quadratic-linear system of equations:

sq. a + sq. b = 12 30 (00), $b = a/7$.

The solution procedure begins in col. *i* with the false values $a = 7$, $b = 1$. These false values satisfy the second equation, but not the first, since sq. $a = 49$, sq. $b = 1$, as noted in col. *ii*, so that sq. a + sq. $b = 50$, instead of 12 30. The false value 50 is noted in the lower right part of the array, and beside it is noted its reciprocal, 1 12 (actually ;01 12). The next step in the solution algorithm is to multiply 12 30 (00) by ;01 12, which is the same as dividing 12 30 (00) by 50. The result is 15 (00). The squareside of 15 (00) is 30, which is the "correction factor" in this application of the rule of false value. The two numbers 15 and 30 are recorded at the bottom of the tabular array.

Next, the false values 7 and 1 are multiplied with the correction factor 30. The result is the true values, $7 \cdot 30 = 3\ 30$ and $1 \cdot 30 = 30$, recorded in col. *iii*. This is the solution to the stated system of equations. However, the procedure continues with a verification of the result. Thus, in col. *iv* are recorded the squares 12 15 (00) and 15 (00) of the computed values. It is now easy to verify that their sum is 12 30, as required. The sum is recorded at the top of the array.

2.3 b. *P.UC 32160* (= *Kahun IV. 3*) # 1: A cylindrical granary

P.UC 32160 (=*Kahun IV.3*) (Griffith, *HPKG* (1898); Imhausen and Ritter, *UCLLP* (2004)) is a papyrus fragment with two mathematical exercises, both in the form of calculations without any accompanying text. The first exercise (Fig. 2.3.3 below) begins with a drawing of a circle, with the numbers 12 and 8 inscribed above and beside it, and with the number 1,365 3' inside it. Under the drawing, the text continues with a series of computations:

a) 1 3' · 12 = 16, b) 16 · 16 = 256, c) 5 3' · 256 = 1,365 3'.

Presumably, the number 5 3' appearing out of nowhere in c) must be understood as 3" · 8. Since the result of the whole computation is the product of a square and a third number, it seems to be motivated to conjecture that the exercise is the computation of a volume. If so, the drawing of a circle suggests that what is computed here is the content of a cylindrical granary with a circular base (cf. Gillings, *MTP* (1982), 151). Now, in *P.Rhind* ## 41-42, the content C of a cylinder of diameter d and height h, expressed in *khar* (= 2/3 cubic cubit), is computed as follows:

C = sq. $(d - \overline{9} \cdot d) \cdot h \cdot 1\ 2'$
(V = sq. $(d - \overline{9} \cdot d) \cdot h$ is the *volume* of the circular cylinder)

The computation in *P.UC 32160* # 1 makes use, instead, of a *modified* equation for the content:

C = sq. $(d + 3' \cdot d) \cdot (3" \cdot h)$.

The two equations for C are mathematically equivalent. Indeed, since sq. 8/9 · 3/2 = 32/27, and also sq. 4/3 · 2/3 = 32/27, both can be replaced by an equation with a single constant:

C = sq. $d \cdot h \cdot 32/27$.

Since fractions of the type 32/27 were not allowed in hieratic mathematical texts, the constant 32/27 had to be replaced by something more legitimate, either sq. $(1 - \overline{9}) \cdot 1\ 2'$ or sq. $(1 + 3') \cdot 3"$. The first alternative had the advantage of letting the volume be computed as an intermediate result, while the second alternative led to slightly less laborious computations.

The unexpected fact that some Egyptian mathematician was capable of manipulating an equation in this nonchalant, yet competent way tends to support the interpretation of *P.Moscow* # 10 suggested above (see Sec. 2.2e), namely that the equation $A = [d \cdot (1 - 1/9) \cdot (1 - 1/9)] \cdot d/2$ for the area of a semicircle was derived from the equation $A = \text{sq.}\ [d \cdot (1 - 1/9)]$ for the area of a circle!

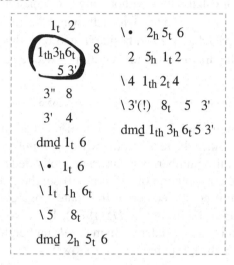

Fig. 2.3.3. *P.UC 32160 (Kahun IV.3)* # 1. The content of a cylindrical granary.

OB mathematicians were capable of manipulating equations in a similar way. An interesting example is offered by the large OB recombination text **BM 85194** (Neugebauer; *MKT 1* (1935),142; see the outline of the text below, in Fig. 2.4.1). This is a text with a rather indiscriminate accumulation of mathematical exercises from a number of different sources. In § 22 (## 34-35) of that text, the content of a circular cylinder is computed in the obvious way, but in § 12 (## 22 and 24) of the same text an alternative method is used. Here is a (corrected) transliteration and translation of BM 85194 # 35:

BM 85194 # 35.

1	1 sìla dug? 10 šu.si *am-ṣu-ur*
2	uš en.nam / *al-li-lik* (*al-li-ik* !)
	za.e
	10 nigin 1 40 *ta-mar*

2.3. Hieratic Mathematical Papyrus Fragments

3	1 40 / *a-na* 5 gúr *i-ši* 8 20 *ta-mar*
4	igi 8 20 du₈.a / 7 12 *ta-mar*
	igi 6 40 *a-na* 7 12 *i-ši* /
5	1 04 48 uš *ta-mar ne-pé-šum*

1 (00) sìla (is) a jar$^?$, 10, fingers, it goes around.
The length, what did I go?
You:
10 square, 1 40 you see.
1 40 to 5, (of) the circle, raise, 8 20 you see.
The opposite of 8 20 resolve, 7 12 you see.
The opposite of 6 40 to 7 12 raise,
1 04 48, the length, you see. The doing.

This is a somewhat difficult text. What it means is that the content of a jar(?) or pipe(?), assumed to be in the form of a circular cylinder, is 1(00) sìla (= 1 barig). The sìla here is what may be called a "cylinder sìla". (See Friberg, *BaM* 28 (1997) § 7 d.) The cylinder sìla is such that 6 40 (00) sìla together have a volume of 1 šar = 1 sq. ninda · 1 cubit. (The constant '6 40' is the storing number for the cylinder sìla.) The circumference a = '10' of the cylinder is also known. Somewhat misleadingly, the word 'finger' is added after the number '10' in order to show that the circumference is ;10 (= 1/6) ninda and not 10 ninda. Actually, ;10 ninda = 2 cubits = 1 00 fingers. The 'length' (or height?) h is computed by use of the following equation for the content C of a cylinder:

C = ;05 · sq. a · h · 6 40 (00) sìla/šar.

Hence, in Babylonian "relative" sexagesimal numbers without zeros,

$h = C \cdot 1/(5 \cdot \text{sq. } a) \cdot 1/6\ 40$.

Consequently, as in the text of BM 85194 # 35,

when a = 10 and C = 1, then
h = 1/(5 · sq. 10) · 1/6 40 = 1/8 20 · 1/6 40 = 7 12 · 9 = 1 04 48.

The precise answer, which is not given in the text, is that h = 1;04 48 cubit.

Now consider, instead, the following exercise from the same clay tablet:

BM 85194 # 22.

1	giš rí.ba.ga (ba.rí.ga!) 4 dal 1_{bg} še bùr! (gúr) ù gúr en.nam /
2	za.e dal nigin 16 *ta-mar*
3	igi 16 du₈.a / 3 45 *ta-mar* 3 45 bùr *ne-pé-šum*

A barig vessel. 4 the transversal, 1(barig) of barley.
The depth and the arc are what?
You:
The transversal square, 16 you see.
The opposite of 16 resolve, 3 45 you see.
3 45 the depth. The doing.

Here, precisely as in # 35, a circular cylinder has the given content 1 (00) sìla, and the object of the exercise is to compute the height of the cylinder, called the 'depth'. Instead of the circumference a, the diameter d is given, $d = $ '4' (meaning ;04 ninda). Neither the circle constant '5' nor the storing number '6 40' for the cylinder sìla is mentioned. That is because here the complicated equation used in exercise # 35 for the content C of a circular cylinder has been transformed into a simpler equation. Indeed, since the circumference $a = $ (appr.) $3 \cdot d$, and since $9 \cdot 6\ 40 = 1\ (00\ 00)$, it follows that

$C = $;05 · sq. $a \cdot h \cdot 6\ 40\ (00)$ sìla/šar $= $;05 · sq. $(3\ d) \cdot h \cdot 6\ 40\ (00)$ sìla/šar
$= $ sq. $d \cdot h \cdot 5\ (00\ 00)$ sìla/šar.

Here, as normally in OB mathematical texts, it is assumed that the cubit is used as the unit of length for the vertical dimension, in particular for the height h of the cylinder. However, in BM 85194, *it is silently assumed that the height is expressed as a multiple of the ninda*, 12 times larger than the cubit. Therefore the mentioned equation for C has now changed into

$C = $ sq. $d \cdot 12 \cdot k \cdot 5\ (00\ 00)$ sìla/šar $= $ sq. $d \cdot k \cdot 1\ (00\ 00\ 00)$ sìla/šar,
where k is the new value for the height.

Hence, in Babylonian relative sexagesimal numbers without zeros,

$k = C \cdot 1/$ sq. d.

Consequently, as in the text of BM 85194 # 22,

when $d = 4$ and $C = 1$, then $k = 1/16 = 3\ 45$ (ninda).

2.3. Hieratic Mathematical Papyrus Fragments

The precise answer, which is not given in the text, is that $h = ;03\ 45$ ninda ($= ;45$ cubits).

2.3 c. *P.UC 32160* # 2: An arithmetic progression with 10 terms

The second exercise on the papyrus fragment *P.UC 32160* ($= Kahun\ IV.\ 3$) is shown in Fig. 2.3.4 below (the photographic detail is borrowed from Griffith, *HPKG* (1898)):

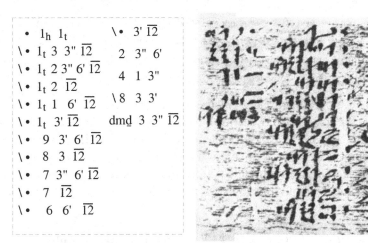

Fig. 2.3.4. *P.UC 32160* ($= Kahun\ IV.\ 3$) # 2. An arithmetic progression.

Just like *P.UC 32160* # 1, this is a computation without any accompanying text. A correct interpretation of the exercise was first published in Cantor, *OLZ* 1 (1898).

In the first column are listed 10 terms of a decreasing arithmetic progression, headed by the numbers 100 and 10. The second column contains the explicit multiplication $9 \cdot 3'\ \overline{12} = 3\ 3"\ \overline{12}$. Apparently, the whole computation is the solution to a question of the following kind:

> 100 (units) are divided among 10 (men). Each one gets 3" 6' more than the next. Which are the shares?

The first step of the solution was probably to compute the average share, $100/10 = 10$. Next, the biggest share could be computed as the average share plus 9 times half the constant difference, that is as

$$10 + 9 \cdot 3'\ \overline{12} = 10 + 3\ 3"\ \overline{12} = 13\ 3"\ \overline{12}.$$

The other shares could then be computed, one by one, through subtraction of the constant difference.

A closely related exercise is **P.Rhind # 64** (see above, Sec. 2.1 c), where 10 *heqat* of barley is divided among 10 men, with a constant difference of 1/8 *heqat*. As mentioned in the discussion of that text, an OB parallel text is Str. 362 # 1.

In the opinion of Imhausen and Ritter, *UCLLP* (2004), 85-86,

"The organization of each calculation into double columns shows that they are the carrying out of mathematical operations in some solution algorithm, now lost. Such 'workings' are common in other mathematical papyri, either individually following each step of the algorithm or collected together, as here at the end of a problem."

However, rather than being "mathematical operations" in some lost solution algorithms, *P.UC 32160* ## 1-2 are probably Egyptian counterparts to the well known OB category of mathematical so called "hand tablets", brief numerical outlines of solution procedures, written either by students listening to a teacher's detailed exposition of the solution algorithm for a mathematical problem, or by students getting their assignments from the teacher. Cf. Fig. 2.4.2 below.

2.3 d. *P.UC 32161* (= *Kahun XLV. 1*): A list of large numbers

P.UC 32161 (= *Kahun XLV.1*) (Griffith, *HPKG* (1898); Imhausen and Ritter, *UCLLP* (2004)) is a small, extremely badly preserved papyrus fragment, inscribed with a list of eight long decimal numbers, all of them ending with at least one fraction. A mirror image conform transliteration of the fragment is shown in Fig. 2.3.5 below. (The conform transliteration is based on the color photo in Imhausen and Ritter (*op. cit.*). Actually, the fragment is even more damaged than what can be shown here.)

No attempt has previously been made to explain the eight numbers, of which none appears to be wholly preserved. Note that if the beginnings of the eight numbers were originally vertically aligned, then the beginning of the first number is missing.

Many examples of OB hand tablets inscribed with one or several very long numbers are known. See Friberg, *MCTSC* (2005), Sec. 1.4-5, a detailed discussion of OB hand tablets with "many-place" sexagesimal numbers. The first step in a successful attempt to understand the meaning

2.3. Hieratic Mathematical Papyrus Fragments

of a given many-place sexagesimal number is normally to look for its factorization into a product of prime numbers. Will the same method work in the case of the "many-digit" *decimal* numbers recorded on *P.UC 32161*?

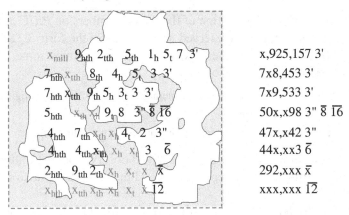

x,925,157 3'
7x8,453 3'
7x9,533 3'
50x,x98 3" $\bar{8}$ $\overline{16}$
47x,x42 3"
44x,xx3 $\bar{6}$
292,xxx $\bar{x}$
xxx,xxx $\overline{12}$

Fig. 2.3.5. *P.UC 32161* (= *Kahun XLVI.1*). Eight many-digit decimal numbers.

The first of the eight number ends in 3' = 1/3. Therefore, the number must be 1/3 of an integer:

x,925,157 3' = 3' · (3 · x,925,157 + 1) = 3' · (x · 3,000,000 + 2,775,472).

The even number 2,775,472 must have 2 as a factor. As it turns out, it has also 2^4 = 16 as a factor:

2,775,472 = 16 · 173,467.

It is not difficult to find out that 7 is a further factor. More precisely,

925,157 3' = 3' · 16 · 7 · 24,781,   where   24,781 is a prime number.

(Actually, 24,781 is the 2,740th odd prime number according to, for instance, Weil's on-line table of the first 28,915 odd primes.)
Obviously, the number 3,000,000 has 16 as a factor, but not 7. Therefore, the final result is that

x,925,157 3' = 3' · (x · 3,000,000 + 2,775,472) = 3' · 16 · (x · 187,500 + 173,467).

Although the second and third numbers are damaged, they can be treated in a similar way:

7x8,453 3' = 3' · (2,125,360 + x · 30,000) = 3' · 16 · 5 · (26,567 + x · 375),
7x9,533 3' = 3' · (2,128.600 + x · 30,000) = 3' · 8 · 25 · (10,643 + x · 150).

Even the fifth number, with two missing digits, yields to some extent to the same kind of analysis:

47,xx42 3" = 3' · (1,410,128 + xx · 100) = 3' · 4 · (352,532 + xx · 25).

The final result of this inspection of the list of numbers on *P.UC 32161* is that the three first and best preserved numbers are of the form 3' · 8 · some integer, while the fifth number is of the form 3' · 4 times some integer, which is even if xx is even. Consequently:

> The first three, and possibly also the fifth number on *P.UC 32161* were originally of the form 3' · 8 · some integer.
>
> At least three of the other four numbers were obviously not of this form.

The second statement follows from the trivial observation that a number of the form 3' · 8 times some integer cannot end in any kind of fractions other than either 3' or 3".

It can hardly be a coincidence that the first three of the eight numbers recorded on the fragment *P.UC 32161* are products of the factor 3' · 8 = 2 3"(= 8/3) and various integers. Apart from that, it is difficult to say anything more about the meaning of those numbers, obviously due to the fact that they are all damaged in one way or another. One can only speculate that they may have been the data for a series of assignments of mathematical exercises.

2.4. Conclusion

New thoughts about the nature of *P.Rhind*

The detailed discussion in sections 2.1-3 above of selected exercises in *P.Rhind*, *P.Moscow*, and the hieratic mathematical papyrus fragments, allow some novel conclusions to be drawn about the nature of these manuscripts, and about the nature of hieratic Egyptian mathematics in general.

Take a look, for instance, at the table of contents for *P.Rhind* (see Sec. 2.1 a above). The 84 exercises of the papyrus can be organized into 36 paragraphs belonging to 12 disparate themes. Compare with the corresponding table of contents (in Fig. 2.4.1 below) for the mathematical text BM 85194, one of several known large mathematical cuneiform texts from Sippar (late OB). Its 35 exercises can be organized into 22 paragraphs

2.4. Conclusion

belonging to various themes, such as

Volumes of large objects	§§ 1-4, 8, 10, 14, 16
Contents of cylindrical vessels	§§ 12, 22
Circle segments	§§ 11, 17
etc., etc.	

BM 85194 is an example of what has here been called a "recombination text". The reason for this designation is the following deliberation. OB mathematical cuneiform texts can be divided into a number of different categories with respect to their interior organization and for what purpose they (probably) were written. It is likely that OB mathematics had its origin in the activities of a relatively small number of particularly gifted but anonymous teachers of mathematics from several important Mesopotamian cities. Building upon old ideas inherited from the pre-Babylonian mathematics of the 3rd millennium, but also exploiting their own new ideas, these clever teachers composed various "theme texts". By that is meant cuneiform texts with long series of closely related exercises (problems with or without explicit solution procedures and answers), often beginning with simple computations and then continuing with permutations of the data and other variations and expansions of the theme. Examples in Neugebauer and Sachs, *MCT* (1945), for instance, are text O (10 problems for 5 types of bricks), text R (22 linear equations for weight stones), texts K and L (40 problems for a small canal), and so on. Text A (the famous table text **Plimpton 322**) is also such a theme text, with its 17 different sets of data for a certain type of quadratic equations (see Friberg, *HM* 8 (1981) and Friberg, *MCTSC* (2005), App. 7). The original combined multiplication table with its up to 42 different head numbers (*op. cit.*, App. 2), and the original combined metrological table with its sub-tables for capacity, weight, area, and length measures (*op. cit.*, App. 5), are other special types of theme texts.

Theme texts were, of course, a great teaching tool, but they seem to have been used frequently also as rich sources of writing exercises or mathematical assignments, by teachers who wanted to give their students *related but non-identical* examples to work with. That is probably why there exist in the various collections of clay tablets so many small cuneiform tablets with single multiplication tables for different head numbers, so many different brief excerpts of metrological tables for measures of capacity, or weight, or area, or length, and that is why there exist so many small clay tablets with only one or a few related mathematical exercises.[18]

Apparently, now and then a teacher who was not in possession of one of the original theme texts wanted to make his own substitute for a theme text, a collection of exercises to be used as a new source of individual assignments. He then collected as many small mathematical cuneiform texts as he could find, sorted them in some way as well as he could, and copied them onto one or several large clay tablets. This is the most likely explanation for the existence of large mathematical cuneiform texts containing many exercises that can be organized into a fairly great number of more or less closely related paragraphs. There are many known Old, and Late, Babylonian examples of such mathematical recombination texts, in addition to BM 85194, the text shown in Fig. 2.4.2.

A look at the table of contents for *P.Rhind* ought to make it clear that that papyrus is a mathematical recombination text of the same general type as BM 85194. Cf. Peet, *BJRL* 15 (1931), 439, footnote 1:

> "The arrangement of Rhind itself is logically far from perfect, and while we may be prepared to find excuses for this in the supposition that the collection was culled, somewhat at random, from other mathematical treatises, we cannot submit to its being held up as a model of consistent and logical arrangements."

---

18. Cf. the detailed discussion in Proust, *TMN* (2004), of the corpus of OB mathematical cuneiform texts from Nippur. According to Proust, at the elementary level of the education in the Old Babylonian scribe schools at Nippur, four series of metrological and mathematical table texts were studied in the following order: a) *metrological lists*, b) *metrological table*s c) *arithmetic tables* (the *table of reciprocals*, the various *multiplication tables*, and the *table of squares*), d) the *tables of square sides and cube sides*. All other mathematical cuneiform texts are relegated by Proust to the category of *exercises at an advanced level*. The students learned to write small sections of the mentioned table texts, one at a time, carefully recording them on the obverse of medium size clay tablets of "type II", and using the reverse of the same tablets to rehearse previously learned sections. Small clay tablets of "type 3", on the other hand, are inscribed on both obverse and reverse with a single section of either a mathematical table text or of a mathematical problem text. Proust does not make it clear what role tablets of type 3 played in the OB mathematical curriculum, but vaguely suggests that they may have been used for examinations of the skills of advanced students. She also points out that tablets of type 3 are particularly wanted by collectors and museums, and that for that reason tablets of type 3 are unproportionately well represented in private collections and certain museum collections. Thus, for instance, the majority of the tablets from the Schøyen collection discussed in Friberg, *MCTSC* (2005) are of type III.

2.4. Conclusion

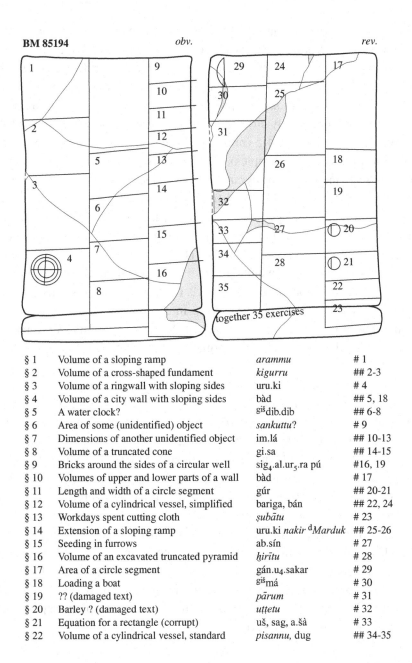

§ 1	Volume of a sloping ramp	*arammu*	# 1
§ 2	Volume of a cross-shaped fundament	*kigurru*	## 2-3
§ 3	Volume of a ringwall with sloping sides	uru.ki	# 4
§ 4	Volume of a city wall with sloping sides	bàd	## 5, 18
§ 5	A water clock?	^{giš}dib.dib	## 6-8
§ 6	Area of some (unidentified) object	*sankuttu*?	# 9
§ 7	Dimensions of another unidentified object	im.lá	## 10-13
§ 8	Volume of a truncated cone	gi.sa	## 14-15
§ 9	Bricks around the sides of a circular well	sig_4.al.ur_5.ra pú	#16, 19
§ 10	Volumes of upper and lower parts of a wall	bàd	# 17
§ 11	Length and width of a circle segment	gúr	## 20-21
§ 12	Volume of a cylindrical vessel, simplified	bariga, bán	## 22, 24
§ 13	Workdays spent cutting cloth	*ṣubātu*	# 23
§ 14	Extension of a sloping ramp	uru.ki *nakir* ^{d}*Marduk*	## 25-26
§ 15	Seeding in furrows	ab.sín	# 27
§ 16	Volume of an excavated truncated pyramid	*ḫirītu*	# 28
§ 17	Area of a circle segment	gán.u_4.sakar	# 29
§ 18	Loading a boat	^{giš}má	# 30
§ 19	?? (damaged text)	*pārum*	# 31
§ 20	Barley ? (damaged text)	*uṭṭetu*	# 32
§ 21	Equation for a rectangle (corrupt)	uš, sag, a.šà	# 33
§ 22	Volume of a cylindrical vessel, standard	*pisannu*, dug	## 34-35

Fig. 2.4.1. BM 85194. An OB mathematical recombination text with mixed content.

In *P.Rhind*, the various paragraphs can be interpreted as longer or briefer excerpts from a great variety of theme texts. The $2/n$ table at the beginning of the papyrus, with its 50 exercises, is an example of a *complete* theme text copied into the recombination text. All other themes or individual paragraphs in *P.Rhind* comprise from only one to a few exercises. Paragraphs with relatively large numbers of exercises are § 3 (n loaves for 10 men; 6 examples), §§ 4-5 (multiplication problems; 14 examples), § 8 (division problems; 8 examples), and the elementary stereometric exercises in § 21 (inclination of pyramids; 5 examples). Well-structured paragraphs are, for instance, §§ 13-14 (round and square granaries), and § 17 (areas of the four basic geometric figures). Many of the problems on the reverse belong to Theme H (baking or brewing numbers). They may all be excerpts from a single large theme text. However, other paragraphs consist of only one or a couple of examples, such as §§ 6-7 (subtraction problems), § 9 (iterated division problems), § 12 (arithmetic progressions), § 20 (applied division problems), and so on. In a couple of cases, the modern numbering of the exercises in the papyrus has been manipulated, to make it appear that the text is more well organized than it really is (## 33-34, 54, and 79).

On a larger scale, at least the obverse of *P.Rhind* is fairly well organized. (In the following discussion of the internal structure of *P.Rhind*, some of the ideas have been borrowed from the close examination of the organization of the papyrus in Spalinger, *SAÄK* 17 (1990).) Thus, it appears that the $2/n$ table in § 1 is accompanied by simple applications in the form of division and multiplication problems in Themes A-B (§§ 2-5). Then follows a blank space, separating these trivial exercises from the completion problems in Theme C. A second blank space separates the completion problems from the division and sharing problems in Themes E-F. Then comes a really big blank space separating the *arithmetic* exercises in Themes A-F from the *geometric* exercises in Theme G, which fill out the rest of the obverse of the papyrus. The text on the reverse seems to be less well organized. In particular, there are no blank spaces separating groups of exercises on the reverse.

The table of contents for *P.Moscow* with its 25 exercises and 9 paragraphs (see Sec. 2.2 a) shows that also *P.Moscow* is an Egyptian mathe-

2.4. Conclusion

matical recombination text. Now, if *P.Rhind* and *P.Moscow* are compared with each other, it is easy to note an obvious and important difference between the two texts, namely that most of the exercises in *P.Rhind,* but none of the exercises in *P.Moscow,* are accompanied by detailed computations. The detailed computations in *P.Rhind* are often preceded by remarks like "The doing as it occurs", "The working out of it", *etc.* Only the exercises ## 57, 59, 71-72, and 78, in *P.Rhind* lack such detailed computations. In Clagett, *AES 3* (1999), 209, the fact that there are no detailed computations in *P.Moscow* is explained as follows:

> "This (the complete lack of system in the Moscow papyrus) leads to the conclusion that the author of the Moscow Papyrus was a student whose training had progressed far enough for the teacher to present various problems to be solved in order to test the skill of the student. The student was apparently not required to present in tabular form, like that found in the Rhind Papyrus, the steps by which the multiplications were carried out, i. e. the doublings, halvings, taking of 2/3 and/or 1/3, and decemplex-multiplications, but rather just to give the results of such multiplications."

A comparison with OB mathematical texts shows that Clagett's explanation misses the point, as there are *never* any detailed computations (such as explicit multiplications) present in mathematical cuneiform texts. The obvious reason for this is that there simply was not space enough on a clay tablet for extensive computations. The situation cannot have been very different for an Egyptian teacher or advanced student of mathematics who recorded collections of mathematical exercises on expensive rolls or sheets of papyrus. Therefore, it is likely that texts like *P.Moscow* were the norm, while *P.Rhind* with its many explicit and detailed computations was an exception. It is also likely that Egyptian school boys wrote their mathematical "brief notes" on small scraps of reused papyrus, or pot shards, or whatever material they could find.

Three of the hieratic mathematical papyrus fragments probably show what mathematical brief notes written by Egyptian school boys normally may have looked like, namely *P.UC 32159*: a $2/n$ table, for n from 3 to 21, *P.UC 32161*: a list of large numbers, and *P.UC 32160* with a) the computation of the content of a cylinder, b) a problem for an arithmetic progression. (Compare with OB mathematical "hand tablets", such as the round clay tablets from Ur discussed in Friberg, *RA* 94 (2000), or the square clay tablets with multiplication, squaring, and division exercises discussed in Friberg, *MCTSC* (2005), Sec. 1.) Three interpolated exercises in *P.Rhind*, possibly late additions to the document, namely # 48 (the area of a circle *vs.* the area of a square), # 53 a-b (a striped triangle and a rectangle with an almost round area number), and # 79 (a "house inventory"), look very much like such school boys' mathematical brief notes. Also # 61 (a general rule for taking 3" of a part $\bar{n}$, followed by a multiplication table for fractions) is a similar brief note at the beginning of the reverse of *P.Rhind*. Cf. the following comment in Chase, *et al.*, *RMP 2* (1929), pl. 83:[19]

> "The table and rule ⋯ are not a part of the text. They are somewhat more hastily written and were evidently jotted down for reference in the blank margin to the right of the point where the text on the *verso* was to begin."

The outline below of *P.UC 31260* is based on a color photo of the text in Imhausen and Ritter, *UCLLP* (2004). The outline clearly suggests that *P.UC 31260* is an example of an Egyptian school boy's mathematical notes, rather than being a "fragment" of a larger mathematical text!

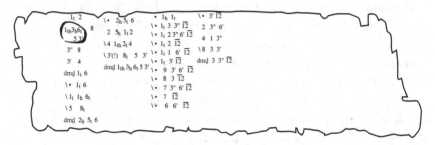

Fig. 2.4.2. *P.UC 32160*. Mathematical brief notes of an Egyptian school boy.

Now, if *P.Rhind* is a recombination text, that is a compilation of a substantial number of students' brief notes or excerpts from longer theme texts, then the fact that almost all the exercises in *P.Rhind* are accompanied by explicit computations can only be explained in the following way: The

---

19. In the preface to *P.Rhind* it is stated that the text is a copy of an older text from the time of the sixth king of the Twelfth Dynasty, in the Middle Kingdom. The interpolated exercises may, of course, have been added to the copy, not to the original. Also other parts of the text may be late additions. See, for instance, Spalinger, *SAÄK* 14 (1987), a paper about the "grain system of Dynasty 18", where it is claimed that some of the exercises in *P.Rhind* must be late additions because they use a late kind of capacity units.

2.4. Conclusion

author of the papyrus wanted to use it *primarily* in order to teach his students how to perform detailed computations. That may be what is meant by the first words of the Introduction: "Accurate (*lit.* Head of) reckoning"! Therefore, he *added* explicit computations to most of the problems he had compiled. How far he went in this respect can be illustrated by, for instance, the solution to the division exercise $f \cdot a = 1$ (*heqat*), $f = 3\ 3'\ (3'\cdot 3')\ \overline{9}$, in exercise # 37. The explicit computations in that exercise are the following:

a) the addition $\quad f = 3 + 3' + 3'\cdot 3' + \overline{9} = 3\ 2'\ \overline{18}\ (= 32/9)$
b) the division $\quad 1/f = 4'\ \overline{32}$ (*heqat*) (= 9/32) $\quad$ the answer as *a sum of parts*
c) the details of the counting in b) $\quad\quad\quad\quad\quad\quad\quad\quad\quad$ summation of parts
d) the multiplication $\ 3\ 3'\ (3'\cdot 3')\ \overline{9}\cdot 4'\ \overline{32}$ (*heqat*) = 1 (*heqat*) $\quad$ verification of b)
e) the details of the counting in d) $\quad\quad\quad\quad\quad\quad\quad\quad\quad$ summation of parts
f) the multiplication $\ 4'\ \overline{32}\cdot 320$ (*ro*) = 90 (*ro*) $\quad$ the answer as *a multiple of the ro*
g) the multiplication $\ 3\ 3'\ (3'\cdot 3')\ \overline{9}\cdot 90$ (*ro*) = 320 (*ro*) $\quad$ verification of f)
h) the conversion $\quad 90$ (*ro*) = $4'_h\ 32'_h$ $\quad\quad\quad$ the answer in *heqat fractions*
i) the multiplication $\ 3\ 3'\ (3'\cdot 3')\ \overline{9}\cdot 4'_h\ 32'_h = 1$ (*heqat*) $\quad$ verification of h)

Note that the answer is given in three different forms: in b) as $4'\ \overline{32}$, a sum of parts, in f) as 90 *ro*, and in h) as $4'_h\ 32'_h$, in binary *heqat* fractions. The details of the computations in d) and i) are not identical. Thus, for instance,

$$3'\cdot 4'\ \overline{32} = \overline{12}\ \overline{96},\ \text{but}\quad 3'\cdot 4'_h\ 32'_h = 16'_h\ 32'_h (1/3 \cdot 90\ ro = 30\ ro = 16'_h\ 32'_h)$$

A detailed outline of the layout of two sections of *P.Rhind* is shown in Figs. 2.4.3-4 below. An effort has been made to demonstrate how most of the exercises begin with the statement of the problem (the question), and then continue with a series of computations (the solution procedure and the answer). This is done by splitting the rectangular outlines of individual exercises into sub-rectangles, one for the statement of the problem, one for the solution procedure, and as many as needed for the various detailed computations associated with the solution procedure.

100   *Unexpected Links Between Egyptian and Babylonian Mathematics*

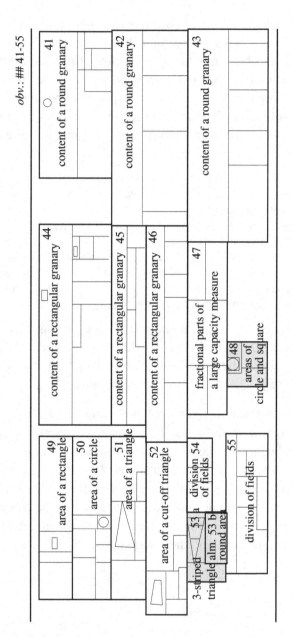

Fig. 2.4.3. The layout of ## 41-55 on the obverse of *P.Rhind*.
Note the different orientations of the triangles in ## 51 and 53 a!
Grey indicates exercises interpolated into available empty spaces.

2.4. Conclusion

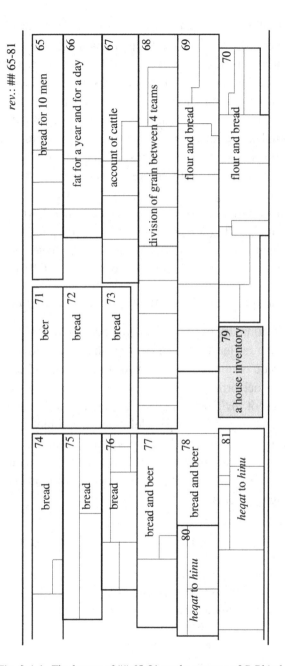

Fig. 2.4.4. The layout of ## 65-81 on the reverse of *P.Rhind*.

From the outline in Fig. 2.4.3, it is evident that in the original layout of the papyrus there was a blank space separating the granary problems in ## 41-47 from the area problems in ## 49-52. Then someone, almost certainly not the author of the main part of the papyrus, interpolated # 48 as a brief explanation of the computations in ## 41-43 of the areas of the circular bases of the round granaries. Similarly, it is clear from the same outline that the incomplete exercises ## 53 a-b were inserted into an empty space to the left of the short text of # 54 by someone who did not really understand what he was doing. Furthermore, it is clear from the outline in Fig. 2.4.4 that the interesting "house inventory" # 79 was inserted, totally out of context, into another empty space, to the left of the short text of the flour and bread exercise # 70. (A much more suitable place for the house inventory exercise with its geometric progression would have been in the large blank space after the arithmetic progression exercise # 40!)

Summary: Comparison of hieratic Egyptian and OB mathematics

Here is a brief summary of the results obtained above, in Sec. 1.1 and Sec. 2.1-3. Altogether, at least 10 clear parallels have been found between exercises in hieratic mathematical papyri or fragments, and exercises in OB mathematical cuneiform texts. The exercises in question have all been chosen as examples of *artificial*, as opposed to *practical*, mathematical problems, because the existence of parallel practical problems is only natural and does not prove anything. (Examples of practical problems are arithmetic exercises of various kinds, sharing problems, computations of areas or volumes, or of inclinations of pyramids, work norm problems, and baking or brewing problems.)

1. *P.Rhind* # 79 // M. 7857.
    Fanciful interpretations of the sum of 5 terms of a geometric progression.
2. *P.Rhind* ## 28-29 // YBC 4652 # 9.
    Repeated division problems.
3. *P.Rhind* # 64 & *P.UC 32160* # 2 // Str. 362 # 1.
    Arithmetic progressions with given common differences.
4. *P.Rhind* # 40 // YBC 9856 & VAT 8522 # 2.
    Arithmetic progressions with given ratios of terms "above" to terms "below".
5. *P.Rhind* # 53 a // Str 364 & IM 43996.
    Complicated problems for striped triangles.

2.4. Conclusion

6. *P.Rhind* # 53 a-b // NCBT 1913 & YBC 7290.
   Almost round area numbers.
7. *P.Rhind* # 72 // YBC 4698 # 6.
   Combined market rate problems.
8. *P.Moscow* # 14 // BM 96954+.
   Correct computations of volumes of truncated pyramids.
9. *P.Moscow* # 17 // YBC 11126 & IM 121613 #1.
   Equations for geometric figures with given area and side ratio.
10. *P.Berlin 6619* # 1 // BM 13901 # 13.
    Quadratic-linear systems of equations for two squares.

The conclusion that can be drawn from the existence of *so many parallels of a non-trivial nature* between hieratic and cuneiform mathematical texts is that Middle Egyptian and OB mathematics must have influenced each other in decisive ways. The many obvious similarities cannot be explained away as due to convergent but independent development. Moreover, there is really no reason to doubt that *the level and extent of mathematical knowledge were nearly identical in Egypt and Mesopotamia at the time when the Egyptian hieratic and OB mathematical texts were written.* The burden of proof ought to be on those who say that this was not the case!

It is unfortunate that the level of difficulty is relatively low in the majority of the problems in *P.Rhind* and *P.Moscow,* in comparison with the level of the most advanced of the OB mathematical texts. However, one must bear in mind 1) that it is natural that the very few known Egyptian mathematical texts cannot be favorably compared with the hundreds of known OB mathematical texts, and 2) that, as was pointed out above, *P.Rhind* seems to have been a mathematical recombination text, compilated and supplied with extremely explicit computations primarily for the purpose of demonstrating elementary calculation techniques, rather than mathematical problem types. Nevertheless, it is clearly indicated by the existence of exercises such as *P.Rhind* ## 40 and 53 a, *P.Moscow* ## 14 and 17, and *P.Berlin 6619* # 1, that in the early part of the second millennium BCE Egyptian mathematics was just as advanced as Babylonian mathematics!

It should further be noted that the implication of this conclusion is that probably *the true extent of Middle Kingdom Egyptian mathematics is not, by far, as well known as it might appear to be*. How can it be, by the way, when even the true extent of the much better documented Old Babylonian mathematics is unknown? Indeed, past experience has shown that each time a new larger cuneiform mathematical text is published, it can be counted on to expand the horizon of the known corpus of Mesopotamian mathematics! For the most recent examples of this phenomenon, see the six new cuneiform problem texts in Friberg, *MCTSC* (2005), Chs. 10-11.

Chapter 3

Demotic Mathematical Papyri and Cuneiform Mathematical Texts

In Parker, *DMP* (1972) were published five Egyptian mathematical papyri written in the demotic script, namely the large *P.Cairo J. E. 89127-30, 89137-43, verso* (consisting of 11 fragments), and the four fragments *P.Br. Museum 10399, 10520, 10794*, and *P.Carlsberg 30*. (A preliminary summary of the contents of *P.Cairo* and the four fragments was published already in Parker, *Cent.* 14 (1969).) Two other fragments published by Parker are *P.Griffith Inst. I E.7*, in *JNES* 18 (1959), and *P.Heidelberg 663*, in *JEA* 61 (1975). The dates given below for the various demotic mathematical papyri are, with one exception, the ones suggested by Parker. All translations of the texts are also, essentially, the ones given by Parker.

In this paper, the mentioned demotic mathematical texts will be searched for traces of possible links with Babylonian mathematics. In the process, some of Parker's interpretations and explanations of the texts will be made considerably more precise and detailed.[20]

3.1. Themes in *P.Cairo* (Ptolemaic, the 3rd C. BCE)

3.1 a. *P.Cairo*, a hieratic mathematical recombination text

---

20. New interpretations are offered in Chapter 3 for the following texts: ***P.Cairo* § 1** (Sec. 3.1 c), **§ 2** (Sec. 3.1 d), **§ 7** (Sec. 3.1 g), **§ 12** (Sec. 3.1 k), ***P.BM 10399* § 1** (Sec. 3.2 a), ***P.BM 10520* § 5** (Sec. 3.3 e), ***P.Carlsberg 30* # 2** (Sec. 3.5 b), ***P.Heidelberg* 663** (Sec. 3.7), **VAT 7531** and **VAT 7621** (Sec. 3.7 c).

The famous legal code of Hermopolis West is inscribed on the obverse (*recto*) of *P.Cairo J. E. 89127-30, 89137-43*. The mathematical text is inscribed on the reverse (*verso*). It is the oldest of all known demotic mathematical texts, dated to the 3rd c. BCE. That means that it is contemporary with some of the Seleucid mathematical cuneiform texts from Mesopotamia. A look at its table of contents, below, shows that *P.Cairo* is a mathematical recombination text of the same kind as *P.Rhind*. It begins with arithmetic exercises, but there are also various kinds of metric algebra exercises (quadratic equations, *etc.*), and geometric exercises for both two- and three-dimensional objects. (A detailed discussion of some interesting aspects of *P.Cairo* can be found in Friberg, *BaM* 28 (1997)/9 b.)

P.Cairo J. E. 89127-30, 89137-43, verso: **Contents.**

...	[Destroyed]	# 1
§ 1 a-b	Division problems:	
	a) $100/(17\ 3") = 5\ 35/53$;	# 2
	b) $100/(15\ 3") = 6\ 18/47$ (binomial fractions)	# 3
§ 2 a-c	Completion problems:	
	a) $6"\ \overline{10}\ \overline{20}\ \overline{120}\ \overline{210} + \overline{280} = 1$	# 4
	b) $6"\ \overline{10}\ \overline{20}\ \overline{120}\ \overline{240}\ \overline{480}\ \overline{510} + \overline{8160} = 1$	# 5
	c) a general rule	# 6
§ 3 a	The area and the side ratio of a sail given.	
	a) $h = \mathrm{sqr.}\ (1000 \cdot 1\ 2')$ c. = appr. $38\ 3"\ \overline{20}$ c., $w = 3" \cdot h$	# 7
§ 4 a-e	Changing the shape of a piece of cloth, keeping the area.	
	a) $7\ \mathrm{c.} \cdot 5\ \mathrm{c.} = (7 - 1)\ \mathrm{c.} \cdot 5\ 6"\ \mathrm{c.}$	# 8
	b) $6\ \mathrm{c.} \cdot 4\ \mathrm{c.} = (6 - 1)\ \mathrm{c.} \cdot 4\ 4/5\ \mathrm{c.} = 5\ \mathrm{c.} \cdot 4\ 3"\ \overline{10}\ \overline{30}\ \mathrm{c.}$	# 9
	c) $6\ \mathrm{c.} \cdot 1\ 2'\ \mathrm{c.} = (6 - 2')\ \mathrm{c.} \cdot (1\ 2' + (1\ 2')/11)\ \mathrm{c.}$	# 10
	d) $6\ \mathrm{c.} \cdot 4\ \mathrm{c.} = 8\ \mathrm{c.} \cdot (4 - 1)\ \mathrm{c.}$	# 11
	e) $6\ \mathrm{c.} \cdot 3\ \mathrm{c.} = 12\ \mathrm{c.} \cdot (3 - 1\ 2')\ \mathrm{c.}$	# 12
§ 5	Pieces of cloth, and silver. Unclear problem.	
	Ratio: $131/60 = 3'\ \overline{15}\ (7\ 2'\ \overline{10})/131$	# 13
§ 4 f	f) $21\ \mathrm{c.} \cdot 5\ \mathrm{c.} = (21 - 1)\ \mathrm{c.} \cdot 5\ 4'\ \mathrm{c.}$	# 14
§ 3 b-c	b) Identical with a);	# 15
	c) $h = \mathrm{sqr.}\ (100 \cdot 1\ 3'\ \overline{15})$ c. = appr. $11\ 6"$ c., $w = 3"\ \overline{30}\ \overline{70} \cdot h$	# 16
§ 4 g	g) $21\ \mathrm{c.} \cdot 5\ \mathrm{c.} = 26\ 4' \cdot (5 - 1)\ \mathrm{c.}$	# 17
§ 3 d	d) $h = \mathrm{sqr.}\ (100 \cdot 10) =$ appr. $31\ 2'\ \overline{10}\ \overline{30}$, $w = h/10 =$ appr. $3\ 6'$	# 18
§ 6	Interest on 50(?) pieces of silver (badly preserved problem)	# 19
...	[Destroyed]	## 20-22
§ 7	A linear equation:	
	$f \cdot a = 2'$, $f = 2' + 2' \cdot 2' + 2' \cdot 2' \cdot 2' = 1\ 6"\ \overline{30}\ \overline{120}$, $a = \overline{5}\ \overline{15}$	# 23

3.1. P.Cairo (Ptolemaic, the 3rd C. BCE) 107

§ 8 a-h	A pole against a wall, d, s given. $(s, h, d) =$	
	a) $2 \cdot (3, 4, 5)$	# 24
	b) $(20, 21, 29)/2$	# 25
	c) $2 \cdot (4, 3, 5)$	# 26
	$d - h$, d given. $(s, h, d) =$	
	d) $2 \cdot (3, 4, 5)$	# 27
	e) $(20, 21, 29)/2$	# 28
	f) $2 \cdot (4, 3, 5)$	# 29
	$d - h$, s given. $(s, h, d) =$	
	g) $2 \cdot (3, 4, 5)$	# 30
	h) $(20, 21, 29)/2$	# 31
§ 9 a-b	The diameter $d =$ sqr. $(A + A/3)$ of a circle with given area	
	a) $A = 100$ sq. c.	# 32
	b) $A = 10$ sq. c.	# 33
§ 10 a-b	The sides of a rectangle with given area and diagonal.	
	a) $A = 60$, $d = 13$	# 34
	b) $A = 60$, $d = 15$	# 35
§ 11 a	An equilateral triangle (side 12 c.) inscribed in a circle.	
	The areas of the triangle and the segments.	# 36
§ 12	A square inscribed in a circle (diameter 30 c.).	
	The areas of the square and the segments.	# 37
§ 11 b	An equilateral triangle (side 10 c.) inscribed in a circle.	
	The areas of the triangle and the segments.	# 38
§ 13	A square pyramid. Height, 300 c., side of the base, 500 c.,	
	find the height of a face.	# 39
§ 14	A square pyramid. Height, 10 c., a side of the base, 10 c.,	
	find the volume.	# 40
§ 15	A quadratic-linear system of equations:	
	sq. $(a + b) = 140$ sq. c., $a/b = 5/2$. Solution: $a = 8\ 3'\ \overline{10}\ \overline{60}$.	# "32"
§ 16	A rectangular-linear system of equations:	
	$a \cdot b = 100$ sq. c., $a - b = 21$ c. Solution: $a = 25$, $b = 4$.	# "33"

3.1 b. *P.Cairo* § 8 (*DMP* ## 24-31). A pole against a wall

In a note about "The position of demotic texts in the history of Egyptian mathematics", Parker makes the important remark (*DMP* (1972), 5) that the existence of a Babylonian influence on Egyptian mathematics in the third century BCE, or earlier, is clearly documented by exercises ## 24-31 (§ 8 a-h) in *P.Cairo*. All these exercises are variants on a single theme, which, in modern notations, can be described as follows:

A pole of length d is leaning against a wall. If its top slides down a certain distance

p, then its base slides out from the wall a corresponding distance s, and a right triangle is formed with the sides s, h, and d, where $h = d - p$. Two of the parameters s, d, and p are known. Find the remaining parameter.

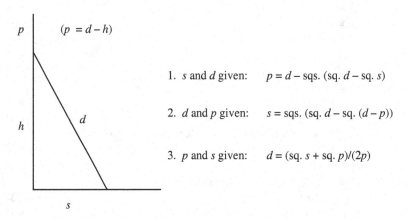

1. s and d given: $p = d - $ sqs. (sq. $d - $ sq. s)

2. d and p given: $s = $ sqs. (sq. $d - $ sq. $(d - p)$)

3. p and s given: $d = ($sq. $s + $ sq. $p)/(2p)$

Fig. 3.1.1. *P.Cairo* § 8 (*DMP* ## 24-31). A pole leaning against a wall.

What Parker had found was that the OB mathematical exercise **BM 85196 # 9** (Neugebauer, *MKT 2* (1935), 53) is essentially identical with *P.Cairo* § 8 d (*DMP* # 27), and that the Seleucid mathematical exercise **BM 34568 # 12** (Neugebauer, *MKT 3* (1937), 22) is essentially identical with *P.Cairo* § 8 g (*DMP* # 30). It is interesting that the demotic papyrus contains all possible variants on the theme, while the OB and Seleucid parallels are incomplete, with only one variant each.

The "pole against a wall" theme is an artificial problem type. Its presence in both a demotic Egyptian and two Babylonian mathematical texts is a first example of possible links between Egyptian and Babylonian mathematics in the late first millennium BCE. However, this is not the only example, as will be shown below, in a systematic survey of the consecutive paragraphs of *P.Cairo*. Cf. the following statement by Høyrup, in *LWS* (2002), 405:

> "In demotic mathematical papyri from the Ptolemaic and Roman period, however, the presence of material with roots in Mesopotamia is indubitable."

In this connection, Høyrup discusses, as examples, ***P.Carlsberg 30* # 1** and ***P.Cairo* § 10** (*DMP* ## 34-35).)

3.1 c. *P.Cairo* § 1 (*DMP* ## 2-3). Two closely related division problems

Here an attempt will be made to describe what is going on in *P.Cairo* § 1 b (*DMP* # 3), which is relatively well preserved. Apparently, the given task is to find out how many times 15 3" goes into 100. The computation begins with the observation that 100 – 6 = 94, and that 6 goes 15 3" times into 96. Next, it is noted that 15 3" = 47 · 3', and that 47 goes 2 6 47 times into 100, where clearly 6 47 stands for (what might be understood as) the common fraction 6/47. (Cf. the discussion in Parker, *DMP*, 8-9.) The final answer is 3 · 2 6 47 = 6 18 47. The result is then verified, as follows:

15 3" · 6 18 47 = (10 + 5 + 3") · 6 18 47 = 63 39 47 + 31 2' 19 2' 47 + 12 47
= 99 2' 23 2' 47 = 100.

In modern terms, the computation in *DMP* # 2 can be explained as follows:

100/(15 3") = 100/(47/3) = (100/47) · 3 = 2 6/47 · 3 = 6 18/47.

P.Cairo § 1 a (*DMP* # 2) is not well preserved but can be fully reconstructed, as pointed out by Parker, because of its close similarity to *DMP* # 3. The given task is to find out how many times 17 3" goes in 100. The computation begins with the observation that 100 + 6 = 106, and that 6 goes 17 3" times into 106. Next, it is noted that 17 3" = 53 · 3', and that 53 goes 1 47 53 times in 100. The final answer is 3 · 1 47 53 = 5 35 53. Again, the result is verified, as follows:

17 3" · 5 35 53 = (2 + 10 + 5 + 3") · 5 35 53
= 11 17 53 + 56 2' 5 2' 53 + 28 16 53 + 3 3" 5 3" 53 = 99 6' 44 6' 53 = 99 6' 6" = 100.

(The signs 6' and 6", meaning 1/6 and 5/6, will be discussed below, in connection with *P.Cairo* § 2 in Sec. 3.1c.)

Nothing like this way of counting is known from any Babylonian mathematical texts. However, the circumstance that in exercise *DMP* # 3 the number 94 is presented as 100 – 6, and that in the damaged exercise *DMP* # 2 the number 106 (probably) is presented as 100 + 6, suggests a geometric explanation of the computations in *P.Cairo* § 1, as in Fig. 3.1.2 below.

A likely interpretation (cf. the discussion below of *P.Cairo* § 4) is that the two division problems in *DMP* ## 2-3 are the computations needed to solve the following pair of metric algebra problems:

a) Consider a rectangle with the area 100 (cubits) and the width 6. If the long side is *increased* by 1 and the short side is decreased so much that the area will still be 100, what is then the new short side?

b) Consider again a rectangle with the area 100 and the width 6. If the long side is *decreased* by 1 and the short side is increased so much that the area will still be 100, what is then the new short side?

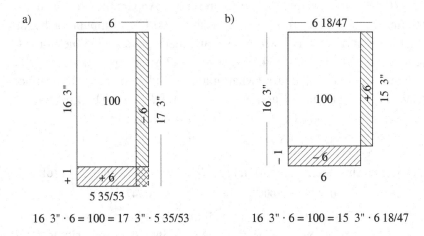

Fig. 3.1.2. *P.Cairo* § 1 (*DMP* ## 2-3). A geometric explanation of two division problems.

Fowler, *MPA* (1987 (1999)), 7.3(e), 7.4(a), contains a detailed discussion of "the evidence for the proposal that early Greek mathematicians conceived and used manipulations of common fractions". In his arguments, which aim to refute that proposal, Fowler also discusses the occurrence of what might be understood as the use of common fractions in a number of demotic mathematical texts, mentioning Parker, *DMP* (1972) ## 2, 3, 10, 13, 51, and 72. These are the exercises *P.Cairo* §§ 1, 4, 5, *P.BM 10399* § 3, and *P.Carlsberg 30*. According to Fowler (*op. cit.*), 264 (262),

> "the evidence of calculations in the papyri displays almost unanimous evidence against this proposal. The very few instances there that can be cited as illustrating notions for common fractions appear, on closer scrutiny, more probably to be abbreviations of unresolved descriptions of divisions that are still conceived as sums of unit fractions, and all can be more naturally explained as relaxations of stylistic conventions about how these divisions should be evaluated and expressed."

It is doubtful that Fowler's conclusion is correct in the case of the mentioned examples from the demotic mathematical texts. (As for the case of Greek mathematical papyri, see the discussion of *P.Akhmîm* in § 4.5 d.) What is clear is that the author of some of the exercises in *P.Cairo* operated smoothly with what may be called "binomial" fractions, precisely *as if* he

3.1. P.Cairo (Ptolemaic, the 3rd C. BCE)

were operating with common fractions. Consider the following operations:

10 · 6 18 47 =	60 180 47 =	63 39 47	P.Cairo § 1 a (DMP # 3)
5 · 6 18 47 =	30 90 47 =	31 2' 19 2' 47	
3" · 6 18 47 =		4 12 47	

sum: 63 39 47 + 31 2' 19 2' 47 + 4 12 47 = 98 2' 70 2' 47 = 99 2' 23 2' 47 = 100

and

2 · 5 35 53 =	10 70 53 =	11 17 53	P.Cairo § 1 b (DMP # 2)
10 · 5 35 53 =	50 350 53 =	56 2' 5 2' 53	(2' · 53 = 26 2')
5 · 5 35 53 =	25 175 53 =	28 16 53	
3" · 5 35 53 =	3 3' 23 3' 53 =	3 3" 5 3" 53	(3' · 53 = 17 3")

sum: 11 17 53 + 56 2' 5 2' 53 + 28 16 53 + 3 3" 5 3" 53 = 99 6' 44 6' 53 = 100
(6" · 53 = 44 6')

The arithmetical operations in these examples are, on one hand *multiplications* of binomial fractions *with integers or basic fractions*, on the other hand *additions* of binomial fractions. Note that while in common fractions of the modern type the numerator always is an integer, in these Egyptian binomial fractions the numerator is allowed to be *an integer plus a basic fraction* (6" = 5/6, 3" = 2/3, 2' = 1/2, 3' = 1/3, 4' = 1/4, or 6' = 1/6).

It is not unlikely that counting with binomial fractions as in *P.Cairo* § 1 was a late development with its roots in the well known counting with "red auxiliaries" in *P Rhind*. An example is **P.Rhind # 23**, a "completion problem" (cf. *P.Cairo* § 2 below). The hand copy below is copied from Chase, *et al, RMP* (1929), pl. 46. The transliteration is a mirror image of the text.

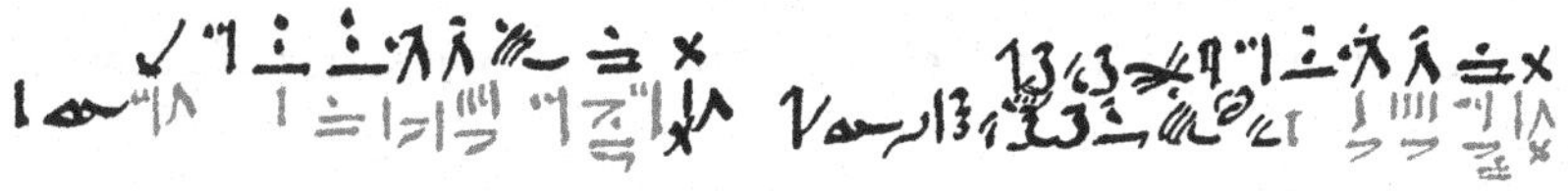

Fig. 3.1.3. *P.Rhind* # 23. A completion problem. An example of the use of red auxiliaries, shown here as grey in the hieratic text and in bold style in the transliteration.

The question in this exercise is formulated as follows:

4' $\overline{8}$ $\overline{10}$ $\overline{30}$ $\overline{45}$. Complete it to 3".

The solution procedure is very brief. It starts by writing red auxiliaries under the given number:

4'	$\overline{8}$	$\overline{10}$	$\overline{30}$	$\overline{45}$
11 4'	5 2' $\overline{8}$	4 2'	1 2'	1

Then it concludes directly, without further motivation, that

Therefore, $\overline{9}$ $\overline{40}$ is to be added to it. It makes 3".

Finally the verification, again by use of red auxiliaries,

4'	$\overline{8}$	$\overline{9}$	$\overline{10}$	$\overline{30}$	$\overline{40}$	$\overline{45}$	3'	
11 4'	5 2' $\overline{8}$	5	4 2'	1 2'	1 $\overline{8}$	1	15	It makes 1.

If the author of *P.Rhind* # 23 had operated with binomial fractions instead of with red auxiliaries, and if he had been more careful with the details, the solution could have proceeded as follows:[21]

a) 4' $\overline{8}$ $\overline{10}$ $\overline{30}$ $\overline{45}$ = 11 4' 5 2' $\overline{8}$ 4 2' 1 2' 1 45 = 23 2' 4' $\overline{8}$ 45,
giving a deficit of 6 $\overline{8}$ 45, since 3" = $\overline{30}$ 45.
b) 6 $\overline{8}$ = 5 + 9 · $\overline{8}$, so that 6 $\overline{8}$ 45 = $\overline{9}$ + $\overline{8}$ · $\overline{5}$ = $\overline{9}$ $\overline{40}$.
c) 4' $\overline{8}$ $\overline{10}$ $\overline{30}$ $\overline{45}$ + $\overline{9}$ $\overline{40}$ + 3' = 3' 4' $\overline{8}$ $\overline{9}$ $\overline{10}$ $\overline{30}$ $\overline{40}$ $\overline{45}$
= 15 11 4' 5 2' $\overline{8}$ 5 4 2' 1 2' 1 $\overline{8}$ 1 45 = $\overline{45}$ 45 = 1.

3.1 d. *P.Cairo* § 2 (*DMP* ## 4-6). Completion problems

(Cf. *P.Rhind* §§ 6-7 (## 21-23).) The question in § 2 b (*DMP* # 5) is posed as follows, according to the correction in Zauzich, *BiOr* 32 (1975):

If it is said to you: 6" $\overline{10}$ $\overline{20}$ $\overline{120}$ $\overline{240}$ $\overline{480}$ $\overline{510}$, what remainder will complete 1?

The given answer can be understood as the following series of computations:

480 = 16 · 30, 510 = 17 · 30, 17 − 16 = 1, $\overline{510}$ · $\overline{16}$ = $\overline{8160}$.

The strangely formulated details of the last step of the computation can be explained as follows:

$\overline{510}$ + $\overline{8160}$ = 16 · $\overline{8160}$ + 1 · $\overline{8160}$ = 17 · $\overline{8160}$ = $\overline{480}$, because 17 · 480 = 8160.

---

21. Compare step b) with *pMoscow* # 20, where 2 3" divided by 20 is given as $\overline{5}$ of 3". The explanation may be that the division was carried out as follows: 2 3" = 4 · 3", so that 2 3" · $\overline{20}$ = 4 · 3" · $\overline{20}$ = $\overline{5}$ · 3".

3.1. P.Cairo (Ptolemaic, the 3rd C. BCE)

The following tentative explanation of what is going on here is based on the observation that while the author of *P.Cairo*, living in Hellenistic Egypt, *nominally counted with traditional sums of parts* (traditional Egyptian), and with what looks very much like *common fractions* (a late Egyptian invention??), he may also have *operated covertly with sexagesimal fractions* (Babylonian)!

Take a renewed look at the given fraction 6" $\overline{10}\ \overline{20}\ \overline{120}\ \overline{240}\ \overline{480}\ \overline{510}$ in *P.Cairo* § 2 b (*DMP* # 5).[22] In this fraction, the basic fraction 6" and the parts $\overline{10}$, $\overline{20}$, $\overline{120}$, $\overline{240}$, and $\overline{480}$ can be equated with the sexagesimal fractions ;50, ;06, ;03, ;00 30, ;00 15 and ;00 07 30, while the final fraction $\overline{510}$ cannot be written as a sexagesimal fraction, for the reason that 510 = 17 · 30, and 17 is a non-regular sexagesimal number. The purpose of the exercise was to 'complete 6" $\overline{10}\ \overline{20}\ \overline{120}\ \overline{240}\ \overline{480}\ \overline{510}$ to 1'. This task was accomplished through the addition of the small "remainder" $\overline{8160}$:

6" $\overline{10}\ \overline{20}\ \overline{120}\ \overline{240}\ \overline{480}\ \overline{510}$ + $\overline{8160}$ = 6" $\overline{10}\ \overline{20}\ \overline{120}\ \overline{240}\ \overline{480}$ + $\overline{480}$ = 1.

In a similar way, in the simpler, but partly destroyed exercise *P.Cairo* § 2 a (*DMP* # 4), the (reconstructed) computation can be explained as follows:

[6" $\overline{10}\ \overline{20}\ \overline{120}\ \overline{210}$] + $\overline{280}$ = 6" $\overline{10}\ \overline{20}\ \overline{120}$ + $\overline{120}$ = 1.

One can carry this analysis one step further by asking: How did the author of the text originally construct the strange numbers 6" $\overline{10}\ \overline{20}\ \overline{120}\ \overline{210}$ and 6" $\overline{10}\ \overline{20}\ \overline{120}\ \overline{240}\ \overline{480}\ \overline{510}$? There is a surprising answer to this question. Remember that in the discussion of the exercises *P.Rhind* ## 53 a-b (§ 2.1 d, Theme G), it was shown that they can be explained with references to "almost round numbers", an idea that can be traced back to the earliest written records, proto-cuneiform texts from the late Uruk period in Mesopotamia. An example is provided by the incomplete exercise

---

22. The initial fraction, meaning 5/6, is written in the text of *P.Cairo* with a special sign, here reproduced as 6". Note that in *hieratic* texts, there were special signs for the "basic fractions" 2/3, 1/2, 1/3, 1/4, 1/6, while 5/6 could be written as a ligature of the signs for 2/3 and 1/6. In *P.Cairo* there are special signs for the mentioned fractions, and for 5/6. See Parker. *DMP* (1972), 8 and 86, and Sethe, *ZZ* (1916), 100, where it is mentioned that the special sign for 5/6 was used already in the Persian period. In Babylonian cuneiform texts there were, in a similar way, special signs for the basic fractions 5/6, 2/3, 1/2, and 1/3. Thus, the use of a special sign for 5/6 in *P.Cairo* may be due to a Babylonian influence. In the present paper, the basic fractions 5/6, 2/3, 1/2, 1/3, 1/4, and 1/6 are transliterated as 5", 3", 2', 4', and 6', respectively, while (*n*-th) *parts* are transliterated with an overbar as $\overline{5}, \overline{6}, \overline{7}$, *etc.*

P.Rhind # 53 b, where the almost round area number 14 2' 8' (*setat*) can be explained as the result of an approximation procedure with the "start number" 12 (*setat*) and the "target number" 15 (*setat*).

In a similar way, it is possible that 6" $\overline{10}\ \overline{20}\ \overline{120}\ \overline{210}$ and 6" $\overline{10}\ \overline{20}\ \overline{120}\ \overline{240}\ \overline{480}\ \overline{510}$ can be understood as "almost sexagesimal fractions", obtained through an approximation procedure where simple sexagesimal fractions are start and target numbers. In both cases, the start and target numbers may have been 6" $\overline{10}\ \overline{20}$ = ;59 and 1, respectively. The successive steps of the construction of the first of the two almost sexagesimal numbers, the one in *P.Cairo* § 2 a, can have been the following:

a) 6" $\overline{10}\ \overline{20}$ = ;59   deficit $\overline{60}$ = ;01,
b) 6" $\overline{10}\ \overline{20}$ + 2' · $\overline{60}$ = 6" $\overline{10}\ \overline{20}\ \overline{120}$ = ;59 30   deficit $\overline{120}$ = ;00 30,
c) 6" $\overline{10}\ \overline{20}\ \overline{120}$ + 4 · 7 · $\overline{120}$ = 6" $\overline{60}\ \overline{120}\ \overline{210}$ deficit   3 · 7 · $\overline{120}$ = $\overline{280}$.

The parallel process possibly used for the construction of the second almost sexagesimal number, the one in *P.Cairo* § 2 b, would have been somewhat more complicated:

a) 6" $\overline{10}\ \overline{20}$ = ;59   deficit $\overline{60}$ = ;01,
b) 6" $\overline{10}\ \overline{20}$ + 2' · $\overline{60}$ = 6" $\overline{10}\ \overline{20}\ \overline{120}$ = ;59 30   deficit $\overline{120}$ = ;00 30,
c) 6" $\overline{10}\ \overline{20}\ \overline{120}$ + 2' · $\overline{120}$ = 6" $\overline{10}\ \overline{20}\ \overline{120}\ \overline{240}$ = ;59 45   deficit $\overline{240}$ = ;00 15,
d) 6" $\overline{10}\ \overline{20}\ \overline{120}\ \overline{240}$ + 2' · $\overline{240}$ = 6" $\overline{10}\ \overline{20}\ \overline{120}\ \overline{240}\ \overline{480}$
               = ;59 52 30   deficit $\overline{480}$ = ;00 07 30,
e) 6" $\overline{10}\ \overline{20}\ \overline{120}\ \overline{240}\ \overline{480}$ + 16 · 17 · $\overline{480}$
               = 6" $\overline{10}\ \overline{20}\ \overline{120}\ \overline{240}\ \overline{480}\ \overline{510}$ deficit   17 · $\overline{480}$ = $\overline{8160}$.

Exercises like these would have taught the students how to find sexagesimal fractions that are close approximations to given non-sexagesimal fractions. This would be useful, for instance, when computing approximations to square roots by the method shown in ***P.BM* 1052 § 6** (below).

For those who cannot readily accept the idea of a hidden use of counting with sexagesimal fractions in *P.Cairo* § 2 a-b, here is an alternative explanation in terms of what looks surprisingly much like counting with *common fractions*. In § 2 b, for instance, the more complicated case, the construction of the given number can have proceeded as follows:

a) 6" $\overline{10}\ \overline{20}$ = (50 + 6 + 3) · $\overline{60}$ = 59 · $\overline{60}$ deficit $\overline{60}$,
b) 6" $\overline{10}\ \overline{20}$ + $\overline{120}$ = (2 · 59 + 1) · $\overline{120}$ = 119 · $\overline{120}$ deficit $\overline{120}$,
c) 6" $\overline{10}\ \overline{20}\ \overline{120}$ + $\overline{240}$ = (2 · 119 + 1) · 240 = 239 · $\overline{240}$ deficit $\overline{240}$,
d) 6" $\overline{10}\ \overline{20}\ \overline{120}\ \overline{240}$ + $\overline{480}$ = (2 · 239 + 1) · $\overline{480}$ = 479 · $\overline{480}$ deficit $\overline{480}$,

e) 6" $\overline{10}$ $\overline{20}$ $\overline{120}$ $\overline{240}$ $\overline{480}$ + $\overline{510}$ = (579 + 16 · $\overline{17}$) · $\overline{480}$ deficit $\overline{17}$ · $\overline{480}$ = $\overline{8160}$.

3.1 e. *P.Cairo* § 3 (*DMP* ## 7, 15-16, 18). A rectangular sail

These exercises are all of the following type:

> A rectangular sail has a given area $A = h \cdot w$ and a given side ratio $h/w = f$.
> Find h and w.

This is a rather simple metric algebra problem, with the solution

> sq. $h = f \cdot A$, h = sqr. $(f \cdot A)$.

The square root approximations appearing in these exercises are:

> *DMP* ## 7 and 15: sqr. 1500 = 38 3" $\overline{20}$ (38;43),
> *DMP* # 16: sqr. 140 = 11 6" (11;50),
> *DMP* # 18: sqr. 1000 = 31 2' $\overline{10}$ $\overline{30}$ (31;38).

A possible reconstruction of the details of the complicated computation of the accurate approximation sqr. 1000 = 31 2' $\overline{10}$ $\overline{30}$ is proposed in Friberg, *BaM* 28 (1997), 38. (The approximation can also have been obtained simply through trial and error.)

Parallel hieratic exercises are ***P.Moscow*** **§ 7** (## 6, 7, 17). Moreover, as mentioned in the discussion of *P.Moscow* § 7 in Sec. 1.4 c, **IM 121613 # 1** and **YBC 11126** are parallel OB exercises.

3.1 f. *P.Cairo* § 4 (*DMP* ## 8-12, 14, 17). Reshaping a rectangular cloth

The exercises in this paragraph are all of the following type (cf. the discussion of *P.Cairo* § 2 in Sec. 3.1 d):

> A rectangular sail has a given height h and a given width w,
> hence a given area $A = h \cdot w$.
> The height (or the width) is decreased by a fraction $1/n$ of its measure.
> By what fraction $1/m$ must the width (or the height) be increased to compensate for the resulting change of area?

Here are, for instance, the successive steps of § 4 a (*DMP* # 8):

> 1. A hair-cloth, 7 cubits in height, 5 cubits in width, hence 35 cloth cubits (in area).
> 2. Take 1 cubit off its height and add (correspondingly) to its width.
> 3. What is then added to its width?
> 4. To make you know it, namely: The height is 7 cubits. Subtract 1 cubit, its $\overline{7}$.
> 5. You shall say: (Its) $\overline{7}$ is taken off, 6 cubits remain. The (new) height is 6 cubits.
> 6. Now, its area taken off makes 5 cloth cubits.

116   *Unexpected Links Between Egyptian and Babylonian Mathematics*

7. You shall say: 5 is what fraction of 6? Result: 6".
8. Add it to 5. Result: 5 6". The width is 5 6" (cubits).
9. You shall count 5 6", 6 times.
   Result 35 cloth cubits, which will make the (given) number.

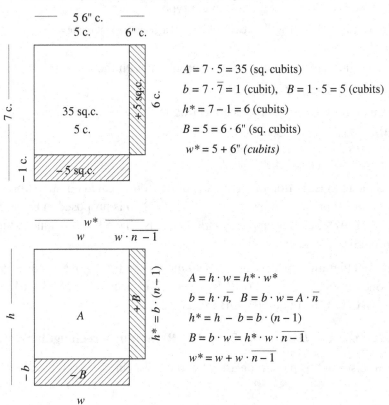

Fig. 3.1.4. Above: *P.Cairo* § 4 a (*DMP* # 8). Reshaping a rectangular cloth.
    Below: The general idea.

In the 7 exercises of § 4, the following values of the parameters appear:

DMP #	h	w	h^* (or w^*)	n	m	w^* (or h^*)
8	7 c.	5 c.	$h^* = h - 1$ c.	7	6	$w^* = 5\ 6"$ c.
9	6 c.	4 c.	$h^* = h - 1$ c.	6	5	$w^* = 4\ 3"\ \overline{10}\ \overline{30}$ c.
10	6 c.	1 2' c.	$h^* = h - 2'$ c.	12	11	$w^* = 1\ 2'\ \underline{1\ 2'}\ 11$ c.
11	6 c.	4 c.	$w^* = w - 1$ c.	3	4	$h^* = 3$ c.

3.1. P.Cairo (Ptolemaic, the 3rd C. BCE)

12	6 c.	3 c.	$w^* = w - 1\ 2'$ c.	1	2	$h^* = 12$ c.
14	21 c.	5 c.	$h^* = h - 1$ c.	21	20	$w^* = 5\ 4'$ c.
17	21 c.	5 c.	$w^* = w - 1$ c.	5	4	$h^* = 26\ 4'$ c.

The following computation in two steps in *DMP* # 10 was observed by Parker *DMP* (1972), 22:

1 2' · 1 $\overline{11}$ = 2' · 3 $\underline{3}$ 11 = 1 2' $\underline{1\ 2'}$ 11.

It is clear from this table that the original from which the exercises in *P.Cairo* § 4 were borrowed must have been a well structured "theme text". This is shown by the way in which the first two exercises are particularly simple, with only 1 (cubit) subtracted from the height. In the third exercise, 1/2 (cubit) is subtracted, and in the fourth and fifth exercises 2 and 6 are *added*. In the last couple of exercises, 1 is subtracted from the *height* in one case but from the *width* in the second case.

The idea behind the exercises in § 4 is illustrated by the diagram in Fig. 3.1.4, below. Apparently, the purpose of the exercises was to use *a geometric model* to demonstrate the following general "reciprocity rule" by a number of explicit examples:

The reciprocal of $1 - \overline{n}$ is $1 + \overline{n-1}$.
The reciprocal of $1 + \overline{n}$ is $1 - \overline{n+1}$.

Cf. the examples in *P.Cairo* § 1, and in *P.BM 10399* § 3 (discussed in Sec. 3.2 c below).

OB mathematicians, too, were familiar with this rule. This is shown by some interesting passages, previously badly understood, in two texts from Uruk, **VAT 7532** (Høyrup, *LWS* (2002), 209) and **VAT 7535** (Neugebauer, *MKT 1* (1935), 303). In VAT 7532, the sides of a trapezoidal field are measured with a measuring stick, a 'reed', of unknown length. After the read has been laid out 1 00 (= 60) times along the length of the field, the top 1/6 of the reed drops off. The broken reed is then laid out 1 12 times along the remainder of the length of the field. And so on. The question is

sag gi en.nam   The head (original length) of the reed is what?

The first step of the solution procedure is to compute the length of the field as a multiple of (the length of) the broken reed:

VAT 7532, obv. 8-11.

> gi ša la ti-du-ú / 1 ḫé.gar
> igi.6.gál-šu ḫu-ṣub-ma 50 te-zi-ib /
> igi 50 du₈-ma 1 12
> a-na 1 šu-ši nim-ma / 1 12
> a-na <1 12> daḫ-ma 2 24 uš lul in.sì

The reed that you do not know, 1 may you set,
its 6th-part break off, 50 you leave.
The opposite of 50 release, then 1 12,
to 1 sixty lift, then 1 12.
To <1 12> join, then 2 24, the false length, it gives.

This means that if the assumed length of the whole reed is 1, then the false length of the broken reed is $1 - 1/6 = ;50$. Since the reciprocal of ;50 is 1;12, the length of the whole reed is 1;12 times the length of the broken reed. Hence, the length of the field is $1\ 12 + 1;12 \cdot 60 = 1\ 12 + 1\ 12 = 2\ 24$ times the (unknown) length r^* of the broken reed. This is what is meant when it is said that 2 24 is the "false length" of the field.

Towards the end of the solution algorithm, the answer is obtained that the length r^* of the broken reed is 25 (meaning ;25 ninda). It remains to find the length r of the whole reed. This time, the computation does not involve the reciprocal of ;50. Instead, it goes like this:

VAT 7532, rev. 6-8.

> aš-šum igi.6.gál re-ša-am iḫ-ḫa-aṣ-bu /
> 6 lu-pu-ut-ma 1 šu-ut-bi 5 te-zi-ib /
> <igi.5 du₈-ma 12>
> <a-na 25 nim 5 in.sì>
> 5 a-na 25 daḫ-ma
> 2'ninda sag gi in.sì

Since a 6th-part at first was broken off,
6 write down, then 1 remove, 5 you leave.
<The opposite of 5 release, then 12.>
<To 25 lift, 5 it gives.>
5 to 25 join, then
1/2 ninda, the original reed, it gives.

Thus, instead of computing the whole reed as $r = 1;12 \cdot r^* = 1;12 \cdot 25 = 30$ (meaning ;30 ninda), the text proceeds as follows:

$(1 - 1/6) \cdot r = r^*$, $6 - 1 = 5$, $1/5 = ;12$, $;12 \cdot r^* = ;12 \cdot 25 = 5$,

$(1 + 1/5) \cdot r^* = 25 + 5 = 30 = r$.

The words in the text within brackets < ··· > were omitted by the scribe, but they are present in parallel instances in VAT 7535, presented below with several improved readings of difficult passages:

VAT 7535, obv. 23-26.

> aš-šum i-na re-ši-in / igi.5.gál.bi-šu iḫ-ḫa-aṣ-bu
> 5 lu-pu-ut-ma / 1 šu-ut-bi 4 si.ni.tum
> igi.4 du₈-ma 15 [in.s]ì /
> a-na 20 [nim 5 in.sì]
> [5 a-na] 20 daḫ 25 sag gi

Since at first a 5th-part was broken off,
5 write down, then 1 remove, 4 the remainder.
The opposite of 4 release, then 15 *it gi*ves.
To 20 *lift, 5 it gives.*
5 to 20 join, then 25, the original reed.

VAT 7535, rev. 21-24.

> aš-šum igi.5.gál.bi-[šu] gaz /
> 5 lu-pu-ut-ma 1 šu-ut-bi 4
> igi.4 du₈-ma / 15
> a-na 20 nim 5 in.sì
> 5 a-na 20 daḫ-ma / 25 sag gi in.sì

Since a 5th-part was broken off,
5 write down, then 1 remove, 4.
The opposite of 4 release, then 15.
To 20 lift, 5 it gives.
5 to 20 join, then 25, the original reed it gives.

Clearly, in these two examples the computation makes explicit use of the observation that

the reciprocal of $1 - 1/5$ is $1 + 1/(5 - 1) = 1 + 1/4$.

Another OB mathematical text related to *P.Cairo* § 4 is **MS 5112 § 9** (Friberg, *MCTSC* (2005), Sec. 11.2 l), an exercise in a large mathematical recombination text with metric algebra as its topic. The question in that exercise is stated as follows:

MS 5112 § 9

> uš sag gu₇.gu₇-*ma*
> 50 a.šà
> *i-na* uš 30 ninda [*zi-ma*]
> *a-n*[*a*] sag / 5 ninda daḫ-*ma* 50 a.šà
> uš sag en.nam
> *etc.*

The length (and) the front (I made) eat (each other), then
50, the field.
From the length 30 ninda (I) *tore out, then*
to the front 5 ninda (I) added, then 50, the field.
The length (and) the front are what?
etc.

This question can be rephrased as follows:

> A rectangle has the area 50 (00 sq. n.).
> The length of the rectangle is diminished by 10 n., and the front augmented by 5 n.
> The new area is again 50 (00 sq. n.). What are the (original and changed) values of the length and the front?

The solution algorithm is based on the observation that if l is the length and s the front (the short side) of the original rectangle, and if l^* is the diminished length, then first a piece of area $30 \cdot s$ is removed from the original rectangle, then a piece of area $5 \cdot l^*$ is added to the shortened rectangle. Since the area of the added piece is assumed to be equal to the area of the removed piece, it follows that $u^* = 30/5 \cdot s$. Therefore,

$$l \cdot l^* = 30/5 \cdot l \cdot s = 30/5 \cdot 50\ (00)\ (\text{sq. n.}) = 5\ (00\ 00)\ (\text{sq. n.}), \quad u - u^* = 30\ \text{n.}$$

This is a rectangular-linear system of equations for the unknowns l and l^*, in the text called uš *pa-nu* 'the earlier length' and uš egir 'the later length'. The solution to this system of equations is

$$u = 2\ 30\ \text{n.}, \quad u^* = 2\ 00\ \text{n.}$$

From this it follows immediately that 'the earlier front' s and 'the later front' s^* must be

$$s = 50\ 00\ \text{sq. n.} / 2\ 30\ \text{n.} = 20\ \text{n.}, \quad \text{and} \quad s^* = 50\ 00\ \text{sq. n.} / 2\ 00\ \text{n.} = 25\ \text{n.}$$

3.1 g. *P.Cairo* § 7 (*DMP* # 23). Shares in a geometric progression

The question in this exercise seems to be phrased more or less in the following way:

> Four ⋯ are what have attained to one broken? (= one half?).
> The question gives the 1st.
> The 2nd gives 2' of the question of the 1st.
> The 3rd gives 2' of the question [of] the 2nd.
> The 4th gives 2' of the question <of> the 3rd.
> The numbers that the fractions have given make its 2'. What is that which has been given to one *tgs*?

The solution procedure goes like this:

> The way of doing it. Namely.
> The first 1, the second 2', the third 4', the fourth $\bar{8}$, total 1 6" $\overline{30}$ $\overline{120}$.
> You shall cause that 1 6" $\overline{30}$ $\overline{120}$ makes the number 8, result 15.
> You shall take $\overline{15}$, its 2' amounts to $\overline{30}$. You shall reckon $\overline{30}$, 8 times, result $\bar{5}$ $\overline{15}$.
> You shall say: The 1st $\bar{5}$ $\overline{15}$, the 2nd $\overline{10}$ $\overline{30}$, the 3rd $\overline{15}$, the 4th $\overline{30}$. The total again.

Parker, who could not explain what is going on here, simply wrote

> "Instead of the usual concrete problem, this appears to be merely some sort of mathematical exercise involving the continuous halving of a given number or fraction. A few key words, which might perhaps explain the purpose of the operation, are either missing or untranslatable with certainty."

In particular, it is not known what the meaning is of the word *tgs*. (Parker points out that it has the 'writing determinative'.) However, a likely interpretation is that *P.Cairo* § 7 is some kind of sharing problem, where 1/2 of some valuable commodity is shared between four *tgs*, in a decreasing geometric progression with the factor 1/2. If the shares are called a, b, c, d, then the problem can be explained (in modern terms) as the following simple system of linear equations:

$a \cdot 2' = b, \quad b \cdot 2' = c, \quad c \cdot 2' = d, \quad a + b + c + d = 2'$.

The solution procedure, which apparently makes use of the rule of false value, starts as follows:[23]

> a) (Let) the 1st = 1. (Then) the 2nd = 2', the 3rd = 4', the 4th = $\bar{8}$.

What then follows is more obscure. The "obvious" way to proceed would

---

23. Cf. Sec. 2.1 b-c, for a number of examples of applications of the rule of false value, both in *pRhind* and in Old and Late Babylonian mathematical texts.

be to divide the given sum 2' of the four shares by the sum 1 2' 4' $\overline{8}$ of the false shares in order to find the necessary "correction factor". Then the false shares times the correction factor would give the true shares.

This goal could have been obtained as follows:

b) The sum of the false shares is 1 2' 4' $\overline{8}$.
c) 1 2' 4' $\overline{8}$ = (8 + 4 + 2 + 1) · $\overline{8}$ = 15 · $\overline{8}$. Therefore, 1 2' 4' $\overline{8}$ · $\overline{15}$ = $\overline{8}$.
d) 2' · $\overline{15}$ = $\overline{30}$, and 8 · $\overline{30}$ = 4 · $\overline{15}$ = 2 · $\overline{10\ 30}$ = $\overline{5\ 15}$.
e) Therefore, 1 2' 4' $\overline{8}$ · $\overline{5\ 15}$ = 2'. Hence, the correction factor is $\overline{5\ 15}$.
f) Consequently, the four shares are
$\overline{5\ 15}$, 2' · $\overline{5\ 15}$ = $\overline{10\ 30}$, 2' · $\overline{10\ 30}$ = $\overline{15}$, and 2' · $\overline{15}$ = $\overline{30}$.
g) The total is correct, since
$\overline{5\ 15}$ + $\overline{10\ 30}$ + $\overline{15}$ + $\overline{30}$ = (6 + 2 + 3 + 1 + 2 + 1) · $\overline{30}$ = 15 · $\overline{30}$ = 2'.

This is close to the actual procedure in the text. However, a complicating factor is the seemingly unmotivated introduction of the sum of parts 1 6" $\overline{30}$ $\overline{120}$ as the sum of the false shares, instead of the obvious sum 1 2' 4' $\overline{8}$. Here is, therefore, an alternative explanation of the solution procedure in *P.Cairo* § 7 (*DMP* # 23), in terms of a *hidden counting with sexagesimal fractions*:

b) The sum of the false shares is
1 2' 4' $\overline{8}$ = 1 + ;30 +;15 + ;07 30 = 1;52 30 = 1 + ;50 + ;02 + ;00 30 = 1 6" $\overline{30}$ $\overline{120}$.
c) 1 6" $\overline{30}$ $\overline{120}$ = 1;52 30 = 15 · ;07 30 = 15 · $\overline{8}$.
Therefore, 1 6" $\overline{30}$ $\overline{120}$ · $\overline{15}$ = ;07 30 = $\overline{8}$.
d) 2' · $\overline{15}$ = $\overline{30}$, and 8 · $\overline{30}$ = 8 · ;02 = ;16 = ;12 + ;04 = $\overline{5\ 15}$.
e) Therefore, 1 6" $\overline{30}$ $\overline{120}$ · $\overline{5\ 15}$ = 2'.
Hence, the correction factor is $\overline{5\ 15}$ = ;12 + ;04 = ;16.
f) Consequently, the four shares are
$\overline{5\ 15}$ = ;16, 2' · $\overline{5\ 15}$ = $\overline{10\ 30}$ = ;08, 2' · $\overline{10\ 30}$ = $\overline{15}$ = ;04, and 2' · $\overline{15}$ = $\overline{30}$ = ;02.
g) The total is correct, since ;16 + ;08 + ;04 + ;02 = ;30 = 2'.

An OB parallel text is **MS 2830,** *obv.* (Friberg, *MCTSC* (2005), Sec. 7.4 a), a small theme text with five exercises, where 4 brothers share given amounts of silver in various kinds of geometric progressions. Another OB parallel text is the round "hand tablet" **MS 1844** (Friberg (*op. cit.*), Sec. 7.4 b) where 7 brothers? share an amount of silver? in a geometric progression, with each brother getting 1/7 times less than his nearest older brother.

3.1 h. *P.Cairo* § 9 (*DMP* ## 32-33). Diameter of a circle with given area

As is well known, the complicated rule for the computation of the area of

3.1. P.Cairo (Ptolemaic, the 3rd C. BCE)

a circle known from *P.Rhind* # 48 is no longer used in the demotic mathematical papyri. An example is shown below:

P.Cairo § 9 a (DMP # 32).
A plot that makes 100 square cubits, being a square.
It is said to you: Let it make a plot that makes 100 square cubits and is round.
What is the diameter? Look, this is what it is like:
You add the 3' of 100 to it, it makes 133 3'.
Let it be a squareside, it makes 11 2' $\overline{20}$.
You shall say: 11 2' $\overline{20}$ is the diameter of the plot that makes 100 square cubits.
Look, this is what it is like:
You count with 11 2' $\overline{20}$, 3 times, it makes 34 2' $\overline{10}$ $\overline{20}$. It is what it is round.
Namely: Its 3' is 11 2' $\overline{20}$, its 4' is 8 2' $\overline{10}$ $\overline{20}$ $\overline{120}$.
You shall count 11 2' $\overline{20}$, 8 2' $\overline{10}$ $\overline{20}$ $\overline{120}$ times, it makes 100 square cubits again.

In this exercise, the area A of a circular field is 100 sq. cubits. The diameter d is computed as

d = sqr. $(A + 3' \cdot A)$ = sqr. $\{(1 + 3') \cdot A\}$

Hence, sq. $d = (1 + 3') \cdot A$. The corresponding rule for the computation of the area of a circle is

$A_{circle} = (1 - 4') \cdot$ sq. d.

Next, the circumference a of the circle is computed as

$a = 3 \cdot d$.

For verification by reversal, the obtained value of a is used to compute the area A:

$A_{circle} = (3' \cdot a) \cdot (4' \cdot a) \; (= 4' \cdot d \cdot a)$.

Consequently, in the latter part of the 1st millennium BCE, *the Egyptians had adopted the Babylonian rule for the computation of the circumference of a circle,* as 3 times the diameter. *They had also adopted two new rules for the computation of the area of a circle,* both essentially identical with the Babylonian rule A = ;05 $\cdot$ sq. a, where ;05 = 1/12 = 1/3 $\cdot$ 1/4.

The square root 11 2' $\overline{20}$ can possibly have been computed in *P.Cairo* § 9 a by use of a simple method like this:

sqr. 133 3' = sqr. (sq. 11 + 37/3) = appr. 11 + 37/66 = 11 2' $\overline{22}$ $\overline{66}$
= appr. 11 2' $\overline{20}$ (= 11;33).

(See Friberg, *BaM* 28 (1997), Sec. 8: "On Babylonian square root approximations" and Sec. 9: "Indian, Egyptian, and Greek square root approximations", for a detailed discussion of ancient accurate methods for the computation of square roots.) On the other hand, 11 2' $\overline{20}$ is a quite *accurate* approximation to sqr. 133 3', since sq. 11 2' $\overline{20}$ = 133 4' $\overline{10}$ $\overline{20}$ $\overline{400}$. Therefore, another possibility is that the square root was computed in the following, more competent way, by use of a very accurate approximation to sqr. 3 (cf. Friberg (*op. cit.*), 324):

sqr. 133 3' = sqr. (400/3) = 20/3 · sqr. 3 = appr. 20/3 · 26/15 = 6;40 · 1;44 = 11;33 20
= 11 2' $\overline{20}$ ($\overline{180}$).

The value used in the text for 4' · *a* is 8 2' $\overline{10}$ $\overline{20}$ $\overline{120}$, probably computed *sexagesimally* as follows:

4' · *a* = 4' · 34 2' $\overline{10}$ $\overline{20}$ = 4' · 34;39 = 8;39 45 = 8;30 + ;06 + ;03 + ;00 30 + ;00 15 =
8 2' $\overline{10}$ $\overline{20}$ $\overline{120}$ ($\overline{240}$).

(Note that without the hidden use of sexagesimal fractions, 4' · *a* could have been computed in a more straightforward manner as

4' · *a* = 4' · 34 2' $\overline{10}$ $\overline{20}$ = 8 2' $\overline{8}$ $\overline{40}$ $\overline{80}$.)

If the final multiplication of 11 2' $\overline{20}$ with 8 2' $\overline{10}$ $\overline{20}$ $\overline{120}$ is actually carried out, the result will be very close to 100, as claimed in the text, since

11 2' $\overline{20}$ · 8 2' $\overline{10}$ $\overline{20}$ $\overline{120}$ = 11;33 · 8;39 30 = 1 40;00 13 30 = 100 ($\overline{300}$ $\overline{3600}$ $\overline{7200}$).

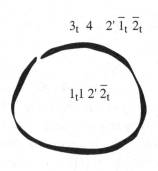 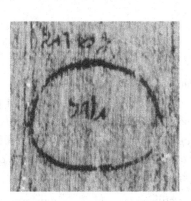

Fig. 3.1.5. *P.Cairo* (*DMP* # 32): circle with diameter 11 2' $\overline{20}$, circumference 34 2' 3$\overline{10}$ $\overline{20}$.

(The photographic detail to the right in Fig. 3.1.5 is copied from Parker, *DMP* (1972), pl. 11.)

3.1 i. *P.Cairo* § 10 (*DMP* ## 34-35). Metric algebra

The two exercises in this paragraph are of the following type:

A rectangular field has a given area $A = h \cdot w$ and a given diagonal d. Find h and w.

This is a metric algebra problem, with a solution in the following form:

$h + w$ = sqr. (sq. $d + 2A$), $h - w$ = sqr. (sq. $d - 2A$), $w = \{(h + w) - (h - w)\}/2$, $h = (h + w) - w$.

In *P.Cairo* § 10 a (*DMP* # 34), the given values are $A = 60$ sq. cubits, $d = 13$ cubits, while the computed values are $w = 5$ and $h = 12$. Consequently, the three rectangle parameters d, h, w together form an *exact* "diagonal triple" 13, 12, 5 of integers. In *P.Cairo* § 10 b (*DMP* # 35), on the other hand, the given values are $A = 60$ sq. cubits and $d = 15$ cubits, so that the computed values are only *approximations*:

$h + w$ = sqr. 345 = appr. 18 2' 12, $h - w$ = sqr. 105 = 10 4', w = appr. 4 6', h = appr. 14 3' $\overline{12}$.

A geometric interpretation of the solution algorithm for the metric algebra problems in *P.Cairo* § 10 is presented in Fig. 3.1.6 below. (It is known that geometric methods similar to this one were employed by both OB and Late Babylonian mathematicians to solve metric algebra problems. See Høyrup, *LWS* (2002) and Friberg, *BaM* 28 (1997) § 1, respectively.)

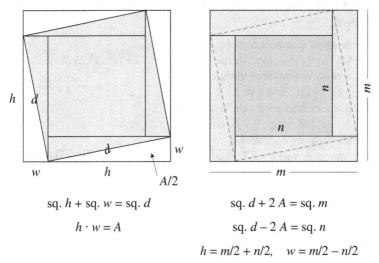

sq. h + sq. w = sq. d sq. $d + 2A$ = sq. m

$h \cdot w = A$ sq. $d - 2A$ = sq. n

                $h = m/2 + n/2$, $w = m/2 - n/2$

Fig. 3.1.6. *P.Cairo* § 10 (*DMP* ## 34-35). A metric algebra problem.

An OB parallel text is **Db$_2$-146 = IM 67118** from Eshnunna (Høyrup, *LWS* (2002), 257, 406; Friberg, *MCTSC* (2005), Sec. 10.1 b), where the length l and front s of a rectangle with given area A and given diagonal d are computed as follows:

$l - s$ = sqr. (sq. $d - 2A$), $(l + s)/2$ = sqr. (sq. $(l - s)/2 + A$),
$l = (l + s)/2 + (l - s)/2$, $s = (l + s)/2 - (l - s)/2$.

In IM 67118, the given values are $d = 1\ 15$, $A = 45\ (00)$, and the computed values are $l = 1\ (00)$, $s = 45$. Consequently, in this text, too, the three rectangle parameters d, l, s together form an exact "diagonal triple" of integers, $d, l, s = 1\ 15,\ 1\ 00,\ 45 = 15 \cdot (5, 4, 3)$.

Another OB parallel text is **MS 3971 § 2** (Friberg (*op. cit.*)), where the length l and front s of a rectangle with given area A and given diagonal d are computed as follows:

$l + s$ = sqr. (sq. $d + 2A$), $(l - s)/2$ = sqr. (sq. $(l + s)/2 - A$),
$l = (l + s)/2 + (l - s)/2$, $s = (l + s)/2 - (l - s)/2$.

Both the given and the computed values are the same as in IM 67118.

3.1 j. *P.Cairo* § 11 (*DMP* ## 36, 38). An equilateral triangle in a circle

In *P.Cairo* § 11 a (*DMP* # 36) an equilateral triangle of side $s = 12$ (divine) cubits is inscribed in a circle. The area of the circle is determined in a number of steps:

1) The height of the equilateral triangle is
   h = sqr. (sq. 12 – sq. 6) c. = √108 c. = appr. 10 3' $\overline{20}$ $\overline{120}$ (10;23 30) c.
2) The area of the equilateral triangle is
   A = 6 c. · sqr. 108 c. = appr. 62 3' $\overline{60}$ (1 02;21) sq. c.
3) The diameter (height) of a circle segment is
   p = 3' · sqr. 108 = appr. 3 3' $\overline{10}$ $\overline{60}$ $\overline{120}$ $\overline{180}$ (3;27 50) c.
4) The area of a circle segment is
   B = appr. $p \cdot (s + p)/2$ = 3 3' $\overline{10}$ $\overline{60}$ $\overline{120}$ $\overline{180}$ c. · 7 3" $\overline{20}$ $\overline{120}$ $\overline{240}$ $\overline{360}$ (7;43 55) c.
   = 26 6" $\overline{10}$ (26;56) sq. c.
5) The area of the circle is
   $A + 3B$ = appr. 143 $\overline{10}$ $\overline{20}$ (2 23;09) sq. c.

In the fourth step, the recorded number 26 6" $\overline{10}$ (26;56) is the result of a small miscalculation. The correct result should be 3;27 50 · 7;43 55 = 26;46 57 20 50, rounded off to 26;46 = 26 3" $\overline{10}$.

3.1. P.Cairo (Ptolemaic, the 3rd C. BCE)

To check the result, the area of the circle is then computed in a different way:

6) The diameter of the circle is
    $d = h + p$ = appr. 13 6" $\overline{45}$ (13;51 20) c.
7) The circumference of the circle is
    a = appr. 3 · d = 41 2' $\overline{15}$ (41;34) c.
8) The area of the circle is
    A = appr. (3' · a) · (4' · a) = 13 6" $\overline{45}$ c. · 10 3' $\overline{20}$ $\overline{120}$ c.
    = 143 6" $\overline{10}$ $\overline{30}$ (2 23;58) sq. c.
9) The difference between the results in steps 5 and 8 is
    143 6" $\overline{10}$ $\overline{30}$ sq. c. − 143 $\overline{10}$ $\overline{20}$ sq. c. = 3" $\overline{10}$ $\overline{20}$ (;49).

The exercise ends with some badly preserved calculations apparently meant to result in an improved value for the diameter of one of the segments:

10) ··· ··· ··· ···
    p = 3 3' $\overline{15}$ $\overline{35}$ $\overline{49}$ c. This is *not* a sexagesimal number!

Here are some pertinent observations:

A. Three drawings in the papyrus illustrate *P.Cairo* § 11 a (*DMP* # 36), showing the equilateral triangle inscribed in a circle, the triangle with its height, and one of the three circle segments. The number recorded inside the drawing of the segment is the height of the segment.

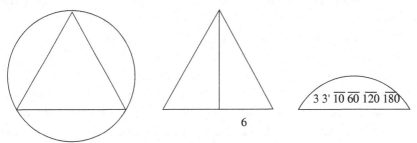

Fig. 3.1.7. Three drawings in the demotic *P.Cairo* § 11 a (*DMP* # 36).

In these three drawings the triangles and the circle segment are all standing upright on their bases. In the drawing in the middle, the height of the triangle is shown at a right angle to the base. This is precisely the way in which isosceles or equilateral triangles and their heights are drawn in the *Late Babylonian* mathematical text **W 23291 §§ 4 a-c** (Friberg, *BaM* 28

(1997), 283-286). In OB mathematical texts, on the other hand, triangles are always drawn with the 'front' pointing to the left. Similarly, circle segments are normally drawn with the arc pointing to the left and the chord vertical. (See the hand copies of **MS 2049, obv.** and **BM 85194 § 11** (## 20-21) in Friberg, *MCTSC* (2005), Sec. 11.1 a. See also the outline of BM 85194 in Fig. 2.4.1.) In Middle Egyptian mathematical texts, like *P.Rhind* and *P.Moscow*, triangles are oriented with the short side pointing either to the left or to the right. Thus, in this respect, *P.Cairo* is more closely related to Late Babylonian mathematics than to OB or hieratic Egyptian mathematics.

Another way in which *P.Cairo* is more closely related to Late Babylonian mathematics than to OB or hieratic Egyptian mathematics is that it measures lengths in cubits, and areas in square cubits, or, in § 4, in "cloth-cubits", which may be the same. In OB mathematical texts, lengths are measured in ninda, and areas in square ninda or related area measures. In hieratic mathematical texts, lengths are measured in *khet* = 100 cubits, and areas in *setat* = square *khet*. In Late Babylonian mathematical texts, on the other hand, such as W 23291 (Friberg *BaM* 28 (1997)) and W 23291-x (Friberg, *et al, BaM 21* (1990)), lengths are normally measured in cubits, and areas either in the related "reed measure", or in "seed measure". Lengths can also be measured in *ninda*, probably in an effort to keep up the tradition from the OB period.[24]

B. In *P.Cairo* § 11 a (*DMP* # 36), just as in *P.Cairo* § 9 a (*DMP* # 32), the circumference a is equal to $3 \cdot d$, where d is the diameter, and the area A of a circle is equal to $(3' \cdot a) \cdot (4' \cdot a)$. Apparently, all calculations are carried out by use of sexagesimal arithmetic, although the results of the computations are expressed in terms of sums of parts. An exception is the improved computation in the last part of the exercise. It is unfortunate that that part of the text is so damaged that it is not clear what is going on there.

C. In *P.Cairo* § 11 a (*DMP* # 36), an equilateral triangle with the side 12 cubits is inscribed in a circle, and it is shown, as the result of a compu-

---

24. In the demotic mathematical text ***P.Heidelberg 663*** (below, Sec. 3.7) and in Greek mathematical papyri (see Sec. 4.1), lengths are measured in *khet* or *schoinia*, both = 100 (or 96) cubits, and areas in *setat* (square *khet*) or *arouras* (square *schoinia*), while in the pseudo-Heronic work *Geometrica*, lengths and areas are measured in feet and (square) feet.

3.1. P.Cairo (Ptolemaic, the 3rd C. BCE)

tation, that the area of that circle is approximately equal to 143 $\overline{10}$ $\overline{20}$ (2 23;09) sq. cubits, or 143 6" $\overline{10}$ $\overline{30}$ (2 23;58) sq. cubits, both values close to 144 (2 24) sq. cubits, that is, to sq. (12 cubits). This is no coincidence, since (in modern notations)

> The height of an equilateral triangle with the side s is $h = \sqrt{3}/2 \cdot s$.
> The diameter of the circumscribed circle is $d = 4/3 \cdot h = 2/3 \cdot \sqrt{3} \cdot s$.
> The area of the circumscribed circle is A = appr. $3/4 \cdot$ sq. d = sq. s.

D. In *P.Cairo* § 11 a (*DMP* # 36), the following "segment area rule" is used to compute the area of one of the three circular segments outside the equilateral triangle:

$A_{\text{segm.}} = p \cdot (s + p)/2$,
where s is the base of the segment, p the height of the segment.

This is *not* the correct equation for the area of a general circle segment, yet it yields the correct result *in this particular case*, as indicated by the fact that the result of the computations in steps 1-5 of the computation agrees almost perfectly with the result of the computations in steps 6-8!

In modern notations, but still with π = appr. 3 and $\sqrt{3}$ = appr. 7/4, the *true* equation for the area of the segment is

$A_{\text{segm.}} = (A_{\text{circle}} - A_{\text{triangle}})/3$
$= ($sq. $s - s/2 \cdot h)/3 = (1 - \sqrt{3}/4)/3 \cdot$ sq. s = appr. $3/16 \cdot$ sq. s.

Since

$p = d - h = 1/3 \cdot h = 1/6 \cdot \sqrt{3} \cdot s$,

the *false* equation for the area of the segment gives instead

$A_{\text{segm.}} = p \cdot (s + p)/2$
$= 1/6 \cdot \sqrt{3} \cdot 1/12 \cdot (6 + \sqrt{3}) \cdot$ sq. $s = (\sqrt{3} + 1/2)/12 \cdot$ sq. s = appr. $3/16 \cdot$ sq. s.

Thus the true and false equations for the area of a segment cut off by an equilateral triangle yield exactly the same result, at least when it is assumed that π = appr. 3 and $\sqrt{3}$ = appr. 7/4.

The only *Late Babylonian* parallels to *P.Cairo* § 11 (*DMP* ## 36, 38) are **W 23291 § 4 c** (Friberg, *BaM* 28 (1997), 286) and **VAT 7848 § 1** (Friberg, *BaM* 28 (1997), 302-304), where the areas of equilateral triangles are computed by use of the very accurate approximation $\sqrt{3}/4$ = appr. ;26, or $\sqrt{3}$ = appr. 1;44 (26/15).

An *OB* parallel is **MS 3051** (Friberg, *MCTSC* (2005), Sec. 8.2 b), a square hand tablet with a drawing and some numbers on the obverse. The reverse is empty, except for a weakly drawn circle. The drawing on the obverse represents an equilateral triangle inscribed in a circle. The equilateral triangle is oriented with one of its sides pointing to the left, in agreement with the OB convention that the 'front' of a triangle should point to the left.

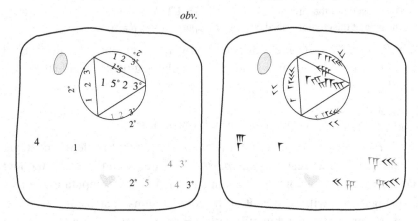

Fig. 3.1.8. MS 3051. An equilateral triangle inscribed in a circle.

The equilateral triangle divides the circumference of the circle in three equal parts. In the drawing on MS 3051, they are all marked with the number '20', obviously meaning '20 ninda'. That means that the whole circumference of the circle is 1 00 (ninda), hence that the diameter is 20 (ninda). It follows that the height of the equilateral triangle is 15. Somewhat misleadingly, the number '15' is recorded along one of the sides of the triangle. Other numbers recorded in the drawing are '1 52 30' inside the triangle and '1 02 30' inside each of the three circle segments.

The student who wrote his solution to an assignment on MS 3051 made himself guilty of a serious error when he tried to compute the area A of the equilateral triangle and the area B of each of the three segments. The numbers he actually recorded on his hand tablet can be analyzed as follows:

"A" = 1 52;30 = 15 · 15/2,   and   "B" = 1 02;30 = (5 00 – 1 52;30)/3.

The mistake he made, absentmindedly thinking about more exciting things than his mathematical assignment, was to use the value '15' both for the height and the side of the equilateral triangle!

3.1 k. *P.Cairo* § 12 (*DMP* # 37). A square inscribed in a circle

In *P.Cairo* § 12 (*DMP* # 37) a square is inscribed in a circle with given diameter $d = 30$ cubits and given area $A = 675$ square cubits (= 3/4 · sq. 30). In the first step of the computation, *the area of the square is computed as half the square of the diagonal of the square* (= *the diameter of the circle*):

1) The area of the square is $A_{\text{square}} = $ (sq. 30 c.)/2 = 450 sq. c.

The computation continues like this:

2) The side of the square is
    $s = $ sqr. 450 sq. c. = 21 $\overline{5}$ $\overline{60}$ (21;13) c.
3) The height of a circle segment is
    $p = (d - s)/2 = $ 4 3' $\overline{20}$ $\overline{120}$ (4;23 30) c.
4) The area of a circle segment is
    $B = $ appr. $p \cdot (s + p)/2$
    $= $ 4 3' $\overline{20}$ $\overline{120}$ c. · 12 3" $\overline{10}$ $\overline{30}$ $\overline{240}$ (12;23 30) c.
    $= $ 56 4' (56;15) sq. c. (actually 56;13 53 52 30, but rounded off).
5) The area of the circle is
    $A + 4 B = $ appr. (540 + 4 · 56 4') sq. c. = 675 sq. c.

Thus, the sum of the areas of the five parts of the circle is *precisely* equal to the area of the circle.

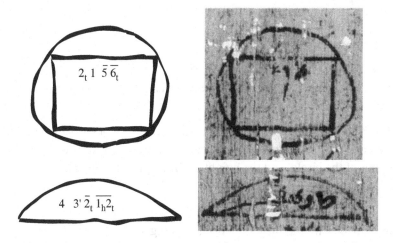

Fig. 3.1.9. *P.Cairo* § 12 (*DMP* # 37). A square inscribed in a circle and a circle segment.

In the drawings illustrating the exercise, two numbers are recorded, 21 $\overline{5}$ $\overline{60}$ = the side s of the square and 4 3' $\overline{20}$ $\overline{120}$ = the height p of one of the segments. (The photographic detail to the right in Fig. 3.1.9 is copied from Parker, *DMP* (1972), pl. 14.)

Note that in *P.Cairo* § 12 the same segment area rule as in *P.Cairo* § 11 is used to compute the area of a circular segment:

$A_{segm.} = p \cdot (s + p)/2$,
where s is the base of the segment, p the height of the segment.

Although this is *not* the correct equation for the area of a general circle segment, it obviously yields the correct result in the case of *P.Cairo* § 12, as indicated by the fact that the result of the computations in step 5 of the computation above agrees with the given value for the area of the circle.

As in the case of the segments cut off by an equilateral triangle, also in the present case the result of an application of the true equation for the area of a segment can be compared with the result of an application of the false equation, using modern notations. Let s again be the base of a segment cut off by the square, and let p be the height of that segment. Then the diameter d of the circle = the diagonal of the square = $\sqrt{2} \cdot s$. Therefore, the true equation gives

$A_{segm.} = (A_{circle} - A_{square})/4 = (3/4 \cdot 2 - 1)/4 \cdot \text{sq. } s = 1/8 \cdot \text{sq. } s$ (if π = appr. 3).

On the other hand, since

$p = (d - s)/2 = (\sqrt{2} - 1)/2 \cdot s$,

the false equation gives

$A_{segm.} = p \cdot (s + p)/2 = (\sqrt{2} - 1)/2 \cdot (\sqrt{2} + 1)/4 \cdot \text{sq. } s = 1/8 \cdot \text{sq. } s$.

Thus, *in this case, too,* the false equation gives the same result as the true equation!

It is interesting to know that segment area rule works just as well also in the case of a semicircle. In that case, the base s of the segment is the diameter of the circle, and the height of the segment is the radius of the circle. Thus, the true equation is

$A_{segm.} = 1/2 \cdot 3/4 \cdot \text{sq. } d = 3/4 \cdot p \cdot d$ (if π = appr. 3),

in agreement with the false equation, which gives

$A_{segm.} = p \cdot (d + p)/2 = 3/4 \cdot p \cdot d$.

Thus, *even in this third case,* the false equation gives the same result as the true equation!

The reason why the segment area rule works so well in the three cases considered is that it can be associated with the idea of approximating the area of a segment with the area of a trapezoid of roughly the same shape, a kind of "quadrature of the circle segment" (cf. the well known explanation of the hieratic circle area rule, as in Fig. 2.1.5 in Sec. 2.1 d).

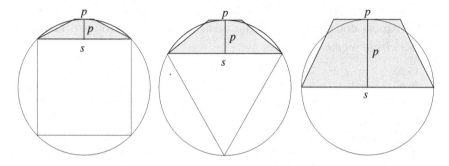

Fig. 3.1.10. Explanation of the apparent accuracy of the segment area rule.

Note that an application of the segment area rule in the case of a semicircle appears explicitly in the Greek mathematical papyrus ***P.Vindob. G 26740 # 5***. (See Sec. 4.3 c and Fig. 4.3.1.) In Babylonian mathematics, the use of the rule is not documented. However, one of the three constants for the area of a semicircle listed in OB tables of constants is

45   *ša* gán u$_4$.sakar ki.3   NSd 54
45, the 3rd (constant) of a crescent-field

(See Sec. 2.1 d.) The meaning of this constant is, obviously, that

$A = 3/4 \cdot p \cdot d$, where d is the 'transversal' (dal) of the circle,
and p is the orthogonal 'cross-line' (*pirkum*).

BM 85194 # 29, is an isolated exercise in a large OB recombination text (outlined in Fig. 2.4.1). There the area of a circle segment is computed in the case when the arc is 1 (00) and the base 50. Unfortunately, the text of that exercise is so corrupt that no conclusion can safely be drawn about what the original author of the exercise tried to do. (See the detailed, inconclusive discussion in Neugebauer, *MKT 1* (1935), 188-190.)

The same segment area rule as in *P.Cairo* can be found also in the early Chinese mathematical manual ***Jiu Zhang Suan Shu* 1:35-36** (Shen,

Crossley, and Lun, *NCMA* (1999), 123-128). In Hero's ***Metrica I*: 30**, the segment area rule is attributed to 'the ancients', and is shown to be inadequate when the improved Archimedean approximation 3 1/7 is used for π. Therefore, Hero suggests various corrected forms of the rule. For details, see Heath, *HGM 2* (1921 (1981)), 330.

***TMS* 3 = BR** (Bruins and Rutten, *TMS* (1961), text 3, is an OB table of mathematical constants. Six of the items are constants for two kinds of *double circle segments*:

13 20	igi.gub	*šà*	a.šà še	constant of a grain-field	BR 16
56 40	dal	*šà*	a.šà še	transversal of a grain-field	BR 17
23 20	*pi-ir-ku*	*šà*	a.šà še	cross-line of a grain-field	BR 18
16 52 30	igi.gub	*šà*	igi.gu$_4$	constant of an ox-eye	BR 19
52 30	dal	*šà*	igi.gu$_4$	transversal of an ox-eye	BR 20
30	*pi-ir-ku*	*šà*	igi.gu$_4$	cross-line of an ox-eye	BR 21

For each one of the two figures the constant stands for the area, the transversal for the longest straight line inside the figure, and the cross-line for the longest straight line inside the figure, at a right angle to the transversal. In both figures, the longest component of the boundary (in cases like these the arc) is supposed to have the length 1 , probably thought of as 1 (00).

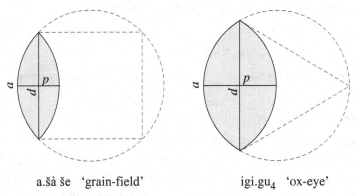

a.šà še 'grain-field'   igi.gu$_4$ 'ox-eye'

Fig. 3.1.11. Two kinds of double circle segments in an OB table of constants.

The given constants are correctly computed when the approximations used are π = appr. 3, √2 = appr. 1;25 = 17/12, and √3 = appr. 1;45 = 7/4.

Indeed, in the figure to the left, if *a* = 1 00, then the circumference of the circle is 4 00, the diameter *D* of the circle is approximately 1 20, and

3.1. P.Cairo (Ptolemaic, the 3rd C. BCE)

$d = 1\ 20/2 \cdot \sqrt{2} =$ appr. $56;40 = 10 \cdot 17/3$. Finally, $p = D - d = 1\ 20 - 56;40 = 23;20$, and the area of the grain-field can be computed as

$A_{\text{grain-field}} = (A_{\text{circle}} - A_{\text{square}})/2 =$ appr. $(1\ 20\ 00 - 53\ 20)/2 = 13\ 20$.

In the figure to the right, if $a = 1\ 00$, then the circumference of the circle is $3\ 00$, the diameter D of the circle is appr. $1\ 00$, and $p = D/2 =$ appr. 30. Furthermore, $d = p \cdot \sqrt{3} =$ appr. $52;30 = 7/8$, and

$A_{\text{ox-eye}} = (A_{\text{circle}} - A_{\text{triangle}}) \cdot 2/3 =$ appr. $(45\ 00 - 26;15 \cdot 45) = 16\ 52;30$.

Thus, it is clear that there existed OB precursors to *P.Cairo* // 11-12 (*DMP* ## 36-38).

A more direct OB parallel to *P.Cairo* § 12 (*DMP* # 37) is **MS 3050** (Friberg, *MCTSC* (2005), Sec. 8.2 c), a thick and round clay tablet with a drawing of a square with diagonals, inscribed in a circle.

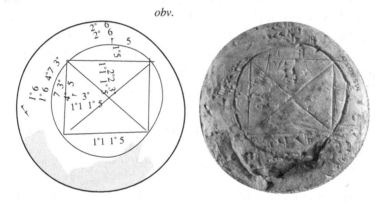

Fig. 3.1.12. MS 3050. An OB round hand tablet with a a square inscribed in a circle.

This is a school boy's hand tablet with various scribbled numbers, apparently giving only the beginning of a correct solution to whatever problem had been assigned to the school boy. It is likely that the diameter of the circle was given as 1(00). In that case, the area of the circle would be approximately 45 (00), and the area of each quarter circle 11 15. This would explain why the number '11 15' is recorded in each of the four quadrants. The number '22 30' can be explained as the area of half the circle, and the number '45' as the area of the whole circle. Similarly, the numbers '15 and '7 30' can be explained as the areas of a quarter square and a half square, respectively. Unfortunately, there is no trace of a number

'3 45' to indicate the area of a segment (3 45 = 11 15 – 7 30), and there is no obvious explanation for the numbers '16 40' and '26'.

3.1.1. *P.Cairo* §§ 13-14 (*DMP* ## 39-40). Pyramids with a square base

Precursors to these exercises exist both in the hieratic *P.Rhind* and *P.Moscow* and in the OB text **BM 96954+**. (See Sec. 2.2 d and Sec. 4.8 e.)

3.1 m. *P.Cairo* §§ 15-16 (## "32-33"). Metric algebra

In the main edition of *P.Cairo*, published in Parker, *DMP* (1972), the text was described as 40 exercises in 19 columns (A-S) on the reverses of 11 papyrus fragments. Parker, *Cent.* 14 (1969) contains a preliminary description of the text, giving brief explanations of 37 exercises in 20 of the columns. A comparison of the two descriptions of the document seems to indicate that exercises *DMP* ## 13-22 in columns H-L were still not understood by Parker in 1969, and therefore not mentioned then. On the other hand, at least two of the exercises mentioned in Parker (1969) are not present in Parker (1972), namely the ones here called §§ 15-16 (## "32" and "33"). Parker describes them quite briefly in the following way:

"32"
"The area of a square is decreased by 40 square cubits to make a rectangle of 100 square cubits.
Determine the sides: result, 11 5/6 by 8 1/3 1/10 1/60."

"33"
"Apparently an area of 100 square cubits has one side larger than the other by 21 cubits.
Determine the sides: result 25 by 4."

These are clearly metric algebra problems, but no details of any likely solution procedures were presented in Parker (*op. cit.*). However, such solution procedures, based on geometric considerations in the Old Babylonian style, are not difficult to find.

Take, for instance, the problem posed in "#32". It can be rephrased (in modern terms) as a couple of rectangular equations (cf. Fig. 3.1.13, left):

$A = h \cdot w = 100$ sq. c.

$h \cdot (h - w) = 40$ sq. c.

This is a simple problem with the following obvious solution:

3.2. P.British Museum 10399 (Ptolemaic, Later than P.Cairo)

sq. $h = (100 + 40)$ sq. c $= 140$ sq. c.

h = sqs. 140 c. = appr. (12 – 4/24) c. = 11 6" c. (= 11;50 c.)

$w/h = 100/140 = 5/7$

w = appr. $5 \cdot 1/7 \cdot 11;50$ c. = appr. $5 \cdot ;08\ 34 \cdot 11;50$ c. = appr. 8;27 c. = 8 3' $\overline{10}\ \overline{60}$ c.

(Without the use of covert counting with sexagesimal numbers, the answer would have been instead that $w = 8\ 3'\ \overline{14}\ \overline{21}$.)

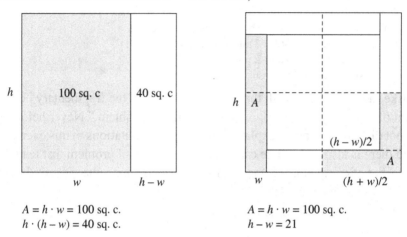

$A = h \cdot w = 100$ sq. c. $A = h \cdot w = 100$ sq. c.
$h \cdot (h - w) = 40$ sq. c. $h - w = 21$

Fig. 3.1.13. *P.Cairo* §§ 15-16 (*DMP* ## "32-33"). Geometric interpretations.

The problem posed in "#33" can be rephrased (in modern terms) as a c rectangular-linear system of equations:

$A = h \cdot w = 100$ sq. c.

$h - w = 21$. c.

This is a problem of a type that appears in both Old and Late Babylonian mathematics. It can be solved as follows (cf. Fig. 3.1.13, right):

sq. $(h + w)/2 = $ (sq. 21/2 + 100) sq. c. = 210 4' c.

$(h + w)/2$ = sqs. 210 4' c. = 14 2' c.

$h = (14\ 2' + 10\ 2')$ c. = 25 c.

$w = (14\ 2' - 10\ 2')$ c. = 4 c.

3.2. *P.British Museum 10399* (Ptolemaic, Later than *P.Cairo*)

The exercises of this text (Parker, *DMP* (1972) ## 42-51) belong to three distinct topics: circle geometry (*DMP* # 41), truncated cones (*DMP* ## 42-

45) and reciprocals of numbers of the type 1 + 1/n (*DMP* ## 46-51).

3.2 a. *P.BM 10399* § 1 (*DMP* # 41). A circle-and-chord problem

There is not much left of the question in this exercise, but more of the solution procedure:

1. The way *of doing* it.
2. You shall carry 7 into 13, result 1 6" $\overline{42}$.
3. You shall count 1 *6" $\overline{42}$ to* 1 6" $\overline{42}$ times, result 3 22 <u>49</u>.
4. *You shall count 3 22 <u>49</u> to* 7 times, result 24 and $\overline{7}$.
5. You shall add 7 *to 24 and* $\overline{7}$, result 31 and $\overline{7}$.
6. You shall take ⋯ 24 $\overline{7}$, result ⋯ (the rest is lost)

Parker admits defeat, with the words "Because of the fragmentary condition of the papyrus it is impossible to state the problem." Nevertheless, it is not difficult to find an explanation for the computations in this exercise, and there is also a plausible candidate for the type of problem that leads to this kind of computations. Indeed, in modern notations, the steps of the solution procedure are:[25]

2. 13/7 = 1 5/6 1/42   (this step is superfluous)
3. sq. 13/7 = 3 22/49
4. sq. 13/7 · 7 = 24 1/7
5. sq. 13/7 · 7 + 7 = 31 1/7 *etc.*

Computations somewhat like these are known to occur, for instance, in OB problems for chords in a circle. An example of such a problem is **TMS 1** (Bruins and Rutten, *TMS* (1961)). See Fig. 3.2.1, left. In Fig. 3.2.1, right, is illustrated a problem where the front *s* and the length *h* of a symmetric triangle inscribed in a circle are given. This is the most likely interpretation of the drawing on *TMS* 1, where *h* (called uš sag.kak *ga-am-ru* 'the whole length of the triangle') = 40, and *s*/2 (called 2' sag '1/2 the front') = 30. The values of *r* (the radius of the circle) and *q* (the distance from the front of the triangle to the center of the circle) can then be found as the solutions to *a quadratic-linear system of equations*:

sq. *r* – sq. *q* = sq. *s*/2,   *r* + *q* = *h*.

This is one of the Babylonian standard forms for a quadratic-linear system

---

25. Knorr, *HM* 9 (1982) observed that the squaring in line 3 can be explained as the result of actually *counting* with common fractions. Thus, sq. 1 6" $\overline{42}$ = sq. 13 <u>7</u> = 169 <u>49</u> = 3 22 <u>49</u>.

3.2. P.British Museum 10399 (Ptolemaic, Later than P.Cairo)

of equations. (See, for instance, Friberg, *RlA* 7 (1990) Sec. 5.7 c, type B3a).[26] The solution, obtained by use of metric algebra, is

$p = r - q = (\text{sq. } s/2)/h, \quad r = (h + p)/2, \quad q = (h - p)/2.$

In the case of *TMS* 1, in particular,

$s/2 = 30, h = 40,$ so that $p = 15\ 00\ /40 = 22;30, r = 31;15, q = 8;45.$

These values for *p* and *q* are explicitly indicated in the drawing on *TMS* 1.

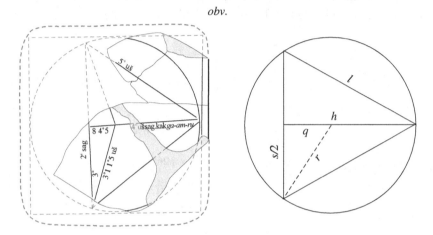

Fig. 3.2.1. *TMS* 1. Metric algebra. A circle-and-chord problem.

In Fig. 3.2.2, left, is shown the opposite case, when $s/2$ (half the length of the chord) and $p = r - q$ (the height of the circle segment cut off by the chord) are known, and *h* has to be computed. This case can be reduced to a quadratic-linear system of equations of the following kind (Friberg, *RlA* 7 (1990) Sec. 5.7 c, type B3b):

$\text{sq. } r - \text{sq. } q = \text{sq. } s/2, \quad r - q = p.$

The solution, obtained through an application of the conjugate rule is:

$h = r + q = (\text{sq. } s/2)/p, \quad r = (h + p)/2, \quad q = (h - p)/2.$

Thus, if the length *s* of the chord, and the height *p* of the circle segment cut

---

26. In the Late Babylonian mathematical recombination text **W 23291 § 1 f** (Friberg, *BaM* 28 (1997)), the "seed measure" and the width of a square band (a region bounded by two concentric and parallel squares) are given. The problem to find the sides of the bounding squares is equivalent to a quadratic-linear system of equations of type B3a.

off by the chord are both known, then the following computation rule can be used to find the length of the diameter:

$d = h + p = (\text{sq. } s/2)/p + p.$

This rule for the computation of the diameter is *close* to the rule applied in *P.BM 10399* § 1, where something is computed as $(\text{sq. } 13/7) \cdot 7 + 7$, which, of course, is equal to $(\text{sq. } 13)/7 + 7$. However, close is not good enough, so one has to find a rule that corresponds *exactly* to the steps of the computation in *P.BM 10399* § 1. How this can be done is shown below:

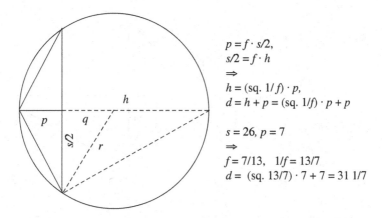

$p = f \cdot s/2,$
$s/2 = f \cdot h$
$\Rightarrow$
$h = (\text{sq. } 1/f) \cdot p,$
$d = h + p = (\text{sq. } 1/f) \cdot p + p$

$s = 26, p = 7$
$\Rightarrow$
$f = 7/13, \quad 1/f = 13/7$
$d = (\text{sq. } 13/7) \cdot 7 + 7 = 31\ 1/7$

Fig. 3.2.2. *P.BM 10399* § 1. Computation of the diameter, using similar triangles.

The crucial idea here is the realization if $h = (\text{sq. } s/2)/p$, as shown above, then it follows that $h/(s/2) = (s/2)/p$, which means (in modern terms) that in Fig. 3.2.2 the right triangle with the sides h and $s/2$ is *similar* to the right triangle with the sides $s/2$ and p. In OB mathematics, this result would have been expressed in terms of side ratio or growth rate, rather than in terms of similarity.[27] Therefore the similarity of the mentioned right triangles in Fig. 3.2.2 would be understood by an OB mathematician as *the equality of the side ratios f of the two triangles*. In modern notations:

$p = f \cdot s/2, \quad \text{and} \quad s/2 = f \cdot h.$

---

27. See, most recently, **MS 3052 § 1 d** (Friberg, *MCTSC* (2005), Sec. 10.2), where the growth rate of the triangular cross section of a mud wall is called *indanum*, probably meaning '(increase in) (n)inda(n) (per cubit of descent)'.

3.2. P.British Museum 10399 (Ptolemaic, Later than P.Cairo)   141

From this follows what may be called the "chord-and-diameter rule".

$h = $ sq. $1/f \cdot p$, and $d = h + p = $ sq. $1/f \cdot p + p$.

Hence, a possible explanation of the computations in the preserved part of *P.BM 10399* § 1 is that the exercise is an application of the chord-and-diameter rule, in the case when the length of the chord is $s = 26$, and when the height of the segment cut off by the chord is 7. What is computed in the preserved first part of the solution procedure is then the diameter

$d = $ sq. $(13/7) \cdot 7 + 7 = 31 \ 1/7$.

MS 3049 (Friberg, *MCTSC* (2005), Sec. 11.1) is a small fragment of a mathematical recombination text. The text appears to be either late OB or post-OB (Kassite). According to a subscript, fortuitously preserved at the end of the inscription, the text originally contained 16 problems, including 6 problems for circles. The first of those 6 problems is preserved. It asks for the length s of a chord (dal an.ta 'the upper transversal'), when the diameter d (dal 'the transversal') and the height p of the segment (ša ur-dam 'that which I went down') are known, $d = 20$ n., and $p = 2$ n. The method used to compute the length of the chord can be explained as follows, in modern notations, and with the symbols introduced in Fig. 3.2.2:

$r = d/2 = 10$, $q = r - p = 8$, $s/2 = $ sqr. (sq. $r -$ sq. q) $= 6$, $s = 2 \cdot 6 = 12$.

It is probably not a coincidence that there were originally precisely 6 "circle-and-chord problems" in the first paragraph of MS 3049. Indeed, in problems of this type, there are 4 primary parameters, namely d, p, h, and s. (Note that the center and the radius of a circle appear to have played only secondary roles in Babylonian mathematics.) In the preserved exercise, the known parameters are d and p, and s is computed. (The trivial computation of h is omitted.) Now, there are 6 ways to choose 2 parameters out of 4. In the case of the circle-and-chord problems the 6 possible choices are (in no particular order):

1. d and p, 2. d and h,
3. d and s, 4. p and h,
5. p and s, 6. h and s.

These 6 possible choices probably correspond to the originally 6 circle-and-chord problems in MS 3049 § 1. The third choice apparently corresponds to the choice of data in *P.BM 10399* § 1.

3.2 b. *P.BM 10399* § 2 (*DMP* ## 42-45). Masts (truncated cones)

In these four exercises, the heights of 4 'masts' (truncated cones) are h =100, 90, 80, and 70 cubits, respectively, The diameter of the mast at its foot is always a = 3 cubits, and the diameter at its top b = 1 cubit. The volume V of the mast is computed using the non-exact equation

$V = (1 - 4') \cdot$ sq. $\{(a + b)/2\} \cdot h$.

The volume is (implicitly) expressed in *cubic cubits*. However, the stated problem is to find out how many *hinu* of water a metal casing for the mast can hold. A *hinu* is defined as a cubic 'palm'. Since 1 (divine) cubit equals 7 palms, it follows that 1 cubic cubit equals $7 \cdot 7 \cdot 7 = 343$ *hinu*. Therefore, the final answer is that the casing for the mast holds $343 \cdot V$ *hinu*, where V is the volume of the mast.

An OB parallel text is **VAT 8522 # 1** (Neugebauer, *MKT 1* (1935), 368). The question in that exercise is the following:

> A log of cedar is 5 ninda long, and
> 1 04 (cylinder) sìla 'thick' at the base, 8 (cylinder) sìla 'thick' at the top.
> The whole log is worth 3 3' minas of silver.
> How much of the log can be cut off (at the top) for 3' mina?

A "cubic" cylinder of diameter d and height d (measured in multiples of ;01 ninda = 6 fingers) contains sq. $d \cdot d$ cylinder sìla. (See the definition of the cylinder sìla in the discussion in Sec. 2.3 a of the granary problem in *P.UC 32160* = *P.Kahun IV. 3* # 1.) Therefore, if a cubic near-cylinder measures 1 04 cylinder sìla at the base of the log and 8 cylinder sìla at the top of the log, then the diameter of the log is a = ;04 ninda at the base and b = ;02 ninda at the top.

In VAT 8522 # 1, the volume of the log (or truncated cone) is computed (incorrectly) as follows:

V = ;05 $\cdot$ sq. $\{(a + b)/2 \cdot 3\} \cdot h$ = ;05 $\cdot$;01 21 sq. n. $\cdot$ 1 00 c. = ;06 45 sq. n. $\cdot$ c.

Then the volume is transformed into capacity measure through multiplication with the constant 6 40 (00), the "storing number" for the cylinder sìla. The result is:

C = ;06 45 sq. n. $\cdot$ c. $\cdot$ 6 40 (00) sìla/sq. n. $\cdot$ c. = 45 (00) sìla.

Since 1/3 mina is 1/10 of 3 1/3 minas, a tenth of this capacity measure can be cut off at the top of the log for the price of 1/3 mina. This is 4 30 cylinder sìla. And so on.

The similarity between VAT 8522 # 1 and the demotic exercises in *P.BM 10399* § 2 is obvious. The encasing of the mast in an envelope of copper in the demotic exercises is only a slightly ridiculous excuse for measuring the size of the mast in capacity measure instead of in volume measure. In VAT 8522 # 1, no such excuse is needed, probably because it was commonplace in Mesopotamia to measure logs in capacity measure. Note, by the way, the similarity between the *hinu* = a cube with the height 7 fingers, and the cylinder sìla = a cubic cylinder with the height 6 fingers.

Another parallel text is **Haddad 104 # 2** (Al-Rawi and Roaf, *Sumer* 43 (1984)), which begins like this:

> If a log, 5 its bottom, 1 40 its top, 30, a reed, it is long, what is the grain it takes?

The 'log' in this text is a truncated cone. Its volume is computed by use of the same incorrect rule as the one that is used in *P.BM 10399* § 2 (*DMP* ## 42-45). Then the volume is multiplied by the storing number 6 (00 00). The result is the capacity measure of the log, in terms of a sìla probably defined as a box with the dimensions 6 fingers · 6 fingers · 5 fingers. (See Friberg, *BaM* 28 (1997), 306, 312.)

3.2 c. *P.BM 10399* § 3 (*DMP* ## 46-51). The reciprocal of 1 + 1/n, etc.

In exercises *DMP* ## 46-51, the stated problem is to compute the reciprocal of $1 + p$, with $p = \overline{5}, 6', \overline{7}, \overline{8}, \overline{9}$, and $6''$, respectively. Here is the text of *DMP* # 50, a relatively well preserved exercise:

> 1. You are told: $\overline{9}$ is the addition, what is the subtraction?
> 2. The way of doing it, namely:
> 3. You add $\overline{9}$ *to 1*, result $1\ \overline{9}$.
> 4. You say: $\overline{9}$ is what of $1\ \overline{9}$? Result, its $\overline{10}$.
> 5. You say: $\overline{9}$ is the addition, $\overline{10}$ the subtraction.
> 6. To make you know it, namely:
> 7. You subtract $\overline{10}$ from 1, remainder $6''\ \overline{15}$.
> 8. The fraction is $6''\ \overline{15}$, namely:
> 9. Its $\overline{9}$ is $\overline{10}$. You add them, result 1 again.

These exercises are clearly of the same type as the exercises in *P.Cairo* § 4 (reshaping a piece of cloth, keeping the area). See Fig. 3.1.4 above. The general rule is here (in modern notations) that

The reciprocal of $1 + \bar{n}$ is $1 - \overline{n+1}$.

This "reciprocity rule" works even in exercise *DMP* # 51, where $1 + p =$ 1 6" = 1 + rec. 1 $\bar{5}$, and where 1 $\bar{5}$ + 1 = 2 $\bar{5}$ (= 11/5), so that, consequently, the answer is

You say: 6" is the addition, 5 $\underline{11}$ is the subtraction (where 5 $\underline{11}$ means 5/11)

OB parallels can be found, as in the case of *P.Cairo* § 4, in **VAT 7532** and **VAT 7535**.

In Fowler, *MPA* (1987 (1999)), 262 (260), the fractions appearing in line 7 of all the exercises in § 3 are examined closely, in an effort to explain how *DMP* # 51 is different from *DMP* ## 46-50. Below, Fowler's examination is repeated, but with different details and a different conclusion.

Essentially, the following reciprocals are computed in line 7 of the six exercises in § 2:

DMP # 46, line 7:	rec. 1 $\bar{5}$	= 1 – 6'	= 6"
DMP # 47, line 7:	rec. 1 6'	= 1 – $\bar{7}$	= 6" $\overline{42}$
DMP # 48, line 7:	rec. 1 $\bar{7}$	= 1 – $\bar{8}$	= 3" $\overline{12}$ $\overline{8}$
DMP # 49, line 7:	rec. 1 $\bar{8}$	= 1 – $\bar{9}$	= 6" $\overline{30}$ $\overline{45}$
DMP # 50, line 7:	rec. 1 $\bar{9}$	= 1 – $\overline{10}$	= 6" $\overline{15}$
DMP # 51, line 7:	rec. 1 6"	= 1 – 5 $\underline{11}$	= 6 $\underline{11}$

Parker, *DMP* (1972), 61 and 63 questioned the procedure in the text in the following two cases:

DMP # 48: For no discernible reason the scribe uses the clumsy fraction
2/3 1/12 1/8, with 1/8 in incorrect order after 1/12, instead of
the more convenient 5/6 1/24.

DMP # 51: That 5/11 is the result is sure from lines 5 and 7, but why did
the scribe use this new type of fraction (see problems 2 and 3)
in just one problem out of six?

Fowler (*op. cit.*), foot notes 83 and 88, tries to explain the first of these obscure points by reference to his observation that "the expression 2' 4' seems to be avoided", suggesting simply without any kind of justification that

"3" 12' may be the standard expression for the 4th of 3".

Actually, also the answers in exercises *DMP* ## 47, 49, and 50 are problematic. If the scribe had counted in the traditional Egyptian way, making use of the 2/*n* table, he would have obtained the following alternative results:

DMP # 47: rec. 1 6' = 1 − $\overline{7}$ = 6 · $\overline{7}$ = (4 + 2) · $\overline{7}$ = 2' $\overline{14}$ + 4' $\overline{28}$ = 2' 4' $\overline{14}$ $\overline{28}$
DMP # 49: rec. 1 $\overline{8}$ = 1 − $\overline{9}$ = 8 · $\overline{9}$ = 4 · 6' $\overline{18}$ = 3" 6' $\overline{18}$ = 6" $\overline{18}$
DMP # 50: rec. 1 $\overline{9}$ = 1 − $\overline{10}$ = 9 · $\overline{10}$ = (8 + 1) · $\overline{10}$ = 3" $\overline{10}$ $\overline{30}$ + $\overline{10}$ = 3" $\overline{5}$ $\overline{30}$

A possible explanation of the mentioned obscure points is that the author of *P.BM 10399* § 3 simply used any method he could think of, including the hidden use of sexagesimal fractions (as often in *P.Cairo*), and the use of some rudimentary idea of a common fraction (as in *P.Cairo* § 1). Thus, he may have counted as follows:

DMP # 46: rec. 1 $\overline{5}$ = 1 − 6'   = 6"
DMP # 47: rec. 1 6' = 1 − $\overline{7}$ = 6 · $\overline{7}$ = 36 · $\overline{42}$ = (35 + 1) · $\overline{42}$ = 6" $\overline{42}$
DMP # 48: rec. 1 $\overline{7}$ = 1 − $\overline{8}$ = 2' 4' $\overline{8}$ = ;30 + ;15 + $\overline{8}$ = ;45 + $\overline{8}$ = 3" $\overline{12}$ $\overline{8}$
                                                                                               (in this order!)
DMP # 49: rec. 1 $\overline{8}$ = 1 − $\overline{9}$ = 1 − ;06 40   = ;53 20 = ;50 + ;02 + ;01 20
                                                                                           = 6" $\overline{30}$ $\overline{45}$
DMP # 50: rec. 1 $\overline{9}$ = 1 − $\overline{10}$ = 1 − ;06   = ;54   = ;50 + ;04   = 6" $\overline{15}$
DMP # 51: rec. 1 6" = 1 − 5 $\underline{11}$ = 6 $\underline{11}$

3.3. *P.British Museum 10520* (Early(?) Roman)

This document (Parker, *DMP* ## 53-65) consists of thirteen exercises on the reverse of a reused papyrus, belonging to seven distinct topics:

§ 1	The iterated sum of the integers from 1 to 10	*DMP* # 53
§ 2	A multiplication table for 64, from 1 to 16	*DMP* # 54
§ 3	A multiplication rule, applied to the product of 13 and 17	*DMP* # 55
§ 4	Expressing 2/35 as a sum of parts	*DMP* # 56
§ 5	Operations with the fractions 3' $\overline{15}$ (2/5) and 3" $\overline{21}$ (5/7)	*DMP* ## 57-61
§ 6	Approximations of square roots; a) sqr. 10, b) sqr. 2'	*DMP* ## 62-63
§ 7	Areas of rectangular fields	*DMP* ## 64-65

3.3 a. *P.BM 10520* § 1 (*DMP* # 53). The iterated sum of 1 through 10

The most interesting of the exercises in *P.BM 10520* is § 1). The initial question there is read and translated in Parker, *DMP* (1972) as follows:

1 is filled twice up to 10.

In Zauzich, *BiOr* 32 (1975), a review of Parker, *DMP*, a modified reading and translation is suggested:

1 is moved back repeatedly, up to 10.

Whichever the correct form of the question is, the form of the solution procedure is clear:

> You count 10, 10 times, result 100.
> You add 10 to 100, result 110. You take 2', result 55.
> You say: 1 up to 10 amounts to 55.
> You add 2 to 10, result 12. You take 3' of 12, result 4.
> You count 4, 55 times, result 220. It is it.
> You say: 1 is filled twice (*or* 1 is moved back repeatedly) up to 10, result 220.

Apparently, what is computed here is

1) $10 \cdot (10 + 1)/2 = (100 + 10)/2 = 55$,
2) $10 \cdot (10 + 1)/2 \cdot (10 + 2)/3 = 55 \cdot 12/3 = 220$.

The first step is, of course, the well known computation of the sum of the first 10 integers. What the second step means can be explained by reference to columns 1-3 in the following table:

1	1	1		
2	3	4		
3	6	10		
4	10	20		
5	15	35	($15 = 3 \cdot 5$,	$35 = 5 \cdot 7$)
6	21	56	($21 = 3 \cdot 7$,	$56 = 7 \cdot 8$)
7	28	84	($28 = 4 \cdot 7$,	$84 = 3 \cdot 4 \cdot 7$)
8	36	120	($36 = 4 \cdot 9$,	$120 = 3 \cdot 4 \cdot 10$)
9	45	165	($45 = 5 \cdot 9$,	$165 = 3 \cdot 5 \cdot 11$)
10	55	220	($55 = 5 \cdot 11$,	$220 = 2 \cdot 10 \cdot 11$)

In *col.* 1 are recorded the first 10 integers. The numbers in *col.* 2 are the sums of the first *n* numbers in *col.* 1, with *n* from 1 to 10. Similarly, the numbers in *col.* 3 are the sums of the first *n* numbers in col. 2, or the "iterated sums" of the first *n* numbers in col. 1, with *n* again from 1 to 10.

The general rule giving the sum $S_1(n)$ of the first *n* integers can very well have been discovered by someone looking at the factorizations of the numbers in *col.* 2 above. Similarly, the general equation for the iterated sum $S_2(n)$ of the first *n* integers can have been discovered by someone looking at the factorizations of the numbers in *col.* 3 above. The two equations are

$S_1(n) = n \cdot (n + 1)/2$ and $S_2(n) = n \cdot (n + 1)/2 \cdot (n + 2)/3$.

Zauzich (*op. cit.*) suggested that the phrase "1 is moved back repeatedly"

3.3. P.British Museum 10520 (Early(?) Roman)

should be understood as referring to some triangular array for the computation of the iterated sums, such as, for instance,

```
1  1  1  1  1  1  1  1  1  1
   2  2  2  2  2  2  2  2  2
      3  3  3  3  3  3  3  3
         4  4  4  4  4  4  4
            5  5  5  5  5  5
               6  6  6  6  6
                  7  7  7  7
                     8  8  8
                        9  9
                          10
```

In any case, clearly the iterated sum of the first 10 integers is what is computed in *P.BM 10520* § 1.

A Late Babylonian (imperfect) parallel is **AO 6484 # 2** (Neugebauer, *MKT 1* (1935), 96), an isolated exercise in a Seleucid mathematical recombination text from Uruk:

AO 6484 # 2.

tam-ḫar-tum šá
ta 1 • 1 : 1 en 10 • 10 : 1 40
ki-i en šid-*tú*
1 • 20 : [3'] / rá-*ma* 20 :
10 • 40 : 2-*ta* šu.2.meš rá-*ma* 6 40 :
6 40 ù 20 7 /
7 • 55 rá-*ma* 6 25
6 25 šid-*tú*

Square-number, that of
from 1 · 1 = 1 to 10 · 10 = 1 40.
Like what is the number?
1 · 20, *3'*, go, then 20.
10 · 40, two-hands (2/3) go, then 6 40.
6 40 and 20 (is) 7.
7 · 55 go, then 6 25.
6 25 is the number.

The given task here is to compute the sum $Q(10)$ of the squares of the integers from 1 to 10. The computation is clearly based on the following general equation for the sum of the first n squares:

$Q(n) = (1 \cdot 1/3 + n \cdot 2/3) \cdot S_1(n) = n \cdot (n + 1)/2 \cdot (2n + 1)/3$.

The general rule may have been found by someone looking closely at a table like the one below:

1	1	1		
2	3	5		
3	6	14		$(14 = 2 \cdot 7)$
4	10	30		$(30 = 5 \cdot 6)$
5	15	55	$(15 = 3 \cdot 5,$	$55 = 5 \cdot 11)$
6	21	1 31	$(21 = 3 \cdot 7,$	$1\,31 = 91 = 7 \cdot 13)$
7	28	2 20	$(28 = 4 \cdot 7,$	$2\,20 = 140 = 4 \cdot 5 \cdot 7)$
8	36	3 24	$(36 = 4 \cdot 9,$	$3\,24 = 204 = 3 \cdot 4 \cdot 17)$
9	45	4 45	$(45 = 5 \cdot 9,$	$4\,45 = 285 = 3 \cdot 5 \cdot 19)$
10	55	6 25	$(55 = 5 \cdot 11,$	$6\,25 = 385 = 5 \cdot 7 \cdot 11)$

In this table, *col.* 1 lists the integers n from 1 to 10, *col.* 2 the sums of the n first integers, and *col.* 3 the sums of the squares of the n first integers.

OB related texts are various tables of "quasi-cube-sides". An example is **MS 3048** (Friberg, *MCTSC* (2005), Sec. 2.4 b):

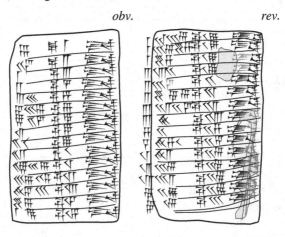

Fig. 3.3.1. MS 3048. A quasi-cube-side table, with 30 entries of the type $n \cdot (n + 1) \cdot (n + 2)$.

The entries in this unusual table text proceed from 6.e 1 íb.si$_8$ '6 makes 1 equalsided' to 8 16.e 30 íb.si$_8$ '8 16 makes 30 equalsided'. It is easy to check that $6 = 1 \cdot 2 \cdot 3$, and that 8 16 (00) = $30 \cdot 31 \cdot 32$.

3.3 b. *P.BM 10520* § 2 (*DMP* # 54). A multiplication table for 64

This is simply a table listing multiples of 64, from 64, 128, 192, ⋯ to ⋯ , 896, 960, 1024. It is interesting only because 64 is the 6th power of 2 and 16 is the 4th power of 2. Therefore, the number 1024 = 16 · 64 in the last line of the table is the 8th power of 2. There are quite a few known examples of computations of high powers of integers, more generally of "many-place regular sexagesimal numbers", in OB school boys' hand tablets. See Friberg, *MCTSC* (2005), Secs. 1.4-1.5. Computations of and with many-place regular sexagesimal numbers were even more popular in Late Babylonian/Seleucid mathematics. See Friberg, (*op. cit.*), App. 8, for various extremely interesting examples.

A survey of all attested "head numbers" for OB multiplication tables is presented in Friberg, (*op. cit.*), Sec. 2.6 e. The smallest attested head number is 1 12 = 72, the head number of the multiplication table **MS 3866**. Thus, the list of allowed OB head numbers for multiplication tables stops just before it reaches the number 1 04 = 64. The reason why 1 04 is not among the attested head numbers is discussed at length in Friberg, *MCTSC* (*op. cit.*), Sec. 2.6 f. There it is also shown that the restriction that all head numbers should belong to a fixed list of such numbers is no longer operative in the case of Late Babylonian multiplication tables.

3.3 c. *P.BM 10520* § 3 (*DMP* # 55). A new multiplication rule

In this exercise, the product of 13 and 17 is computed as follows:

$17 \cdot 13 = (10 + 7) \cdot (10 + 3)$
$= 10 \cdot 10 + 10 \cdot 3 + 10 \cdot 7 + 7 \cdot 3 = 100 + 30 + 70 + 21 = 221$.

This is interesting because it is a departure from the use a binary expressions for the multiplicator in the Egyptian hieratic mathematical papyri, where the product of 13 and 17 would have been computed as

$17 \cdot 13 = (1 + 16) \cdot 13 = 13 + 208 = 221$, or
$13 \cdot 17 = (1 + 4 + 8) \cdot 17 = 17 + 68 + 136 = 221$.

The only known example of an explicit multiplication in a Babylonian cuneiform text is **BM 34601** = *LBAT* 1644, a fragment of a Late Babylonian text, where the square of 3^{46} = is computed in the same way as we would do it. (See Friberg, *MCTSC* (2005), App. 8 b.)

3.3 d. *P.BM 10520* § 4 (*DMP* # 56). 2/35 expressed as a sum of parts

This exercise is an explanation of the identity $2 \cdot \overline{35} = \overline{30}\,\overline{42}$. The same expression is used for $2 \cdot \overline{35}$ in **the 2/n table in *P.Rhind***, although it breaks the pattern of the other 2/n expansions. In his systematic survey of the various expansion methods used in the 2/n table, Neugebauer, *ARÄ* (1930), 366, Table 9, characterizes the expansion of $2 \cdot \overline{35}$ as "systemlos" (unsystematic).

The importance of this exercise is that it seems to point to an intimate connection between a demotic and a hieratic mathematical text, *P.BM 10520* on one hand and the 2/n table in *P.Rhind* on the other.

3.3 e. *P.BM 10520* § 5 (*DMP* ## 57-61). Operations with fractions

Fowler's discussion of "the evidence for the proposal that early Greek mathematicians conceived and used manipulations of common fractions" in *MPA* (1987 (1999)), 7.3(e), 7.4(a), was mentioned above, in Sec. 3.1 c, in connection with the analysis there of the computations in *P.Cairo* § 1. Fowler reinforces his arguments, which aim to refute that proposal, with the following emphatic statement:

> "Of course, the manipulations of fractions expressed as unit fractions are (arithmetically) equivalent to the same manipulations when expressed as common fractions; but they will be conceived differently in the two systems. ⋯ ⋯ Just one example of some operations such as the addition, subtraction, multiplication, or division of two fractional quantities, expressed as something like 'the *n*th of *m* multiplied by the *q*th of *p* gives the *nq*th of *mp*' and *clearly unrelated, by context, to any conception in terms of simple and compound parts*, could be fatal to my thesis that we have no good evidence for the Greek use or conception of common fractions. I know of no such example."

Immediately after that passage follows Fowler's footnote 91:

> "For a recent, useful, detailed study of Greek and Egyptian fractional techniques, see Knorr, TFAEG ((1982]). But what are represented there as manipulations of common fractions *m/n* are clearly manipulations of the descriptions 'the *n*th of *m*', which are conceived throughout the texts under discussion in terms of unit fractions. ⋯ ⋯ (The answer here is surely that the scribes were actually conceiving, teaching, and developing unit fraction techniques, and had little or no conception of common fractions.)"

This categorical dismissal of Knorr's results is so much more surprising as

3.3. P.British Museum 10520 (Early(?) Roman)

Knorr, *HM* 9 (1982) specifically mentions a systematic list of five examples of operations with fractions in *P.BM 10520* § 5, a fact which in some way escaped Fowler's attention. Those examples are worth a closer look:

P.BM 10520 § 5 a (*DMP* # 57): Multiplication of fractions

This is the text of the exercise:
1. Count 3' $\overline{15}$ 3" $\overline{21}$ times.
2. You shall count 5 7 times, result 35.
3. You shall bring 3' $\overline{15}$ (to the) number 5, result 2.
4. You shall bring 3" $\overline{21}$ (to the) number 7, result 5.
5. You shall count 2 5 times, result 10.
6. You shall <cause> that 10 make part of 35, result 4' $\overline{28}$.
7. You shall say: Result 4' $\overline{28}$.

In terms of sums of parts, this example can only be interpreted in the following way:

3' $\overline{15}$ · 3" $\overline{21}$ = 4' $\overline{28}$.

However, this simplistic interpretation does not make any sense at all of the following five steps of the actual computation:

1. 3' $\overline{15}$ · 3" $\overline{21}$ = ?
2. 5 · 7 = 35.
3. 3' $\overline{15}$ · 5 = 2   (3' $\overline{15}$ · 5 = 1 3" + 3' = 2)
4. 3" $\overline{21}$ · 7 = 5   (3" $\overline{21}$ · 7 = 4 3" + 3' = 5)
5. 2 · 5 = 10.
6. 10 as part of 35 is 4' $\overline{28}$.
7. Answer: 4' $\overline{28}$.

(Note that there are two different terms for multiplication in this text, 'count *a*, *b* times' when *a* is an integer, and 'bring *a* to the number *b*' in other cases!) The multiplication in line 2 makes sense only if the author of the text was thinking of the two given fractions as

3' $\overline{15}$ = (3' · 15 + 1) · $\overline{15}$ = 6 · $\overline{15}$ = 2 · $\overline{5}$ or $\overline{5}$ of 2   (the 5th part of 2),

and

3" $\overline{21}$ = (3" · 21 + 1) · $\overline{21}$ = 15 · $\overline{21}$ = 5 · $\overline{7}$ or $\overline{7}$ of 5   (the 7th part of 5).

The text of *P.BM 10520* § 5 a can be interpreted a *proof by numerical example* of a "multiplication rule for (common) fractions". The proof is, essentially, of the following form:

a. $P = (3'\,\overline{15}) \cdot (3''\,\overline{21})$ is silently understood as $(2 \cdot \overline{5}) \cdot (5 \cdot \overline{7})$.
b. $5 \cdot 7 = 35$, $3'\,\overline{15} \cdot 5 = 2$, $3''\,\overline{21} \cdot 7 = 5$.
c. $P \cdot 35 = (3'\,\overline{15} \cdot 5) \cdot (3''\,\overline{21} \cdot 7) = 2 \cdot 5 = 10$.
d. $P = 10 \cdot \overline{35} = 2 \cdot \overline{7} = 4'\,\overline{28}$.

Clearly, what is going on here is that the product P of the two given fractions is first multiplied by $5 \cdot 7 = 35$, then divided by the same factor. Knorr (*op. cit.*) calls this method "raising the terms". In modern notations, the proof can be understood as follows:

$m/n \cdot q/p = (m/n \cdot n) \cdot (q/p \cdot p)/(n \cdot p) = (m \cdot q)/(n \cdot p)$.

It is important to note that *both the question, the answer, and much of the solution procedure in § 5 a are expressed in terms of sums of parts*. Knorr describes the situation by saying that the sums of parts "perform merely a notational, rather than a computational, role". In other words, just as there seems to be a hidden use of sexagesimal fractions in *P. Cairo*, there seems to be a similar *hidden use of common fractions* in *P.BM 10520*. This suspicion is strengthened by the following explanations of the remaining exercises in *P.BM 10520 § 5*.

P.BM 10520 § 5 b (*DMP # 58*): Division of fractions

1. Cause that $3'\,\overline{15}$ make part of $3''\,\overline{21}$.
2. You shall count 5 7 times, result 35.
3. Find its $3''\,\overline{21}$, result 25.
4. Find its $3'\,\overline{15}$, result 14.
5. You shall cause that 14 make part of 25, result $2'\,\overline{25}\,\overline{50}$.
6. You shall say: Result $2'\,\overline{25}\,\overline{50}$.

These are the steps of the computation:

1. $3'\,\overline{15} / 3''\,\overline{21} = ?$
2. $5 \cdot 7 = 35$.
3. $3''\,\overline{21} \cdot 35 = 25$, $(3''\,\overline{21} \cdot 35 = 23\,3' + 1\,3'' = 25$
4. $3'\,\overline{15} \cdot 35 = 14$ $(3'\,\overline{15} \cdot 35 = 11\,3'' + 2\,3' = 14)$
5. $14 \cdot \overline{25} = 2'\,\overline{25}\,\overline{50}$.
6. Answer: $2'\,\overline{25}\,\overline{50}$.

The text of *P.BM 10520 / 5 b* can be interpreted as a *proof by numerical example* of a division rule for (common) fractions. The proof is, essentially, of the following form:

a. $Q = 3'\,\overline{15} / 3''\,\overline{21}$ is silently understood as $2 \cdot \overline{5} / 5 \cdot \overline{7}$.
b. $5 \cdot 7 = 35$, $3''\,\overline{21} \cdot 35 = 25$, $3'\,\overline{15} \cdot 35 = 14$.

3.3. P.British Museum 10520 (Early(?) Roman)

c. $Q = (3'\,\overline{15} \cdot 35) / (3''\,\overline{21} \cdot 35)) = 14/25 = 2'\,\overline{25}\,\overline{50}$.

Clearly, what is going on here is that the quotient Q of the two given fractions is first multiplied by $5 \cdot 7 = 35$, then divided by the same factor. This is another application of what Knorr called the method of *raising the terms*. In modern notations, the proof can be understood as follows:

$(m/n) / (q/p) = (m/n \cdot n \cdot p) / (q/p \cdot n \cdot p) = (m \cdot p) / (n \cdot q)$.

P.BM 10520 § 5 c (DMP # 59): Fractions of fractions

1. Take the 3' $\overline{15}$ of 3" $\overline{21}$.
2. You shall count 5 7 times, result 35.
3. Find its 3" $\overline{21}$, result 25.
4. You shall take the 3' $\overline{15}$ to 25, result 10.
5. You shall cause that 10 make part of 25, ⋯ (the rest is corrupt).

These are the steps of the computation, according to Parker's reconstruction of the text:

1. 3' $\overline{15}$ of 3" $\overline{21}$ = ?
2. $5 \cdot 7 = 35$.
3. 3" $\overline{21} \cdot 35 = 25$,
4. 3' $\overline{15} \cdot 25 = 10$.
5. 10 as part of 35 is 4' $\overline{28}$.
6. Answer: 4' $\overline{28}$.

The question in § 5 c is a differently formulated variant of the question in § 5 a. *The solution procedure is also different*, but the answer is, of course, the same. Thus, in § 5 c the *proof by numerical example* of a "composition rule for (common) fractions" goes as follows:

a) $C = (3'\,\overline{15})$ of $(3''\,\overline{21})$ is silently understood as $(2 \cdot \overline{5})$ of $(5 \cdot \overline{7})$.
b) $5 \cdot 7 = 35$, 3" $\overline{21}$ of $35 = 25$, 3' $\overline{15}$ of $25 = 10$.
c) C of $35 = (3'\,\overline{15})$ of $(3''\,\overline{21})$ of $35 = (3'\,\overline{15})$ of $25 = 10$.
d) $C = \overline{35}$ of $10 = \overline{7}$ of $2 = 4'\,\overline{28}$.

In modern terms, this proof can be understood as follows:

m/n of $q/p = (m/n$ of q/p of $n \cdot p) / (n \cdot p) = (m/n$ of $q \cdot n) /(n \cdot p) = (m \cdot q) / (n \cdot p)$.

P.BM 10520 § 5 d (DMP # 60): Subtraction of fractions

1. Subtract 3' $\overline{15}$ from 3" $\overline{21}$.
2. You shall count 5 7 times, result 35.
3. Find its 3" $\overline{21}$, result 25.
4. Find its 3' $\overline{15}$, result 14.
5. You shall subtract 14 from 25, result 11.

6. You shall cause that 11 make part of 35, result 4' $\overline{28}$ $\overline{35}$.
7. You shall say: Result 4' $\overline{28}$ $\overline{35}$.

This time, the steps of the computation are as follows:

1. 3" $\overline{21}$ – 3' $\overline{15}$ = ?
2. 5 · 7 = 35.
3. 3" $\overline{21}$ · 35 = 25,
4. 3' $\overline{15}$ · 35 = 14.
5. 25 – 14 = 11.
6- 11 / 35 = 4' $\overline{28}$ $\overline{35}$.
7. Answer: 4' $\overline{28}$ $\overline{35}$.

This is a *proof by numerical example* of a "subtraction rule for (common) fractions". As in the three preceding examples, the method used is *raising the terms*:

$D \cdot 35 = (3" \overline{21} - 3' \overline{15}) \cdot 35 = 3" \overline{21} \cdot 35 - 3' \overline{15} \cdot 35 = 25 - 14 = 11,$
$D = 11 / 35 (= (10 + 1) / 35) = 2/7 + 1/35 = 4' \overline{28}\,\overline{35}.$

In modern terms,

$m/n - q/p = (m/n \cdot n \cdot p - q/p \cdot n \cdot p)/(n \cdot p) = (m \cdot p - q \cdot n)/(n \cdot p).$

P.BM 10520 § 5 e (*DMP #* 61): **Addition of fractions**

1. Choose 3' $\overline{15}$ to 3" $\overline{21}$.
2. You shall count 5 7 times, result 35.
3. Find its 3" $\overline{21}$, result 25.
4. Find its 3' $\overline{15}$, result 14.
5. You shall add 14 to 25, result 39.
6. You shall carry 35 into 39, result 1 $\overline{10}$ $\overline{70}$.
7. You shall say: Result 1 $\overline{10}$ $\overline{70}$.

The steps of the computation are:

1. 3' $\overline{15}$ + 3" $\overline{21}$ = ?
2. 5 · 7 = 35.
3. 3" $\overline{21}$ · 35 = 25,
4. 3' $\overline{15}$ · 35 = 14.
5. 25 + 14 = 39.
6. 39 / 35 (= 1 + 8/70) = 1 $\overline{10}$ $\overline{70}$.
7. Answer: 1 $\overline{10}$ $\overline{70}$.

This is a *proof by numerical example* of an "addition rule for (common) fractions". Again, the method used is *raising the terms*:

$S \cdot 35 = (3" \overline{21} + 3' \overline{15}) \cdot 35 = 3" \overline{21} \cdot 35 + 3' \overline{15} \cdot 35 = 25 + 14 = 39,$
$S = 39 / 35 = 1 \overline{10}\,\overline{70}.$

In modern terms,

$m/n + q/p = (m/n \cdot n \cdot p + q/p \cdot n \cdot p)/(n \cdot p) = (m \cdot p + q \cdot n)/(n \cdot p)$.

3.3 f. *P.BM 10520* § 6 (*DMP* ## 62-63). The square side rule

P.BM 10520 § 6 a (*DMP* # 62).

1. Let 10 be reduced to its squareside.
2. You shall count 3 3 times, result 9, remainder 1. (Its) 2', result 2'.
3. You let 2' make part of 3, result 6'.
4. You add 6' to 3, result 3 6'. It is the squareside.
5. Let it be known, namely:
6. You shall count 3 6', 3 6' times, result 10 $\overline{36}$.
7. Its difference of square side $\overline{36}$.

The computations in the text above can, of course, be explained as follows:

1) sqr. 10 = sqr. (sq. 3 + 1) = appr. 3 + 2' · 1 · 3' = 3 6'.
2) sq. 3 6' = 10 $\overline{36}$. Error: $\overline{36}$.

This is a simple application of the "square side rule", which appears to have been widely used also in both Old and Late Babylonian mathematical texts (see Friberg, *BaM* 28 (1997) § 8), although there are no known Babylonian examples of *explicit* applications of the rule, such as the one above.

P.BM 10520 § 6 b (*DMP* # 63).

1. Let 2' be reduced to its squareside.
2. You shall count 6, 6 times, result 36. Namely, its half, 18.
3. Let 18 be reduced to its square side, result 4 4'.
4. Let 4 4' be <part of 6>.
5. The way of doing it, namely:
6. You shall bring 4 4' (to) <4'>, result 17.
7. You shall bring 6 (to) the <same> number, result 24.
8. You shall let 17 make <part> of 24, result 3" $\overline{24}$.
9. You shall count 3" $\overline{24}$ 3" $\overline{24}$ times, result 2' $\overline{576}$.
10. Its difference of square side $\overline{576}$.

The object of the exercise is to compute the square side (the square root) of 2' = 1/2. This is done in a roundabout way. The idea is to multiply 2' with a suitable square number, compute the square side of the product, and then divide the result by the square side of the square number.[28] Apparently, the computation proceeds (essentially) as follows, in a series of simple steps:

1) 2' · sq. 6 = 18.
2) sqr. 18 = sqr. (sq. 4 + 2) = appr. 4 + 2' · 2 · 4' = 4 4'.
3) What is 4 4' as part of 6? (In other words, what is 4 4' / 6?)
4) Now, 4 4' = 4' · 17, and 6 = 4' · 24.
5) Therefore, the answer is: 4 4' / 6 = 17 / 24, namely 3" $\overline{24}$.
6) Test: sq. 3" $\overline{24}$ = sq. (17 · $\overline{24}$) = 289 · $\overline{576}$ = 2' $\overline{576}$. Error: $\overline{576}$.

In this connection, it is interesting to note that

if sqr. 2' = appr. 3" $\overline{24}$ = 17 · $\overline{24}$, then sqr. 2 = appr. 2 · 3" $\overline{24}$ = 1 3' $\overline{12}$ = 17 · $\overline{12}$.

It is also interesting to note that the method used in *P.BM 10520* § 6 b can be used to find improved approximations of square roots, as in the following example:

The OB standard approximation to sqr. 3 was:

sqr. 3 = sqr. (4 − 1) = appr. 2 − 1/4 = 1;45 (7/4).

However, in two consecutive exercises in the Late Babylonian mathematical recombination text **W 23291/4 b-c** (Friberg, *BaM* 28 (1997), 286), *two different approximations* to (sqr. 3)/4 are used in order to compute the area of an equilateral triangle. (Compare with the discussion above, in Sec. 3.1 j, of *P.Cairo/11* . See also the discussion of *P.Vindob. G. 19996* **# 10** in Sec. 4.8 b.) The first approximation is ;26 15, obtained from the OB standard approximation: (sqr. 3)/4 = appr. 1;45/4 = ;26 15. The second, approximation is ;26, corresponding to the approximate value 1;44 for sqr. 3. That value can have been obtained as follows:

sqr. (3 · sq. 3) = sqr. 27 = sqr. (25 + 2) = appr. 5 1/5 = 5;12,
sqr. 3 = appr. 5;12 /3 = 1;44 (= 26/15).

It is interesting to compare the accuracy of the Old and Late Babylonian approximations to sqr. 3:

sq. 1;45 = 3;03 45, and sq. 1;44 = 3; 00 16.

---

28. The procedure is the opposite of the OB factorization method used to compute the square side of many-place semi-regular sexagesimal numbers. The idea of that method is to first multiply all regular square factors of the given number by their reciprocals, so that after a number of steps there remains only the non-regular "factor-reduced core" of the given number. The square side of the given number can then be computed as the product of the square side of the factor-reduced core and the square sides of all the removed square factors. For explicit examples, see, for instance, the round hand tablet **UET 6/2 222**, Friberg, *RA* 94 (2000) § 2 d, and examples related to the well known table text **Plimpton 322**, Friberg, *MCTSC* (2005), App. 7, § A7 a.

3.3 g. *P.BM 10520* § 7 (*DMP* ## 64-65). The quadrilateral area rule

P.BM 10520 § 7 a (*DMP* # 64) is a computation of the area of a rectangle. The text, which is accompanied by a drawing (Fig. 3.3.2 below), is as follows:

1. A piece (of land).
2. You shall add the south and the north, result 20. Namely, its half, 10.
3. You shall add the east and the west, 24. Namely, its half, 12.
4. You shall count <it> 10 times, result 120.
5. You shall carry 100 into 120 in order to bring another (formulation), result 1 $\overline{5}$ cubit(-strips).
6. You shall say: In order to bring another (formulation), 1 $\overline{5}$ cubit(-strips).

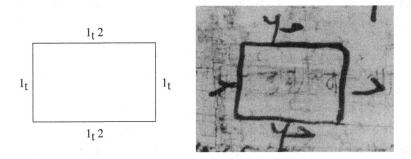

Fig. 3.3.2. *P.BM 10520* § 7 a. Data for an application of the quadrilateral area rule.

Here, the area of a rectangle with the sides 12 and 10 is computed according to the rule that

the area A = the half-sum of the short sides (south and north) times the half-sum of the long sides (east and west).

Thus, in this particular case,

$A = (10 + 10)/2 \cdot (12 + 12)/2 = 10 \cdot 12 = 120$.

The result is divided by 100, and the answer is given in the following alternative form:

$A^* = 120/100 = 1\ \overline{5}$ 'cubits'.

It is silently understood that the sides of the rectangle are measured in *cubits*, and the area A in *square cubits*. The alternative value of the area, here called A^*, is expressed in 'cubits', actually *cubit-strips*, where 1 cubit-strip = 1 cubit · 100 cubits. Compare with the hieratic text **P.Rhind**

 # 51 (Sec. 2.1 e), where a triangle has the sides 10 *khet* and 4 *khet*, with 1 *khet* = 100 cubits. The area is computed as 2,000 (cubit-strips) = 2 (thousand-cubit-strips).

It is interesting that in *P.BM 10520* § 7a, the area of the rectangle is not computed simply as the short side times the long side, but as the half-sum of the two (equal) short sides times the half-sum of the two (equal) long sides. This is obviously a step towards teaching the students to use what may be called the "quadrilateral area rule" in order to compute areas of non-rectangular quadrilaterals. (The rule is approximatively correct in the case of *nearly* rectangular figures.)

P.BM 10520 § 7 b (*DMP # 65*) is another computation of the area of a rectangle. The text is accompanied by a simplified drawing (Fig. 3.3.3). Here is the brief text:

1. A piece (of land). Its plan.
2. You shall add the south and the north, 3. Namely, its half, 1 2'$_s$.
3. You shall add the east and the west, 4 2'$_s$. Namely, its half, 2 4'$_s$.
4. You shall count 2 4'$_s$ 1 2'$_s$ times, result 3 4'$_s$ 8'$_s$.
5. It is its specification of field.

In this exercise, the area of a rectangle is again computed by use of the quadrilateral area rule. The lengths of the sides are written with special number notations (see below, Fig. 3.7.2), indicating that the unit of length measure is 1 *khet* (Greek: *schoinion*). Consequently, the area is measured in sq. *khet* = 1 *setat* (Greek: *aroura*). Thus, the computed area in § 7 b is

1 2'$_s$ · 2 4'$_s$ = 3 4'$_s$ 8'$_s$ = 3 1/4 1/8 *setat*.

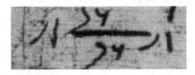

```
              2 4'_s
1 2'_s   ─────────────   1 2'_s
              2 4'_s
```

Fig. 3.3.3. *P.BM 10520* § 7 b. Data for another application of the quadrilateral area rule.

The two applications of the quadrilateral area rule in *P.BM 10520* § 7 a-b with the associated drawings, copied in Figs. 3.3.2-3.3.3, are embarrassingly simple, since the quadrilaterals in both examples are *rectangles*, for which the use of the rule is superfluous.

As noted by Parker, *DMP* (1972), 72, there are more interesting exam-

3.3. P.British Museum 10520 (Early(?) Roman)

ples in the *demotic* ***Theban Ostracon D 12*** in Thompson, *TO 2* (1913), 42-44 and pl. 3, dated by a superscript to 'year 11, Khoiak day 20'.

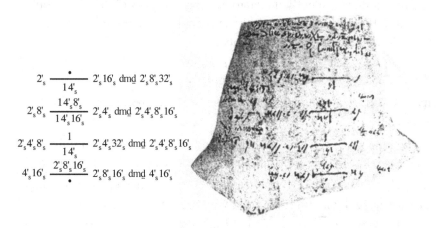

$2'_s \quad \dfrac{\cdot}{14'_s} \quad 2'_s\,16'_s \text{ dmd } 2'_s\,8'_s\,32'_s$

$2'_s\,8'_s \quad \dfrac{14'_s\,8'_s}{14'_s\,16'_s} \quad 2'_s\,4'_s \text{ dmd } 2'_s\,4'_s\,8'_s\,16'_s$

$2'_s\,4'_s\,8'_s \quad \dfrac{1}{14'_s} \quad 2'_s\,4'_s\,32'_s \text{ dmd } 2'_s\,4'_s\,8'_s\,16'_s$

$4'_s\,16'_s \quad \dfrac{2'_s\,8'_s\,16'_s}{\cdot} \quad 2'_s\,8'_s\,16'_s \text{ dmd } 4'_s\,16'_s$

Fig. 3.3.4. *Theban O. D 12*. A land survey with the dimensions and areas of four fields.

The ostracon contains four simplified field plans of the same type as the one in *P.BM 10520* § 7b. One of these, the one mentioned by Parker (*op. cit.*), has unequal opposite sides:

north $2'_s\,8'_s$, south $2'_s\,4'_s$, west $1\ 4'_s\,16'_s$, east $1\ 4'_s\,8'_s$.

The area of the field is given on the ostracon as $2'_s\,4'_s\,8'_s\,16'_s$. No details of the actual computation of this value are recorded. However, if the unit of length used here is a *khet* (Greek: *schoinion*) of 96 cubits (*mḥ*), so that $32'_s = 3$ cubits (see below), then it is likely that the computations were carried out in the following way:

$A = (2'_s\,8'_s + 2'_s\,4'_s)/2 \cdot (1\ 4'_s\,16'_s + 1\ 4'_s\,8'_s)/2$
$= 2'_s\,8'_s\,16'_s \cdot 1\ 4'_s\,16'_s\,32'_s$
$= 2'_s\,8'_s\,32'_s\ mḥ\ 1\ 2' + 8'_s\,32'_s\ mḥ\ 1\ \overline{8} + 16'_s\ mḥ\ 2\ \overline{16}$
$= 2'_s\,4'_s\,8'_s\,32'_s\ mḥ\ 1\ 2'\ \overline{8}\ \overline{16} = $ appr. $2'_s\,4'_s\,8'_s\,16'_s$.

There are many known examples of applications of the quadrilateral area rule in Mesopotamian cuneiform texts, from the oldest to the youngest. Outstanding among the oldest examples is **W 19408, 76** (Friberg, *AfO* 44/45 (1997/98), Fig. 2.1; Nissen/Damerow/Englund, *ABk* (1993), Fig. 50), a *proto-cuneiform* school text from around 3000 BCE. On the reverse of that text, for instance, are given (according to the most plausible reconstruction of damaged parts) two lengths 'along', 16(60) 30 and 23(60) 30, and two lengths 'across', 21(60) 20 and 8(60) 40, all in non-positional sexagesimal numbers, silently understood to be multiples of the ninda (appr. 6 meters). If the quadrilateral area rule is applied to these numbers, the result is the large and round area number

(16(60) 30 + 23(60) 30)/2 · (21(60) 20 + 8(60) 40)
= 20(60) · 15(60) = 5(60 · 60 · 60) sq. n. = 10 šár.

Here 10 šár is an unrealistically large area number, 38.8 square kilometers.

The rule is used also, quite frequently, in Sumerian administrative texts. More surprisingly, the simplistic rule is still used in **IM 121565**, a relatively sophisticated OB mathematical text with the topic "metric algebra". In that text, one finds the following badly preserved exercise:

IM 121565, *obv.* **iii.** (Friberg and Al-Rawi *(to appear)*. From Tell Haddad.

```
... ... ... ...
za.e
1 uš an.ta ù 40 uš ki.ta x x x
[1 40] / ḫe-pé-ma [?] 50 ta-mar
tu-úr
20 sag / an.ta ù 6 40 sag ki.ta x x x
[26 40] / ḫe-pé-ma 13 20 ta-mar
tu-úr /
50 a-na 13 20 i-ta-aš-ši-ma [11] 06 40 /
igi 11 06 40 du₈ 5 24 ta-mar
etc.
```

... ...

You:
1, the upper length, and 40, the lower length, x x x
1 40 break, then 50 you see.
Return.
20, the upper front, and 6 40, the lower front, x x x
26 40 break, then 13 20 you see.
Return.
The 50 to the 13 20 always lift, then 11 06 40.
The opposite of the 11 06 40 resolve, 5 24 you see.
etc.

Before and after the lines copied above, there are some additional lines of text, less well preserved but at least complete enough for the reconstruction of both the statement of the problem and the final part of the solution procedure. Thus, let u_a and u_k be the unknown upper and lower lengths, and let s_a and s_k be the unknown upper and lower fronts. Then, apparently, these four unknowns were supposed to be the solution to the following rectangular-linear system of four equations:

$(u_a + u_k)/2 \cdot (s_a + s_k)/2 = A = 2\;46;40$ sq. n.
$u_k = 2/3 \cdot u_a$
$s_a = 1/2 \cdot u_k$
$s_k = 1/3 \cdot s_a$

The solution procedure started by choosing the following false values for the four sides:

$u_a{}^* = 1\;00,\quad u_k{}^* = 40,\quad s_a{}^* = 20,\quad s_k{}^* = 6;40.$

With these values, the linear equations are satisfied, but the equation for the area is not, since

$(u_a{}^* + u_k{}^*)/2 \cdot (s_a{}^* + s_k{}^*)/2 = A\;^* = 50 \cdot 13;20 = 11\;06;40,$
which means that $A = A\;^* \cdot\;;15$ sq. n.

To get the right value for the area, the false values have to be multiplied by ;30 n. (1 reed). Thus,

$u_a = 1\;00 \cdot\;;30$ n. $= 30$ n.
$u_k = 40 \cdot\;;30$ n. $= 20$ n.
$s_a = 20 \cdot\;;30$ n. $= 10$ n.
$s_k = 6;40 \cdot\;;30$ n. $= 3;20$ n.

A second example is **IM 52301**, a single-column tablet from Shaduppum (Tell Harmal) inscribed with two metric algebra exercises and a brief table of constants, with an additional brief text in two(!) columns on the left edge. The text on the edge, not understood before, was shown in Friberg, *RA* 94 (2000) § 2 f, to be a general, non-numerical formulation of the quadrilateral area rule:

IM 53 301 (Baqir, *Sumer* 6 (1950). Group 7b, Shaduppum.

šum-ma a.šà uš *la mi-it-ḫa-ru-ti at-ta* /

> na-ap-ḫa-ar uš li-iq-bu-ni-kum-ma /
> 4 ša-ar er-bé-tim lu-<pu>-ut-ma /
> i-gi 4 pu-ṭú-ur-ma /
> a-na na-ap-ḫa-ar sag i-ši-ma /
> ma-la i-li-ku
> tu-uš-ta-ka-al-ma
> i-na li-ib-bi / a.šà ta-tam-sa-aḫ

If a field, the lengths not equal.
You:
The sum of the lengths may they say to you,
4 of the four winds may you touch,
the opposite of the 4 resolve, then
to the sum of the fronts raise (it), then
whatever came up for you,
you let (them) eat each other, then
(that) inside the field you will have measured.

The awkwardly formulated rule should probably be understood as saying that

$$A = (u_a + u_k) \cdot (s_a + s_k) \cdot 1/4,$$

where

u_a, u_k are the two 'lengths' and s_a, s_k the two 'fronts'.

Several *Late Babylonian* examples can be found, for instance, in Nemet-Nejat, *LBFP* (1982), a survey of Late Babylonian "field plans" in the British Museum. Such a field plan is typically a small clay tablet, inscribed with a schematic drawing of one or several rectangular fields, together with numerical data and other specifications. A particularly well preserved field plan is **BM 47437** (*op. cit.*, text 24; p. 91, pl. 13), with a drawing of two fields, one of which has the following side lengths: north 9 cubits 8 fingers, south 19 cubits 18 fingers, west 13 cubits 15 fingers, and east 11 cubits. In the Late Babylonian system of "common linear measure", the cubit was divided into 24 fingers, and 7 cubits made 1 reed, as shown by the following "factor diagram":

$$\text{clm (LB):} \quad \leftarrow \text{gi} \xleftarrow{7} \text{kùš} \xleftarrow{24} \text{šu.si} \xleftarrow{7} \text{še} \leftarrow$$
$$\qquad\qquad\quad \text{reed} \quad \text{cubit} \quad\; \text{finger} \;\; \text{barley-corn}$$

Fig. 3.3.5. Factor diagram for the Late Babylonian "common linear measure".

3.3. P.British Museum 10520 (Early(?) Roman)

Hence the area of the mentioned field could be computed as

(19 c. 8 f. + 19 c. 18 f)/2 · (13 c. 15 f. + 11 c.)/2 = 19 c. 13 f · 12 c. 7 1/2 f.

However, this result had to be converted to Late Babylonian "common reed measure", one of several Late Babylonian replacements for the traditional, Sumerian/OB system of area measure. In this new system, the main units were the 'reed', the 'cubit', and the 'finger', as in the system of common linear measure. However, just as the Egyptian "cubit-strip" = 1 cubit · 1 *khet* (100 or 96 cubits), so the mentioned units of common reed measure have to be understood as

1 'reed' =	1 reed-strip =	1 reed · 1 reed,
1 'cubit' =	1 cubit-strip =	1 cubit · 1 reed,
1 'finger =	1 finger-strip =	1 finger · 1 reed.

Actually, in the Late Babylonian mathematical recombination text **W 23291-x** (Friberg, *et al, BaM 21* (1990)), § 11 can be described as a combination of

a) a "multiplication table from common linear measure to common reed measure",
b) a "structure table" for common reed measure.

See (Friberg, *et al.* (*op. cit.*) Sec. 10, and Friberg, *GMS 3* (1993), text 9):

rev.

§ 11
```
1 gi a.rá 1 gi 1 gi 1 gi a.rá 1 kùš 1    kùš
1 gi a.rá 1 šu.si 1 šu.si 1 kùš a.rá 1 gi 1 kùš
1 kùš a.rá 1 kùš 1 kùš tur-tú 1 kùš a.rá 1 šu.si 1 še
1    šu.si 1 gi 1    šu.si 1 šu.si a.rá 1 kùš 1 še
1 šu.si a.rá 1    šu.si tur-tú 2°4 šu.si.meš tur.meš
1 še 7 še.meš 1 šu.si 2°4 šu.si.meš 1 kùš 7 kùš.meš 1 gi
3 šu.si 3 še 1 kùš tur-tú 7 kùš    tur.meš    1 kùš
```

Fig. 3.3.6. W 23291-x § 11. A multiplication table for common linear measure followed by a structure table.

Reorganized into a proper tabular form, the multiplication table and its associated structure table can be displayed as follows:

1 gi	a.rá	1 gi	1 gi
1 gi	a.rá	1 kùš	1 kùš
1 gi	a.rá	1 šu.si	1 šu.si
1 kùš	a.rá	1 gi	1 kùš
[1] kùš	a.rá	1 kùš	1 kùš tur-*tú*
1 kùš	a.rá	1 šu.si	1 š[e]
[1 š]u.si	<a.rá>	1 gi	1 šu.si
1 šu.si	a.rá	1 kùš	1 [še]
[1 š]u.si	a.rá	1 šu.si	<1 šu.si> tur-*tú*
24 šu.si.meš tur.meš			[1] še
7 še.meš			1 šu.si
24 šu.si.meš			1 kùš
7 kùš.meš			1 gi
3 šu.si 3 še			1 kùš tour-*tú*
7 kùš tur.meš			1 kùš

1 reed	times	1 reed	1 reed
1 reed	times	1 cubit	1 cubit
1 reed	times	1 finger	1 finger
1 cubit	times	1 reed	1 cubit
1 cubit	times	1 cubit	1 small cubit
1 cubit	times	1 finger	1 barley-corn
1 finger	times	1 reed	1 finger
1 finger	times	1 cubit	1 barley-corn
1 finger	times	1 finger	1 small finger
24 small fingers			1 barley-corn
7 barley-corns			1 finger
24 fingers			1 cubit
7 cubits			1 reed
3 fingers 3 barley-corns			1 small cubit
7 small cubits			1 cubit

Thus, the common reed measure R of the mentioned field in the Late Babylonian field plan BM 47437 can have been computed as follows:

R = (19 c. 8 f. + 19 c. 18 f)/2 · (13 c. 15 f. + 11 c.)/2
   = 19 c. 13 f · 12 c. 7 1/2 f.
   = 228 small cubits 298 1/2 barley-corns 97 1/2 small fingers
   = 4 reeds 4 cubits 4 small cubits
      + 1 cubit 18 fingers 4 1/2 barley-corns

+ 4 barley-corns 1 1/2 small fingers
= 4 reeds 5 cubits 4 small cubits 18 fingers 9 barley-corns
(− 1/2 barley-corn + 1 1/2 small fingers)
= appr. 4 reeds 6 cubits 9 fingers.

Inside the contour of the mentioned rectangular field in BM 47437 is recorded precisely this value for the common reed measure of the field, 4 reeds 6 cubits 9 fingers.

Another Late Babylonian example of an application of the quadrilateral area rule is offered by **BM 67314**, a computation of the volume and seed measure of a symmetric trapezoid:

BM 67314 § 3 (Friberg, *BaM* 28 (1997), 296).

```
30 uš gíd.da 24 uš lúgud.da /
1 sag an.ta 1 sag ki.ta x x /
uš gíd.da ù uš lúgud.da gar.gar / 54 2'-šú 27
sag an.ta / ù sag ki.ta gar.gar 2 2'-šú 1 /
27 a.rá 1 27
27 a.rá 18 / 8 06
8 06 še.numun x / xxx
```

30 is the long length, 24 the short length,
1 is the upper front, 1 the lower front. x x
The long length and the short length heap, (it is) 54. 1/2 of it is 27.
The upper front and the lower front heap, (it is) 2. 1/2 of it is 1.
The 27 steps of 1 is 27.
The 27 steps of 18 is 8 06.
The 8 06 is the seed xxxx.

3.4. *P.British Museum 10794* (Date Uncertain)

Multiplication tables for $\overline{90}$ and $\overline{150}$ (*DMP* ## 66-67)

This small fragment contains only the first ten lines of two multiplication tables, one for $\overline{90}$, the other for $\overline{150}$. Note that 90 and 150 both can be expressed as regular sexagesimal numbers, 90 = 1 30 and 150 = 2 30. Their reciprocals are the sexagesimal fractions ;00 40 and ;00 24. The computation of the two tables probably made use of sexagesimal arithmetic.

In his paper about Greek and Egyptian techniques of counting with fractions, Knorr, *HM* 9 (1982), 156, confessed that he (like Parker before him) was puzzled by "idiosyncrasies" in the computational procedure in *P.BM 10794*. Why was $2 \cdot \overline{150}$ given as $\overline{90}\ \overline{450}$ and not as $\overline{75}$, and why was $3 \cdot \overline{150}$ given as $\overline{60}\ \overline{300}$ and not as $\overline{50}$, and so on? Knorr was even moved to suggest (*ibid.*), footnote 32, that

> "the scribe might hold in view a base of 360 for its association with the number of days in a year. Then 90 corresponds to the days in a 3-month period, 150 to those in a 5-month period, and the entries in the tables give the fraction of the base period which each of the days in a decanal (10-day) period amounts to. Such tables might be useful for the computation of interest on short-term loans."

The assumption that the scribe used sexagesimal arithmetic leads to a much simpler explanation, with, for instance, the entries in the multiplication table for $\overline{150}$ analyzed as follows:

The way of taking $\overline{150}$ to 10:		
1	to	$\overline{150}$
2	to	$\overline{90}\ \overline{450}$
3	to	$\overline{60}\ \overline{300}$
4	to	$\overline{45}\ \overline{225}$
5	to	$\overline{30}$
6	to	$\overline{30}\ \overline{150}$
7	to	$\overline{30}\ \overline{90}\ \overline{450}$
8	to	$\overline{20}\ \overline{300}$
9	to	$\overline{30}\ \overline{45}\ \overline{225}$
10	to	$\overline{15}$

$$
\begin{array}{rcl}
\overline{150} & = & ;00\ 24 \\
2 \cdot ;00\ 24 & = & ;00\ 48 \quad = \quad ;00\ 40 + ;00\ 08 \quad = \quad \overline{90}\ \overline{450} \\
3 \cdot ;00\ 24 & = & ;01\ 12 \quad = \quad ;01 + ;00\ 12 \quad = \quad \overline{60}\ \overline{300} \\
4 \cdot ;00\ 24 & = & ;01\ 36 \quad = \quad ;01\ 20 + ;00\ 16 \quad = \quad \overline{45}\ \overline{225} \\
5 \cdot ;00\ 24 & = & ;02 \quad = \quad \overline{30} \\
6 \cdot ;00\ 24 & = & ;02\ 24 \quad = \quad ;02 + ;00\ 24 \quad = \quad \overline{30}\ \overline{150} \\
7 \cdot ;00\ 24 & = & ;02\ 48 \quad = \quad ;02 + ;00\ 40 + ;00\ 08 \quad = \quad \overline{30}\ \overline{90}\ \overline{450} \\
8 \cdot ;00\ 24 & = & ;03\ 12 \quad = \quad ;03 + ;00\ 12 \quad = \quad \overline{20}\ \overline{300} \\
9 \cdot ;00\ 24 & = & ;03\ 36 \quad = \quad ;02 + ;01\ 20 + ;00\ 16 \quad = \quad \overline{30}\ \overline{45}\ \overline{225} \\
10 \cdot ;00\ 24 & = & ;04 \quad = \quad \overline{15}
\end{array}
$$

The only remaining, slightly puzzling feature is why, in line 4, the sexagesimal fraction ;01 36 was not simply resolved as ;01 + ;00 36 = $\overline{60}\ \overline{100}$, and why similarly, in line 9, the sexagesimal fraction ;03 36 was not resolved as ;03 + ;00 36 = $\overline{20}\ \overline{100}$. Maybe, in the first case, the answer is that the given resolution of 4 times $\overline{150}$ is to be understood as "approximately $\overline{45}$, with the much smaller remainder $\overline{225}$", clearly a better result than the alternative "approximately $\overline{60}$, with the not much smaller remainder $\overline{100}$". Similarly in the second case.

3.5. *P.Carlsberg 30* (Probably 2nd Century BCE)[29]

3.5 a. *P.Carlsberg 30* # 1 (*DMP* # 69). The diagonal of a square, *etc.*

On this fragment, only three drawings and a few lines of text are preserved. Apparently, the three drawings are meant to represent a) a trapezoid, divided into two triangles and a central rectangle or square, b) a square, and c) a second square with a diagonal. The photographic detail in Fig. 3.5.1 below is copied from Parker, *DMP* (1972), pl. 25.

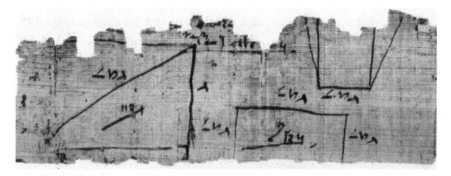

Fig. 3.5.1. *P.Carlsberg 30, obv.* Three drawings illustrating an exercise.

One of the squares has all sides equal to 14 $\overline{7}$ and the area 200. Here 14 $\overline{7}$ is an approximation to the square root of 200, obtained by the method exemplified in *P.BM 10520* § 6 (*DMP* ## 62-63), a method used also in Old and Late Babylonian mathematical texts, in *P.Cairo*, and possibly in the hieratic text *P.Berlin 6619* # 1. (Since 200 = sq. 14 + 4, the square side rule shows that sqr. 200 = appr. 14 + 4/(2 · 14) = 14 $\overline{7}$).

---

29. For the corrected date, see the review of Parker, *DMP* (1972) by Kaplony-Heckel, *OLZ* 76 (1981), 118 footnote 1.

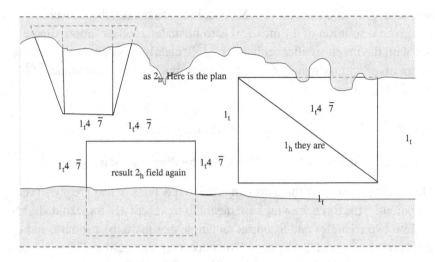

Fig. 3.5.2. P.Carlsberg 30, obv., conform transliteration.

The second square has all sides equal to 10 and the diagonal equal to sqr. 200 = 14 $\overline{7}$. The trapezoid, finally, has its base equal to 14 $\overline{7}$, but the indications of the lengths of its other sides are not preserved. A rather likely conjecture is that the trapezoid was meant to be symmetric, and that its height and the three parts of its top were all equal to 14 $\overline{7}$. In that case, the trapezoid would be divided in three parts, a square in the middle with the area 200, and two flanking triangles, each with the area 100. (Cf. the further examples of divided trapezoids in *P.Heidelberg 663* in Sec. 3.7.)

Unfortunately, there is not enough left of the text of this exercise to make it possible to understand what is really going on. Note that all recorded numbers are abstract, with no mention of any metrological units for length or area measure.

3.5 b. *P.Carlsberg 30* # 2 (*DMP* ## 71-72). A system of linear equations

On the reverse of the papyrus fragment *P.Carlsberg 30*, there is, according to Parker, *DMP* (1972), 75-77, the brief end of one problem (*DMP* # 71), and the beginning of another (*DMP* # 72). However, a more likely interpretation is that the reverse of the fragment contains the end of one very long exercise, and a few lines of the beginning of another exercise. Here is the text of the end of the long exercise, as given by Parker, with recon-

3.5. P.Carlsberg 30 (Probably 2nd Century BCE)

structed parts of the text in italics:

... ...
... ... amounting to 2'.
... ... in it.
... ...

The number of the silver (is) 42 $\underline{76}$. Namely, the 6' (is) 7, the $\overline{7}$ (is) 6, 13 again.
The number of the gold (is) 72 $\underline{76}$. Namely, the $\overline{8}$ (is) 9, the $\overline{9}$ (is) 8, 17 again.
The number of the copper (is) 110 $\underline{76}$. *Namely, the $\overline{10}$ (is) 11, the $\overline{11}$ (is) 10, 21 again.*
The number of the lead (is) 156 $\underline{76}$. Namely, <the> $\overline{12}$ (is) 13, the $\overline{13}$ (is) 12, total 25 again.
The 6', the $\overline{7}$ of the silver, 13. The $\overline{8}$, the $\overline{9}$ of the gold, 17.
The $\overline{10}$, the $\overline{11}$ of the copper, 21. The $\overline{12}$, <the> $\overline{13}$ of *the* lead, 25. Four numbers.
Their specification: 13, 17, 21, 25, amounting to 76 again.

Four 'numbers' for *silver, gold, copper, and lead* are mentioned in this text fragment, namely:

silver:	$s =$	42 $\underline{76}$	(42/76)
gold:	$g =$	72 $\underline{76}$	(72/76)
copper:	$c =$	110 $\underline{76}$	(110/76)
lead:	$l =$	156 $\underline{76}$	(156/76).

There are also some simple computations, namely:

$$6' \overline{7} \cdot 42 = 7 + 6 = 13,$$
$$\overline{8} \, \overline{9} \cdot 72 = 9 + 8 = 17,$$
$$\overline{10} \, \overline{11} \cdot 110 = 11 + 10 = 21,$$
$$\overline{12} \, \overline{13} \cdot 156 = 13 + 12 = 25.$$

In the final line of the exercise, it is simply stated that

$$13 + 17 + 21 + 25 = 76.$$

Parker, (*op. cit.*), 76-77, sees that there is a clear pattern with regard to the form of the four fractions, but is unable to suggest an explanation.

This fragmentary exercise can be compared with the "combined price exercise" **P.Rhind** § **24 (# 62)**, a problem about buying a bag filled with <equal> amounts of *gold, silver, and lead*, for a given total price of 84 *sha'ty*, when the price of gold is 12 *sha'ty* per *deben*, that of silver 6 *sha'ty* per *deben*, and that of lead 3 *sha'ty* per *deben*.[30] This amounts to a rather trivial division problem.

---

30. Cf. Imhausen, *ÄA* (2003), footnote 373, concerning the relative value of gold and silver in Egypt at different times.

In Sec. 2.1 f above, an indirect OB parallel to *P.Rhind* § 24 was discussed, namely **YBC 4698 § 3 a** (# 4), where it is stated that *iron and gold* are 90 (sic!) and 9 times more valuable than silver, and that 1 shekel of iron and gold together is bought for 1 mina of silver. This leads to a system of two linear equations for two unknowns, the amounts of gold and iron, respectively. The solution to this system of equations given in Sec. 2.1 f is in imitation of the explicit solutions to problems of the same kind in the OB mathematical texts **VAT 8389** and **VAT 8391**. (See Friberg, *MCTSC* (2005), Sec. 11.2 m, Fig. 11.2.14 left.)

Is it possible that *P.Carlsberg* 30 # 2 is an exercise of the same general type as YBC 4698 § 3 a, but with *four unknowns* instead of two? If that is the case, a system of *four linear equations* is needed in order to fix the values of the four unknowns. Actually, it is not difficult to find *two* linear equations satisfied by the numbers mentioned in the exercise, namely:

$$42\ \overline{76} + 72\ \overline{76} + 110\ \overline{76} + 156\ \overline{76} = (42 + 72 + 110 + 156)\ \overline{76} = 380\ \overline{76} = 5,$$

and

$$6'\ \overline{7} \cdot 42\ \overline{76} + \overline{8}\ \overline{9} \cdot 72\ \overline{76} + \overline{10}\ \overline{11} \cdot 110\ \overline{76} + \overline{12}\ \overline{13} \cdot 156\ \overline{76}$$
$$= (13 + 17 + 21 + 25)\ \overline{76} = 76\ \overline{76} = 1.$$

The simple forms of these equations can hardly be a coincidence, in particular as it is explicitly mentioned in the last line of the exercise that the sum of 13, 17, 21, and 25 is 76.

For the remainder of the discussion it is convenient to introduce, in addition to the notations s, g, c, and l for the four 'numbers' for silver, gold, copper, and lead, mentioned in the fragment, also the following notations for the four associated numbers:

$S = \ 6'\ \overline{7} \cdot s\ (= 13/42 \cdot 42/76 = 13/76),$
$G = \ \overline{8}\ \overline{9} \cdot g\ (= 17/72 \cdot 72/76 = 17/76),$
$C = \ \overline{10}\ \overline{11} \cdot c\ (= 21/110 \cdot 110/76 = 21/76),$
$L = \ \overline{12}\ \overline{13} \cdot l\ (= 25/156 \cdot 156/76 = 25/76).$

A search for two further simple linear relations between some of the mentioned numerical values immediately gives a positive result, namely the observation that

$$C + G = 38/76 = L + S, \quad \text{and} \quad C - G = 4/76 = 3' \cdot (L - S).$$

Therefore, it is likely that the object of the exercise, of which only the last column is preserved in the fragment *P.Carlsberg 30* # 2, was the following

3.5. P.Carlsberg 30 (Probably 2nd Century BCE)

system of linear equations:

a) $s + g + c + l = 5$
b) $S + G + C + L = 1$, where $S = 6' \; \overline{7} \cdot s, \; G = \overline{8} \; \overline{9} \cdot g, \; C = \overline{10} \; \overline{11} \cdot c, \; L = \overline{12} \; \overline{13} \cdot l$
c) $C + G = L + S$
d) $C - G = 3' \cdot (L - S)$.

This is a cleverly devised system of *four linear equations for the four unknowns s, g, c,* and *l*. It can be solved by use of a method similar to the method used in VAT 8389 and VAT 8391 to solve a system of two linear equations for two unknowns. The basic idea is to start by looking for a "partial solution" satisfying all but one of the linear equations. In the present case, *a simple partial solution is the one for which C = G, and which satisfies the three equations b), c), and d)*. Indeed,

Suppose that $C = G$.
Then equation d) is also satisfied if, in addition $L = S$.
Then equation c) is also satisfied if, in addition $G = S$, so that $C = G = L = S$.
Then equation b) is also satisfied if, more precisely $C = G = L = S = 4'$.

However, this is not the correct solution. Indeed, since

$S = 13 \; \underline{42} \cdot s, \; G = 17 \; \underline{72} \cdot g, \; C = 21 \; \underline{110} \cdot c, \; L = 25 \; \underline{156} \cdot l$,

it follows that, conversely,

$s = 42 \; \underline{13} \cdot S, \; g = 72 \; \underline{17} \cdot G, \; c = 110 \; \underline{21} \cdot C, \; l = 156 \; \underline{25} \cdot L$.

Hence, if $C = G = L = S = 4'$, then

$s + g + c + l = (42 \; \underline{13} + 72 \; \underline{17} + 110 \; \underline{21} + 156 \; \underline{25}) \cdot 4'$.

This cannot be equal to 5, as required by equation a). In fact, since

$13 \cdot 17 \cdot 21 \cdot 25 = 116{,}025$,

it follows that

$s + g + c + l = (42 \; \underline{13} + 72 \; \underline{17} + 110 \; \underline{21} + 156 \; \underline{25}) \cdot 4'$
$= (374{,}850 + 491{,}400 + 607{,}750 + 723{,}996) \; \underline{116{,}025} \cdot 4'$
$= 2{,}197{,}996 \; \underline{116{,}025} = 549{,}499 \; \underline{116{,}025}$.

On the other hand, equation a) requires that sum of the four unknowns *s, g, c,* and *l* to be equal to

$5 = 5 \cdot 116{,}025 \; \underline{116{,}025} = 580{,}125 \; \underline{116{,}025}$.

This means that the mentioned partial solution gives a *deficit* in equation a) which is equal to

$(580{,}125 - 549{,}999) \; \underline{116{,}025} = 30{,}626 \; \underline{116{,}025}$.

In the next step of the proposed solution procedure, which tries to imitate one of the two different procedures in VAT 8389 and VAT 8391, an effort is made to *eliminate the calculated deficit* in equation a) by adding a small amount a to C and subtracting a from G. What then happens is the following:

> Suppose that $C = 4' + 1 \cdot a$ and $G = 4' - 1 \cdot a$.
> Then equation d) is satisfied if $L - S = 3 \cdot (a + a) = 6 \cdot a$,
> and equation c) is satisfied if $L + S = 2 \cdot 4' = 2'$.
> Therefore, equations c) and d) are both satisfied if
> $C = 4' + 1 \cdot a$, $G = 4' - 1 \cdot a$, $L = 4' + 3 \cdot a$, and $S = 4' - 3 \cdot a$.
> With these values for C, G, L, and S, equation b) is also satisfied.

Simultaneously, in equation a) the following amount is *subtracted from the deficit*:

$3 \cdot a \cdot (723{,}996 - 374{,}850) \; \underline{116{,}025} + a \cdot (607{,}750 - 491{,}400) \; \underline{116{,}025}$
$= a \cdot (1{,}047{,}438 + 116{,}350) = a \cdot 1{,}163{,}788 \; \underline{116{,}025}.$

Therefore, in order to eliminate the deficit, a must satisfy the equation

$a \cdot 1{,}163{,}788 = 30{,}626.$

Since $1{,}163{,}788 = 38 \cdot 30{,}626$, the deficit $30{,}626 \; \underline{116{,}025}$ is precisely eliminated if $a = \overline{38}$. Consequently, the four equations a), b), c), and d) are all satisfied if

$C = 4' + 1 \cdot \overline{38} = 21 \; \underline{76},$
$G = 4' - 1 \cdot \overline{38} = 17 \; \underline{76},$
$L = 4' + 3 \cdot \overline{38} = 25 \; \underline{76},$
$S = 4' - 3 \cdot \overline{38} = 13 \; \underline{76}.$

Ordered by size,

$S = 13 \; \underline{76}$, $G = 17 \; \underline{76}$, $C = 21 \; \underline{76}$, and $L = 25 \; \underline{76}$.

Hence the final result is that

$s = 42 \; \underline{13} \cdot 13 \; \underline{76} = 42 \; \underline{76},$
$g = 72 \; \underline{17} \cdot 17 \; \underline{76} = 72 \; \underline{76},$
$c = 110 \; \underline{21} \cdot 21 \; \underline{76} = 110 \; \underline{76},$
$l = 156 \; \underline{25} \cdot 25 \; \underline{76} = 156 \; \underline{76}.$

These are precisely the four numbers for silver, gold, copper, and lead mentioned in the preserved column of text on *P.Carlsberg 30, rev*. Actually, that last column of the exercise can be understood as the *verification of the result*, namely the demonstration that with departure from the com-

3.5. P.Carlsberg 30 (Probably 2nd Century BCE) 173

puted values for s, g, c, and l, the corresponding values for S, G, C, and L can be computed 'again', and that the sum of S, G, C, and L will then 'again' be equal to 1.

There is no obvious interpretation of the four coefficients $\overline{6}\,\overline{7}$, $\overline{8}\,\overline{9}$, $\overline{10}\,\overline{11}$, and $\overline{12}\,\overline{13}$ as meaningful constants associated in some way with silver, gold, copper, and lead. They can definitely not, in any case, be interpreted as the values per weight unit, or as the weights per volume unit, *etc.*, of the four metals. A more likely alternative is that *the names of the four metals are used as arbitrary names for four unknowns*, and that the four mentioned coefficients are arbitrarily chosen numbers.[31] Compare with the way in which a symbol for *drachma* and the same symbol crossed-over are used in ***P.Mich. 620 # 1*** (see Sec. 4.4 a below) to denote known and unknown units, respectively.

Additional surprises in the proposed interpretation of *P.Carlsberg 30 # 2* are the counting with many-place decimal numbers, and the extensive manipulation of binomial fractions. Note however that a list of many-place decimal number appears already in the hieratic mathematical fragment *P.UC 32161* (Sec. 2.3 c). In Mesopotamia, manipulations of many-place regular sexagesimal numbers was a popular topic in Old and Late Babylonian mathematical texts, and many-place sexagesimal numbers were indispensable for Late Babylonian/Seleucid mathematical astronomy.

As for the proposed manipulation of binomial fractions in *P.Carlsberg 30 # 2*, this feature is not an essential part of the suggested interpretation of the exercise, it just makes the computations so much easier. It is, by the way, entirely possible that the use of binomial fractions was introduced in demotic mathematics precisely in order to make it easier to deal with complicated computations of the kind that apparently were necessary for the solution of systems of linear equations with complicated coefficients like the one in *P.Carlsberg 30 # 2*.

---

31. Cf. the well known passage about education in *Plato, Laws vii*, 817 E - 820 D (Thomas, *SIHGM* (1939, 21)), where it is said, among other things, that " ⋯ the freeborn ought to learn as much of these things as a vast multitude of boys in Egypt learn along with their letters ⋯ the boys should play with bowls containing gold, bronze, silver and the like mixed together, or the bowls may be distributed as wholes. For, as I was saying, to incorporate in the pupils' play the elementary applications of arithmetic will be of advantage to them later ⋯ "

3.6. *P.Griffith Inst. I. E. 7* (Late Ptolemaic or Early Roman)

A theme text with linear equations

This fragment (Parker, *JNES* 18 (1959)) contains what is apparently part of a theme text, with a semi-systematically arranged series of 3 linear equations of the following type (in modern notations)

$(2' \cdot a + p) + (3' \cdot a + q) = r, \quad a = ?$

followed by 2 equations of the type

$(2' \cdot a - p) + (3' \cdot a + q) = r, \quad a = ?$

Since $2' + 3' = 6''$ (5/6), all the equations are reduced after some steps to the simpler form

$6'' \cdot a = s, \quad a = ?$

Next, the reciprocal of $6''$ is computed, in a phrase of the following form:

You say, $6''$, what customarily goes with it to 1?
Result 1 $\bar{5}$.

Finally, the solution is obtained as 1 $\bar{5}$ times *s*. As pointed out by Parker, solving a division problem $f \cdot a = s$ in this way by first computing the reciprocal of the coefficient f and then multiplying the right hand side *s* with that reciprocal is the method used in OB mathematical texts, and also in the hieratic *P.Moscow*, but not in *P.Rhind*.

In the first group of equations, $p = q = 3, 4, 5$. In the second group, $p = 3$, and $q = 2, 3$. In four of the five equations $r = 10$, but in the third equation $r = 15$. The reason is obvious, since if r had been equal to 10 also in the third equation, then the equation for *a* would have been reduced to $6'' \cdot a = 0$. Presumably, the author of the text did not know how to formulate or handle a case like that.

3.7. *P.Heidelberg 663* (Ptolemaic, 2nd or 1st C. BCE)

This document (Parker, *JEA* 61 (1975)) consists of two narrow papyrus fragments, containing parts of four geometric exercises illustrated by four drawings of divided trapezoids. The three first of these exercises were briefly, though essentially correctly, interpreted by Parker. So little is preserved of the fourth exercise that no detailed interpretation can be attempt-

3.7. P.Heidelberg 663 (Ptolemaic, 2nd or 1st C. BCE)

ed. The photographic details in Figs. 3.7.1, 3.7.4, 3.7.6 are copied from Parker (*op. cit.*).

3.7 a. P.Heidelberg 663 # 1. A vertically striped trapezoid

Presumably, *P.Heidelberg 663* # 1 is concerned with a symmetric (i. e. isosceles) trapezoid divided symmetrically in six parts by five transversals. Just like the equilateral triangles in **P.Cairo** § **11** *(DMP ##* 36, 38), the trapezoid is shown standing on its base, in this case the longer of the two parallel sides. The transversals are drawn at right angles to the base.

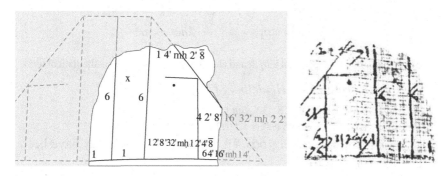

Fig. 3.7.1. *P.Heidelberg 663* # 1. A fragment of a drawing of a vertically striped trapezoid.

In the mirror image conform transliteration of *P.Heidelberg 663* in Fig. 3.7.1, left, and in the reconstruction of the trapezoid in Fig. 3.7.3, the numbers present in the drawing are interpreted as the lengths a_1, a_2, a_3 of the three segments of the right half of the base, and the lengths h_1, h_2 of the two transversals to the right. The length numbers are written as multiples and fractions of an unnamed major length unit, followed by multiples and fractions of a cubit (*mḥ*). In the following, this unnamed major length unit will be assumed to be a *khet* of 96 cubits, that is a slight modification of the *khet* of 100 units used in the hieratic mathematical texts. The correctness of this assumption will be demonstrated in the course of the analysis of the computations in the text.

In *P.Heidelberg 663*, ordinary notations for fractions are used for fractions of 1 cubit, but special notations for fractions of 1 *khet*. The same special notations for fractions of 1 *khet* are used, for instance, in **P.BM 10520** § **7 b** (Fig. 3.3.3 above), and in **Theban O. D 12** (Fig. 3.3.4 above). In

Theban O. D 12, the same special notations are used also for fractions of an unnamed area unit, presumably equal to 1 sq. *khet* = 1 *setat*. Following Parker, *JEA* 61 (1975), the form of the number signs used in demotic texts for fractions of the *khet* (and the *setat*) can be displayed in a diagram, such as the one below. The signs for the combinations 1/2+1/4 and 1/16+1/32 (*khet* or *setat*) are not ligatures but independent constructions.

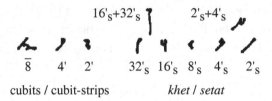

Fig. 3.7.2. Notations for fractional length and area numbers in demotic mathematical texts.

Thus, for instance, in *P.Heidelberg 663* # 1,

the length number a_2 = 1 2'$_s$ 8'$_s$ 32'$_s$ *mḥ* 1 2' 4' $\overline{8}$

stands for   1 1/2 1/8 1/32 · 96 cubits + 1 1/2 1/4 1/8 · 1 cubit.

The question in *P.Heidelberg 663* # 1 is not preserved. It may have been of the following form:

> A trapezoid with the base 18 (*khet*), the top 2 (*khet*), and the height 6 (*khet*) is divided symmetrically in five parts by four vertical transversals. The area of the two outer parts, of the two middle parts, and of the central part, are proportional to 5, 3, and 2, respectively,. Find the lengths of all line segments in the divided trapezoid.

The solution procedure probably started with the computation of the areas:

> The whole area is 60 (*setat*), hence the area of each one of the outer parts is A_1 = 15, the area of each one of the middle parts is A_2 = 9, and the area of each half of the central part is A_3 = 6.

Consequently, two of the vertical transversals must pass through the upper corners of the trapezoid, dividing the trapezoid into a central rectangle and two flanking triangles. (See Fig. 3.7.3.) The inclination of the diagonal of each flanking triangle is f = 8/6 = 4/3 = 1 3'. Therefore (cf. *P.Moscow* § 7, Sec. 2.2 c above), the base of the triangle can be computed as follows:

$a_1 \cdot h_1 = 2 A_1$, and $a_1 = f \cdot h_1$, hence sq. $a_1 = 2 f \cdot A_1 = 2\ 3'' \cdot 15 = 40$ (*setat*).

The square root was probably computed by repeated use of the square side rule (see **P.BM 10520 § 6** in Sec. 3.3 f above):

3.7. P.Heidelberg 663 (Ptolemaic, 2nd or 1st C. BCE)

sqr. 40 = appr. 6 3',   sq. 6 3' = 40 $\overline{9}$,   hence
sqr. 40 = appr. 6 3' – $\overline{9}$ / (2 · 6 3') = 6 37/114 (*khet*).

The division 37/114 was then performed according to the rules, possibly in the following way:

37/114 · 32 = 10 22/57,   and 22/57 · 3 = 1 3/19 = appr. 1 $\overline{8}$.

Therefore,

a_1 = sqr. 40 *setat* = appr. 6 10/32 *khet* 1 $\overline{8}$ cubit-strips = 6 4'$_s$ 16'$_s$ *mḥ* 1 $\overline{8}$.

Next, the base of the middle trapezoid was computed as follows:

a_2 = 8 – a_1 = 8 – 6 4'$_s$ 16'$_s$ *mḥ* 1 $\overline{8}$
= 1 2'$_s$ 8'$_s$ 32'$_s$ *mḥ* 1 2' 4' $\overline{8}$.

In the drawing, a dot indicates that the rectangle with the base a_2 and the height h_1 (see Fig. 3.7.3) also has the width a_2 at the top. The length of h_1 is computed as follows:

h_1 = 1/f · a_1 = 2' 4' · 6 4'$_s$ 16'$_s$ *mH* 1 $\overline{8}$
= 3 8'$_s$ 32'$_s$ *mH* 2 $\overline{16}$ + 1 2'$_s$ 16'$_s$ *mH* 1 2' 4' $\overline{32}$
= 4 2'$_s$ 8'$_s$ 16'$_s$ 32'$_s$ *mH* 2 4' $\overline{16}$ $\overline{32}$.

Finally, the height of the triangle on top of the rectangle is computed as

k_1 = h_2 – h_1
= 6 – 4 2'$_s$ 8'$_s$ 16'$_s$ 32'$_s$ *mH* 2 4' $\overline{16}$ $\overline{32}$
= 1 4'$_s$ *mH* 2' $\overline{8}$ $\overline{32}$.

All these complicated calculations were performed without mistakes.

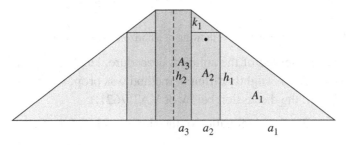

A_1 = 15, A_2 = 9, A_3 = 6 a_1 = sqr. 40 = 6 4'$_s$ [16'$_s$ *mḥ* 1 $\overline{8}$]
h_2 = 6 a_2 = 8 – a_1 = 1 2'$_s$ 8'$_s$ 32'$_s$ *mḥ* 1 2' 4' $\overline{8}$
a_1 + a_2 = 8, a_3 = 1 h_1 = 2' 4' · a_1 = 4 2'$_s$ 8'$_s$ [16'$_s$ 32'$_s$ *mḥ* 2 4' $\overline{16}$ $\overline{32}$]
                                          k_1 = h_2 – h_1 = 1 4'$_s$ *mḥ* 2' $\overline{8}$ [$\overline{32}$]

Fig. 3.7.3. *P.Heidelberg 663* # 1. The divided trapezoid and its parameters.

3.7 b. *P.Heidelberg 663* # 2. A horizontally striped trapezoid

The lengths of the transversals

In *P.Heidelberg 663* # 1 a trapezoid was divided by four *vertical* transversals in such a way that the areas of the outer, middle, and central parts were 30, 18, and 12 (*setat*), respectively. In # 2, the same trapezoid is apparently divided by two *horizontal* transversals in such a way that the areas of the lower, middle, and upper parts of the divided trapezoid are, again, 30, 18, and 12 (*setat*). See Figs. 3.7.4-5 below. Apparently, the aim of the exercise was to compute the lengths of the transversals.

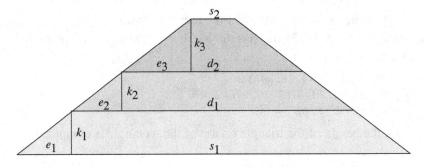

$e_1 = (s_1 - d_1)/2 = 2\ 2'_s\ 16'_s\ 32'_s$ $k_2 = 2'\ 4' \cdot e_2 = 1\ 2'_s\ 8'_s\ 16'_s\ mh\ 2\ 4'$
$k_1 = 2'\ 4' \cdot e_1 = 1\ 2'_s\ 4'_s\ 8'_s\ 16'_s\ mh\ 2'\ 4'$ $e_3 = (d_2 - s_2)/2 = 3\ 8'_s$
$e_2 = (d_1 - d_2)/2 = 2\ 4'_s\ 32'_s$ $k_3 = 2'\ 4' \cdot e_3 = 2\ 4'_s\ 16'_s\ 32'_s$

Fig. 3.7.4. *P.Heidelberg 663* # 2. A drawing of a horizontally striped trapezoid.

Very little is preserved of the solution procedure. Here is one way of solving the problem. (A slightly different method was proposed in Parker, *JEA* 61 (1975). Cf. the discussion below of VAT 7621.)

Let $f = 4/3 = 1\ 3'$ be the inclination of the sloping sides of the trapezoid. Then, in Fig. 3.7.5,

$A_1 = (s_1 + d_1)/2 \cdot (s_1 - d_1)/f = (\text{sq. } s_1 - \text{sq. } d_1)/(2f)$,
$A_2 = (\text{sq. } d_1 - \text{sq. } d_2)/(2f)$,
$A_3 = (\text{sq. } d_2 - \text{sq. } s_2)/(2f)$.

On the other hand, it is clear that, with the given values for the areas,

$A_1 = A_2 + A_3$,

3.7. P.Heidelberg 663 (Ptolemaic, 2nd or 1st C. BCE)

and

$A_1 + A_2 = 4 A_3$.

Therefore,

(sq. s_1 − sq. d_1) = (sq. d_1 − sq. s_2),

and

(sq. s_1 − sq. d_2) = 4 (sq. d_2 − sq. s_2).

This means that the lengths of the transversals can be computed as follows. First

sq. d_1 = (sq. s_1 +sq. s_2)/2 = (324 + 4)/2 = 164 (*setat*),

so that

d_1 = sqr. 164 = appr. 12 2' 4' $\overline{20}$ (*khet*) = appr. 12 2'$_s$ 4'$_s$ 16'$_s$.

Then

sq. d_2 = (sq. s_1 + 4 sq. s_2)/5 = (324 + 4 · 4)/5 = 68 (*setat*),

so that

d_2 = sqr. 68 = appr. 8 4'$_s$.

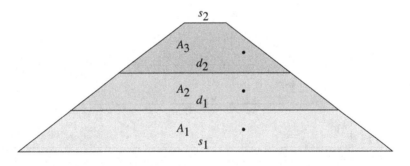

$A_1 = 30$, $A_2 = 18$, $A_3 = 12$ sq. d_1 = (sq. s_1 + sq. s_2)/2 = 164
$s_1 = 18$, $s_2 = 2$ sq. d_2 = (sq. s_1 + 4 sq. s_2)/5 = 68

d_1 = sqr. 164 = appr. 12 2'$_s$ 4'$_s$ 16'$_s$
d_2 = sqr. 68 = appr. 8 4'$_s$

Fig. 3.7.5. *P.Heidelberg 663* # 2. A proposed explanation of the computation of the lengths of the transversals.

The lengths of the partial heights

P.Heidelberg 663 # 3 is a continuation of # 2. The aim of the exercise is to compute the partial heights k_1, k_2, k_3 and the corresponding base lines e_1, e_2, e_3 (see Figs. 3.7.6-7). The computations are relatively simple and straightforward.

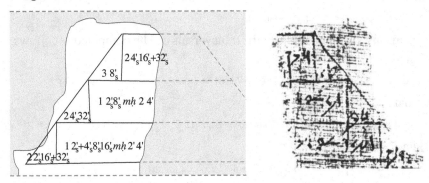

Fig. 3.7.6. *P.Heidelberg 663* # 3. A horizontally striped trapezoid.

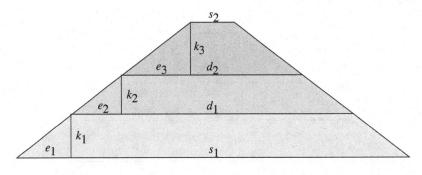

$e_1 = (s_1 - d_1)/2 = 2\ 2'_s\ 16'_s\ 32'_s$ $k_2 = 2'\ 4' \cdot e_2 = 1\ 2'_s\ 8'_s\ 16'_s\ mh\ 2\ 4'$

$k_1 = 2'\ 4' \cdot e_1 = 1\ 2'_s\ 4'_s\ 8'_s\ 16'_s\ mh\ 2\ 4'$ $e_3 = (d_2 - s_2)/2 = 3\ 8'_s$

$e_2 = (d_1 - d_2)/2 = 2\ 4'_s\ 32'_s$ $k_3 = 2'\ 4' \cdot e_3 = 2\ 4'_s\ 16'_s\ 32'_s$

Fig. 3.7.7. *P.Heidelberg 663* # 3. Computation of the partial heights, *etc.*

The lengths of the sloping sides

P.Heidelberg 663 # 4 is a continuation of ## 2-3. Very little is preserved of the text, but in the associated drawing the length '10' of the sloping side of the trapezoid is recorded. It is likely, therefore, that the object of this exercise was to compute various lengths by use of the diagonal rule.

3.7 c. *P.Heidelberg 663*. Parallel texts

For the history of mathematics, *P.Heidelberg 663* is an important text. The earliest known example of a divided trapezoid of the kind appearing in *P.Heidelberg 663* ## 2-4 is a drawing on **IM 58045**, an Old Akkadian school boy's round hand tablet from around 2200 BCE (Friberg, *MCTSC* (2005), App. 6 c). The topic was very popular in OB mathematics. See Friberg, *RlA 7* (1990) Sec. 5.4 k. No Late Babylonian example have yet been found, but the topic reappears in this Ptolemaic papyrus fragment and in books attributed to Euclid and Hero, both from Egyptian Alexandria.

The only OB parallels to the divided trapezoid with *vertical* transversals in *P.Heidelberg 663* # 1 can be found in the OB text **VAT 7531**, which will be discussed below. It is much easier to find parallels to the divided trapezoid with *horizontal* transversals in *P.Heidelberg 663* ## 2-4, since divided trapezoids in OB mathematical texts normally have transversals parallel with the parallel fronts of the trapezoid. The similarity is imperfect, however, for several reasons. First, of course, because the trapezoid in *P.Heidelberg* stands on its base, while OB trapezoids are orientated to the left. Secondly, because OB divided trapezoids usually are divided into an even number of sub-trapezoids, while the trapezoid in *P.Heidelberg* is divided into three parts. Finally, because the lengths of the transversals in the OB examples almost always are finite sexagesimal numbers (and therefore rational numbers), while the lengths of the transversals in *P.Heidelberg* are approximations to square roots of non-square numbers. In this respect, the divided trapezoids in *P.Heidelberg* is more like the striped triangles in the hieratic exercise ***P.Rhind* # 53 a** and the OB exercise **IM 43996**. (See Figs. 2.1.8-9 above.)

A likely explanation is that the exercises in *P.Heidelberg* are close parallels to a kind of *Late* Babylonian geometric exercises of which no examples have yet been found. Typically "Late Babylonian" traits in *P.Heidelberg* are the orientation of the figures, the computation of heights as in *P.Heidelberg* # 3, and the elaborate computations of approximations to square roots.

The mentioned OB parallels to *P.Heidelberg 663* # 1 are exercises ## 1-4 in the Old Babylonian text **VAT 7531**, probably from Uruk. The first accurate reading of this text was published in Thureau-Dangin, *TMB* (1938), 98. A correct (although incomplete) interpretation was presented in Vaiman, *SVM* (1961), 258-263, where it was noticed that the trapezoids in these exercises have the unusual property that the *long sides* (the 'lengths') are parallel, not the short sides (the 'fronts') as in most other trapezoids in OB texts. The missing details in Vaiman's interpretation will be filled in below, in an improved interpretation inspired by the example of *P.Heidelberg 663* # 1! Here is the text of VAT 7531:

VAT 7531 ## 1-4.

2 35 50 uš gíd.da 1 54 10 [uš lugúd.da] /   50 sag an.na 41 40 sag ki.[ta] /   a.šà.bi *ma-la ma-ṣú-ú a-mu-[ur-ma]* /   *a-na* 3 šeš-*a-ni mi-it-ḫa-r[i-iš zu-uz]* /   *ù* uku.uš *sí-ka-a[s-sú ku-ul-li-im-š]u*
10 37 30 uš gíd.da 2 17 30 uš lugúd.da /   10 sag an.na 8 20 sag ki.ta /   a.šà.bi *ma-la ma-ṣú a-mu-ur-ma* /   *a-na* 5 *šu-ši* erín.ḫá 1$_{èše}$ aša$_5$.[ta].àm *pu-lu-uk* /   *ù* uku.uš *sí-ik-ka-as-sú ku-ul-li-im-šu* /   *si-ta-at* a.šà.*ka a-n[a]* 2 *šu-ši* er n /   *mi-it-ḫa-ri-iš i-di-in ù* uku.uš /   *sí-ik-ka-as-sú ku-ul-li-im-šu* /   šu.nigin 7 *šu-ši* erim a.šà.bi en.nam
2 58 30 uš gíd.da 1 16 30 uš lugúd.da /   2 sag an.na 1 36 sag ki.ta /   a.šà.bi *ma-la m[a-ṣú a]-mu-ur-ma* /   *a-na* 1 *li-im* 8$^?$ *me* [erín.ḫá *mi-it*]-*ḫa-ri-iš* / *zu-uz*   *ù* [uku.uš *sí-ka-as-sú*] / *ku-ul-l[i-im-šu]*
2 43 30 uš gíd.da 1 56 30 uš lugúd.da /   1 37 30 sag an.na 1 30 30 sag ki.ta /   a.šà.bi *ma-la ma-ṣú a-mu-ur-ma* /   *a-na* 5 šeš-*a-ni mi-it-ḫa-ri-iš* / *zu-uz*   *ù* uku.uš *sí-ka-as-sú* / *ku-ul-li-im-šu*

3.7. P.Heidelberg 663 (Ptolemaic, 2nd or 1st C. BCE)

#1 2 35 50 the long length, 1 54 10 *the short length,*
    50 the upper front, 41 40 the low*er* front.
    Its area, how much it is, find o*ut, then*
    to 3 brothers equal*ly divide it,*
    and (each) soldier s*how him his stake*

#2 10 37 30 the long length, 2 17 30 the short length,
    10 the upper front, 8 20 the lower front.
    Its area, how much it is, find out, then
    to 5 sixties of men 1 èš e of field each delimit
    and (each) soldier show him his stake.
    The remainder of your field to 2 sixties of men
    equally give, and (each) soldier
    show him his stake.
    Altogether 7 sixties of men. Its area is what?

#3 2 58 30 the long length, 1 16 30 the short length,
    2 the upper front, 1 36 the lower front.
    Its area, how much *it is, f*ind out, then
    to 1 thousand 8? hundred *men equ*ally divide it,
    and *(each) soldier* sho*w him his stake.*

#4 2 43 30 the long length, 1 56 30 the short length,
    1 37 30 the upper front, 1 30 30 the lower front.
    Its area, how much it is, find out, then
    to 5 brothers equally divide it,
    and (each) soldier show him his stake.

Properly speaking, VAT 7531 ## 1-4 is a collection of assignments rather than exercises, since the questions are not followed by solution procedures. An attempt is made below to show what the solution procedures would have been like.

In VAT 7531 # 1, the given figure can be interpreted as a trapezoid composed of a central rectangle, and two flanking non-equal triangles (Fig. 3.7.8). If the rectangle is removed, what remains is a rotated symmetric triangle, with two sides equal to 41;40 and the third side equal to 50. The height against the side 50 can be computed by use of the diagonal rule. It is 33;20, so that the triangle can be reinterpreted as the composition of two equal right triangles with the sides 8;20 · (3, 4, 5).

The height h against the side 41;40 can be computed as follows:

$h \cdot 41;40/2 = A_{triangle} = 33;20 \cdot 50/2$,

$h = 33;20 \cdot 50/41;40 = 40$.

After h has been computed, the two components a and b of the base of the triangle can be computed by use of the diagonal rule. They are 30 and 11;40. Hence, the triangle has an alternative composition as two right triangles with the sides $10 \cdot (3, 4, 5)$ and $1;40 \cdot (7, 24, 25)$ joined together.

The aim of exercise # 1 is to divide the given trapezoid equally between three brothers. Now, it is clear that the area of the trapezoid is equal to

$(2\ 35;50 + 1\ 54;10)/2$ n. $\cdot$ 40 n. = $2\ 15$ n. $\cdot$ 40 n. = $1\ 30\ 00$ (sq. ninda) = 3 bùr.

One third of that area is 30 00 (sq. ninda) = 1 bùr, which is equal to the area of a rectangle with the length 45 n. and the height 40 n. Hence, the middle brother gets a central rectangle with these sides, while the first brother gets the left triangle plus a rectangle with the sides 30 and 40, and the third brother gets the right triangle plus a rectangle with the sides 39;10 and 40. In this way, each brother gets a field with the area 1 bùr. See again Fig. 3.7.8.

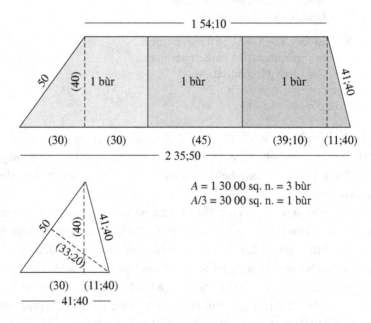

Fig. 3.7.8. VAT 7531 # 1. Three brothers sharing a trapezoidal field.

3.7. P.Heidelberg 663 (Ptolemaic, 2nd or 1st C. BCE)

The given trapezoid in VAT 7531 # 2 is, just like the one in # 1, composed of a central rectangle and two flanking, non-equal triangles. See Fig. 3.7.9 below. It is easy to check that the flanking triangles in # 2 are similar to the flanking triangles in # 1, but with sides that are 12 times larger. The area of the whole trapezoid is

(10 37;30+ 2 17:30)/2 n. · 8 00 n. = 12 55 /2 n. · 8 00 n. = 51 40 00 sq. n.

This is surprising, since 51 40 00 (sq. ninda) = 1 43 bùr 1 èše, which is not a round area number, in contrast to the 1 30 00 (sq. ninda) = 3 bùr in # 1. As a matter of fact, it is likely that the author of the text made a mistake here. Suppose that his intention was to construct a trapezoid with the mentioned flanking triangles and with a total area of precisely 1 00 00 00 (sq. ninda) = 2 00 bùr. Since the sum of the areas of the two flanking triangles is 8 20 n. · $h/2$ = 8 20 n. · 4 00 n. = 33 20 00 sq. n., he *should* have computed the "short length" of the trapezoid as follows:

(1 00 00 00 − 8 20 · 4 00) sq. n./8 00 n. = 26 40 00 sq. n./8 00 n.= 3 20 n.

By mistake, he computed the short length instead as

(1 00 00 00 − 8 20 · 5 00) sq. n./8 00 n.= 18 20 00 sq. n./8 00 n. = 2 17;30 n.

The corrected lengths 3 20 and 3 20 + 8 20 = 11 40 are shown in Fig. 3.7.9.

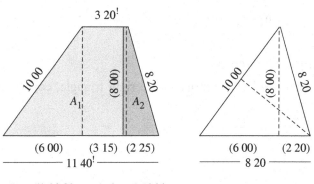

A_1 = 50 00 00 sq. ninda = 1 40 bùr
A_2 = 10 00 00 sq. ninda = 20 bùr

Fig. 3.7.9. VAT 7531 # 2. Two teams sharing a trapezoidal field. *Corrected data.*

The assignment in # 2 (corrected) is to divide the trapezoid between two teams. In the first team, each one of 5 00 men gets 10 00 (sq. ninda) = 1 èše. This means that the total share of the first team is 5 00 · 10 00 (sq.

ninda) = 50 00 00 (sq. ninda). In the second team, 2 00 men divide the remaining part of the field in equal shares. Hence the total share of the second team is 10 00 00 (sq. ninda), and each one of the 2 00 men in the second team gets 5 00 (sq. ninda) = 1/2 èše = 3 iku. The share of the second team is slightly more than the area of the right triangle, which is 2 20/2 · 8 00 = 9 20 00 (sq. n.). The difference is 40 00, to be taken out of the central rectangle, for a length of 40 00/8 00 = 5 (ninda).

Theoretically, each man in each of the two teams can be allotted his own narrow strip, bounded by two parallel vertical lines.[32] The computation of the width of the individual strips is particularly simple in the case of strips falling within the central rectangle. The first team gets 3 15 · 8 00 = 26 00 00 (sq. ninda) in the central rectangle, enough for 6 · 26 = 2 36 men. The width of each one of the 2 36 strips is 3 15 / 2 36 = 1;15 (ninda).

It is more complicated to compute the widths of the strips falling within the left triangle for the remaining 2 24 men of the first team. The method that has to be used in this case is the same as the method used for a similar purpose in *P.Heidelberg 663* # 1. Thus, let a_n be the sum of the widths of the first n strips, and let h_n be the length of the nth bounding line. Then

$a_n \cdot h_n/2 = 10\ 00 \cdot n,$ and $a_n = ;45 \cdot h_n.$

Consequently,

sq. $a_n = ;45 \cdot 2 \cdot 10\ 00 \cdot n = 15\ 00 \cdot n,$

so that

$a_n = 30 \cdot \text{sqr.}\ n,$ and $h_n = 40 \cdot \text{sqr.}\ n.$

Similarly for the individual strips in the right triangle, counted from the right end:

sq. $b_n = ;17\ 30 \cdot 2 \cdot 10\ 00 \cdot n = 5\ 50 \cdot n,$

so that

$b_n = \text{sqr.}\ (5\ 50 \cdot n),$ and $h_n = 24/7 \cdot b_n.$

---

32. Compare with the ring wall problem **BM 85194 # 4**, Thureau-Dangin, *TMB* (1938), text 48. There a circular ring-shaped mud-wall around a "city" has the length 1 30 ninda and the volume 1 07 30 šar. Since the prescribed work norm is ;10 šar/man-day, 6 45 00 man-days are needed for the construction of the ring wall. The length of the ring wall marked out for each worker to build in a day is then 1 30 ninda/ 6 45 00 = ;00 13 20 ninda = appr. 2 centimeters, a completely unrealistic figure!

3.7. P.Heidelberg 663 (Ptolemaic, 2nd or 1st C. BCE)

In VAT 7531 # 3, the trapezoid can be shown to be composed of a rectangle and a *single* right triangle with its sides equal to 24 · (3, 4, 5). The area, 3 00 00 (sq. ninda) = 6 bùr = 18 èše, is divided equally between 1 thousand [8 hundred]? men, so that each man gets 1/100 èše, a surprisingly small share.

VAT 7531 # 4 is a much more interesting text. The trapezoid here, with the lengths 2 43;30 and 1 56;30, and the fronts 1 37;30 and 1 30;30, is divided equally between 5 brothers.

In this assignment, the trapezoid is again composed of a central rectangle and two unequal flanking triangles. The two triangles together form a triangle with the sides 1 37;30, 1 30;30, and 47. Since there is no pair of equal sides, this case is more difficult than the two cases of isosceles triangles in ## 1-2. Evidently, the OB author of this text was confident that his students knew how to compute the height of a non-isosceles (scalene) triangle! The way they did it was probably as follows:

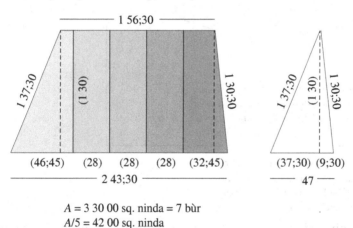

$A = 3\ 30\ 00$ sq. ninda = 7 bùr
$A/5 = 42\ 00$ sq. ninda

Fig. 3.7.10. VAT 7531 # 4. Five brothers sharing a trapezoidal field.

Let a, b, c be the sides of the triangle, and suppose that the height against the side b divides b into the segments p and q, where p is greater than q. Then,

$p + q = b$ and
sq. a − sq. p = sq. c − sq. q (by the diagonal rule, since both are equal to sq. h).

This leads to a *quadratic-linear system of equations* for p and q:

$p + q = b$ and sq. p − sq. q = sq. a − sq. c.

This quadratic-linear system of equations can be solved by use of metric algebra. (For details, see the discussion of exercise # 2 in the Greek mathematical papyrus fragment *P.Chic. litt 3* in Sec. 3.7 below.) The solution can take several forms, for instance the following:

$p = \{b + (\text{sq. } a - \text{sq. } c)/b\}/2, \quad q = \{b - (\text{sq. } a - \text{sq. } c)/b\}/2$.

With a, b, c = 1 37;30, 1 30;30, 47, these equations show that p = 37;30, q = 9;30. It is then easy to compute h = 1 30. The remaining part of the solution algorithm is straightforward.

An OB parallel to the trapezoid with horizontal transversals in *P.Heidelberg 663* ## 2-4, can be found in **VAT 7621 # 1** (Thureau-Dangin, *TMB* (1938), 99):

VAT 7621 # 1.

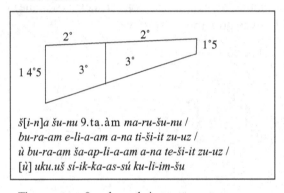

š[i-n]a šu-nu 9.ta.àm ma-ru-šu-nu /
bu-ra-am e-li-a-am a-na ti-ši-it zu-uz /
ù bu-ra-am ša-ap-li-a-am a-na te-ši-it zu-uz /
[ù] uku.uš sí-ik-ka-as-sú ku-li-im-šu

They are two, 9 each are their sons.
The upper b r in nine divide,
and the lower b r in nine divide,
and (each) soldier show him his stake.

The divided trapezoid in this example is of the standard type, with a transversal parallel to the two fronts. The lengths s_a, d, s_k = 15 · (7, 5, 1) of the upper front, the transversal, and the lower front, respectively, form a solution to the "equipartitioned trapezoid equation"

sq. s_a + sq. s_k = 2 · sq. d.

Consequently, the transversal divides the trapezoid in two parts of equal area. The normalized length of the trapezoid is 1 00 (ninda). Since 5 − 1 = 2 · (7 − 5), the transversal divides the length in two parts, 20 and 40, in the ratio 1: 2. The factor 15 multiplying the triple (7, 5, 1) is chosen so that the two equal sub-areas are both equal to the round area number 30 00 (sq. ninda) = 1 bùr.

An unusual feature of this text is that the two sub-trapezoids are further divided in nine parts each, allotted to the 2 · 9 sons of the two brothers, Thus, each son gets 1/9 bùr = 1/3 èše = 2 iku. To 'show each soldier his stake' means, as usual, to determine the lengths and positions of the transversals separating the 18 lots from each other. This can be done by use of the method suggested by Parker, *JEA* 61 (1975) as an explanation of the given length numbers in *P.Heidelberg 663* # 2.

Let the sum of the first n sons' lots, counted from the left, be the trapezoid with the fronts s_a, d_n, and the length $(s_a - d_n)/f$, where f is the inclination of the sloping side of the trapezoid. Let the corresponding area be A_n. Then

$A_n = (s_a + d_n)/2 \cdot (s_a - d_n)/f = ($sq. $s_a -$ sq. $d_n)/(2f)$, so that sq. $d_n =$ sq. $s_a - 2f \cdot A_n$.

In VAT 7621, $s_a = 1\ 45$, $f = (1\ 45 - 15)/1\ 00 = 1;30$, and $A_n = 1\ 00\ 00/18 \cdot n = 3\ 20 \cdot n$. Hence, the length d_n of the n-th transversal can be computed as

$d_n =$ sqr. (sq. $s_a - 2f \cdot A_n) =$ sqr. $(3\ 03\ 45 - 10\ 00 \cdot n)$, for $n = 1, 2, \cdots, 18$.

When the lengths of the transversals are known, it is easy to find also their positions.

3.8. Conclusion

The importance of *P.Cairo* for the history of mathematics

The rather detailed analysis in Sec. 3.1 a above of the contents of *P.Cairo* clearly shows that *in the third century BCE*, if not sooner, Egyptian mathematics had become deeply influenced by Babylonian mathematics.[33] This is shown, to some extent, by the fact that almost all the topics treated in *P.Cairo* are known also from Old or Late Babylonian/Seleucid mathematical texts. More important is the circumstance that the equation for the area of a circle that was used in hieratic mathematical papyri is replaced in *P.Cairo* by the kind of equations for the area of a circle that one meets in mathematical cuneiform texts. Above all, the Babylonian influence is evident in *the hidden use of sexagesimal fractions* as a convenient computational tool, side by side with the Egyptian traditional use of sums of parts and the apparent novel use of binomial fractions. (Note that there are no detailed, *explicit* computations in *P.Cairo* of the kind that one meets so frequently in the hieratic *P.Rhind*.)

The hidden use of sexagesimal fractions in *P.Cairo*, a text from the *third century BCE*, is so much more remarkable in view of the following remark in Fowler, *MPA* (1987, 222; 1999, 223):

> "In scientific computations from the time of Hipparchus and Hypsicles (c. 150 BC), and especially in Ptolemy's *Almagest* and its commentaries, the Greeks employed an alphabetic version of the sexagesimal system found earlier in Babylonian cuneiform clay tablets that date back to almost 2000 BC. But there is no trace whatsoever of sexagesimal numbers in any Greek text from before the second century BC ⋯ "

The orientation of the equilateral triangle in *P.Cairo* § 11 a (# 36),

---

33. Cf. the following commentary in Parker, *DMP* (1972), 5: "In the third century B.C. we are admittedly on very shaky ground if we regard anything that is new in demotic texts as a purely Egyptian development, free of any influence from Greece or Babylonia. ⋯ Elsewhere I have shown that the content of a demotic papyrus of the Roman period concerned with celestial omina can be definitely ascribed to Babylonia, the transmission of such literature having taken place during the Persian rule of Egypt, the late sixth and fifth centuries B.C. It is very likely, indeed, that some amount of Babylonian mathematical literature also came to Egypt at the same time ⋯ in the third century B.C. Egypt, after all, was a part of the Hellenistic world, in which both Greek and Babylonian science was widely diffused." — Knorr, *HM* 9 (1982), footnote 4, concurs, with the words: "On the basis of the technical evidence from Egypt, Mesopotamia, and Greece, in the light of literary sources, I have proposed a transmission of Mesopotamian mathematical techniques to the Greeks through Egypt after the Persian occupation of the late sixth and early fifth centuries B.C. The details of this argument are given in an unpublished paper, 'The Greeks learn geometry' ⋯ " (Unfortunately, that paper was never published.) — Jones, *GMS* 3 (1993), 88, writes: "The general pattern of transmission of Babylonian astronomy seems to be a gradual trickle of basic concepts and the occasional parameter from about 500 B.C., followed by a sudden flood of detailed information in the second century B.C." Something similar may have happened in the case of the transmission of Babylonian mathematics. — The following passage from Jones, *UOS* (2002]) may also be of some interest here: "Two further points deserve to be mentioned. One is the conservative character of the tradition. The System B tables surviving in cuneiform fall within an interval of two centuries, from about 250 B.C. to about 50 B.C. Now after a documentary gap of two hundred years or more, in a different region and a different language, we encounter tables not only based on the same astronomical theory and mathematical methods, but even preserving such details of format as placing the names of zodiacal signs following the degrees and the indications of subtractive quantities following the quantities. This fact demonstrates, overriding all the historians' *a priori* assumptions about the difficulty of deep cultural contacts between Mesopotamia and the classical world, that an unbroken practical tradition at a high technical level led from the scribes of Babylon and Uruk to those of provincial Egypt."

3.8. Conclusion

depicted as standing on one of its sides, suggests an influence from *Late Babylonian*, rather than OB mathematics. On the other hand, it is known that many of the traditions from the OB school of mathematics were still kept alive in Late Babylonian mathematics. (See examples in Friberg, *et al.*, *BaM* 21 (1990): W 23291-x §§ 1-4, and Friberg, *BaM* 30 (1999).) Therefore, the claim that Egyptian mathematics in the third century BCE was influenced by Babylonian mathematics should, perhaps, be replaced by the more precise claim that it was influenced by Late Babylonian mathematics. What makes it difficult to say anything more precise about the matter is that very little is known about the true character and extent of Late Babylonian mathematics. After all, the number of known Late Babylonian mathematical texts is relatively small. (See the survey in Friberg, *BaM* 28 (1997), 356-357], to which must now be added the fragment N 2873 in Robson, *SCIAMVS* 1 (2000), 44.)[34] The truth may be that the mathematical content of the Egyptian demotic papyri can tell us something that we otherwise would not know about the kind of problems that were considered in Late Babylonian mathematics!

There are no traces of influence from high level "academic" Greek mathematics in *P.Cairo*. Instead, texts like *P.Cairo* may have had a decisive influence on Greek mathematics. Indeed, the impact of the existence of a Ptolemaic demotic mathematical text like *P.Cairo* on the history of the origin of Greek mathematics must be considerable. *P.Cairo* tells us that at the time when Euclid was working in Alexandria in Egypt (or not much later), Egyptian mathematicians were familiar with many of the traditions and methods of Babylonian mathematics. So, when Greek historians like Herodotus claim that the Greeks got their mathematics from Egypt (see, for instance, Heath, *HGM* (1921 (1981)), I: 4, 121, and II: 440, or Peet, *RMP* (1923), 31), what that means may very well be that they got it from the Babylonians, by way of Egypt! However, the situation is complicated by the fact that it may be difficult to distinguish between ideas that the demotic mathematical texts *borrowed* from OB mathematics and ideas

---

34. See also the discussion of NB mathematical cuneiform texts (c. 750-500 BCE) in Robson, *SCIAMVS* 5 (2004), in particular the remark on p. 62 that "The Kish and Babylon evidence taken together thus suggest that the mathematical elements of elementary scribal education in the mid-first millennium BCE consisted *only* of metrological lists and tables of squares."

that they *inherited* directly from Middle Kingdom Egyptian mathematics, since, as was argued in Chapter 2 of this work, the mathematics of the Egyptian hieratic mathematical texts was not very different from OB mathematics.

The other demotic mathematical papyri

There is no other demotic mathematical papyrus obviously covertly using sexagesimal arithmetic like *P.Cairo*, although *P.BM 10794* comes close. Cf. Fowler, *MPA* (1987 (1999)), 7.1(b):

> "Among the thousands, possibly tens of thousands, of examples of fractions to be found in *contemporary* Egyptian (hieroglyphic, hieratic, and demotic), Greek, and Coptic texts, all but a few isolated examples in five texts (P.Lond. ii 265; M.P.E.R., N.S. i 1; and three demotic papyri published in Parker, *DMP*), all to be described in detail in Section 7.3 below, use throughout the following 'Egyptian' system of expressing fractions: (sums of parts)."

(It is regrettable that Fowler did not think of using the opportunity to look for the *hidden* use of sexagesimal fractions in some of those thousands of examples!)

Nevertheless, all the other demotic mathematical papyri show other clear signs of being influenced by Babylonian mathematics, with the exclusion of *P.Griffith Inst.* with its simple theme of linear equations.

There is, on the other hand, no demotic mathematical papyrus showing any influence from high level Greek mathematics, not even the ones dating from the Roman period.

The final conclusion must be that there can be little doubt that there were no significant differences between the general level and extent of the knowledge of mathematics in Egyptian demotic mathematical texts and in Mesopotamian cuneiform mathematical texts towards the end of the first millennium BCE, and that there are no signs of influence on either from high-level Greek mathematics. That Greek mathematics was inspired by Babylonian mathematics is another matter, but to (begin to) make that claim precise is left to Chapter 4 below.

Chapter 4

Greek-Egyptian Mathematical Documents and Cuneiform Mathematical Texts

What is usually meant by the term "Greek mathematics" is mainly known from late Byzantine or Islamic sources, and consists of copies or translations of important works written by ancient mathematicians and commentators, most of them known by name. However, if one wants to consider original, "contemporary" texts, rather than late copies and translations, what is then available is a small number of *anonymous* mathematical ostraca, papyrus leaves, and papyrus fragments. These can be divided into two discrete groups of documents.

One group of contemporary sources consists of texts closely associated with Euclid's *Elements*. A survey of published texts belonging to this group, with an extensive commentary, can be found in Fowler, *MPA* (1987 (1999)), Sec. 6.2, and plates 1-3. The documents mentioned by Fowler are

a) *six fragmentary ostraca* containing text and figures, found on Elephantine Island (***Pack no. 2323***), dating to the third quarter of the 3rd century BCE, and dealing with the results found in *Elements XIII*, 10 and 16 concerning the pentagon, hexagon, decagon, and icosahedron,

b) a papyrus roll from Herculaneum, (***P.Herc. 1061***), thus dating to 79 CE, and containing a hardly legible essay on *Elements I* by Demetrius Lacon,

c) a papyrus fragment from Oxyrhynchus (***P.Oxy. i 29***), dated to the end of the 1st century CE, and possibly part of a manuscript of notes by someone working through *Elements II*,

d) a fragment from Fayûm (***P.Fay. 9***), assigned to the latter half of the 2nd century CE, and containing parts of *Elements I*, 39 and 41,

e) the fragment ***P.Mich. iii 143***, containing the first ten definitions of *Elements I*.

The second group of contemporary sources consists of anonymous mathematical ostraca, papyri, and papyrus fragments *having no or insignificant associations with Euclidean mathematics*. They are all from the Ptolemaic and Roman periods in Egypt. Since also all the demotic mathematical papyri discussed in Chapter 3 above are from the Ptolemaic and Roman periods, the two kinds of documents are to some extent contemporary with each other. It can therefore be expected that there should be no noticeable difference in form and content between demotic and (non-Euclidean) Greek mathematical papyri. Below, an effort will be made to show, by means of a survey of examples from various Greek mathematical papyri, how little difference there really is between mathematical exercises in Greek and demotic texts.[35] To make the point even more obvious, the discussion will begin with a discussion of a Greek land survey ostracon.

4.1. *O.Bodl. ii 1847* (30 BCE), an Ostracon with Schematic Field Plans

Fowler, *MPA* (1987 (1999)), Sec. 7.1(d), is a fairly detailed discussion of Greek-Egyptian "land surveys" in the Ptolemaic period, exemplified by the Greek ostracon ***O. Bodl. ii 1847***(photo: (*op. cit.*), pl. 6), which is from Thebes like the demotic ***Theban Ostracon D 12*** (Thompson, *TO 2* (1913); Fig. 3.3.4 above). The two look very much alike. The Greek ostracon has eight schematic drawings of the same type as the one in *P.BM 10520* § 7 b (Fig. 3.3.3), and the ones on *Theban Ostracon D 12*, although with Greek letters to denote fractions of the *schoinion* and the *aroura*. Cf. Fowler (*op. cit.*), 233-234:

> "The earliest surviving land registers such as I have described here date from the second century BC ⋯ ⋯ ⋯ After the conquest of Egypt by Alexander in 331 BC, the Greeks took over the existing Egyptian administration, including the land survey, merely imposing Greek as the official language for virtually all administrative and financial documents that were of concern to them. On the other hand, the Greeks never seemed to have operated a land survey in Greece itself ⋯ ⋯ ⋯ ."

---

35. Several new or improved interpretations are offered here for both Greek-Egyptian and Babylonian mathematical texts. Thus, new interpretations are offered in Chapter 4 for the following texts: *O.Bodl.ii 1847* (Sec. 4.1), **Michael. 62 # 2, YBC 4698 § 4, MLC 1842** (Sec. 4.6), *P.Chicago litt. 3* (= *pAyer*) and *P.Cornell 69* (Sec. 4.7 b-c), **BM 96954 + BM 102366 + SÉ 93** (Sec. 4.8 f).

Fowler tries to explain (*op. cit.*), 232 (233), how the computation in line 3 of *O.Bodl. ii 1847* can have been carried out by "long multiplication". The result of the proposed long multiplication is a sum of parts, where the smallest part is 1024' (1/1024). Fowler admits that he has no idea how "the taxman" performed the computation and rounded off the result. However, it is not likely that Fowler's reconstruction of the computation is correct, for the simple reason that the smallest fraction of both the *schoinion/khet* and the *aroura/setat* was $32'_s$ (1/32).[36] (See Fig. 3.7.2.) Smaller lengths were expressed as multiples or fractions of the cubit, and smaller areas as multiples or fractions of the cubit(-strip) = 1 cubit · 1 *schoinion*.

Compare the following two computations. First Fowler's proposed long multiplication:

2' 4' 8' 16'				(the half-sum of 2' 4' 8' and 1)
× 4' 16'		64'	(the half-sum of 4' 8' 16' 32' and 8' 16')

$$
\begin{array}{llll}
8'\ 16'\ 32'\ 64' & & & \\
\quad\ \ 32'\ 64'\ 128'\ 256' & & & \\
\qquad\quad\ 128'\ 256'\ 512'\ 1024' & & &
\end{array}
$$

8' 16' 16' 32' 64' 128' 512' 1024' = appr. 4'   16' (??)

Then the same example, with small lengths or areas expressed in terms of cubits and cubit-strips, and with the *left* factor as the multiplier (the multiples or fractions of the cubit(-strip) are written *after* the sign c.):

$2'\,4'\,\overline{8}\,\overline{16} \cdot 4'\,\overline{16}\,\text{c.}\ 1\,2'$

$= \overline{8}\,\overline{32}\ \text{c.}\ 2'\,4' + \overline{16}\ \text{c.}\ 1\,2'\,4'\,\overline{8} + \overline{32}\ \text{c.}\ 2'\,4'\,\overline{8}\,\overline{16} + \text{c.}\ 1\,2'\,4'\,\overline{8}\,\overline{16}\,\overline{32}$

$= 4'\,\overline{32}\ \text{c.}\ 2\,2'\,\overline{32} = 4'\,\overline{16} - \text{c.}\ 4'\,\overline{8}\,\overline{16}\,\overline{32} = \text{appr.}\ 4'\,\overline{16}$ (*arouras*).

4.2. *P.Oxyrhynchus iii 470* (3rd C. CE). A Water Clock

P.Oxyrhynchus iii 470 (Greenfell and Hunt, *OP 3* (1903)) is a single leaf

---

36. It is known that in Greek-Egyptian mathematical papyri two common units of length measure were the *schoinion* and the cubit. According to a metrological table in one such papyrus (Bell, *GPBM* (1917), *Pap. 1718*, lines 85-86, (6th c. CE) there were (at least) two kinds of *schoinion*. One, called the "hieratic" *schoinion* was equal to 100 cubits, while another, the "geometric" *schoinion*, was equal to 96 cubits. It is likely that the hieratic *schoinion* was identical with the *khet* of 100 cubits occurring in the hieratic mathematical papyri, while the geometric *schoinion* ('geometric' in the literal sense of 'used for land measurement') was a modified *khet*, with 1/32 of a geometric *schoinion* = 3 cubits.

from a papyrus codex. Lines 31 to the end of this manuscript are concerned with the construction of a water clock(?), shaped like a flower pot. More precisely, the object is an inverted truncated circular cone, in which the diameter of the circular top is $a = 24$ fingers and the diameter of the circular base $b = 12$ fingers. The depth is $h = 18$ fingers. The volume is computed as the depth times the area of the circle whose diameter is the average of the diameters at the top and the base. This rough approximation of the volume of a truncated cone can be compared with the exercises in *P.BM 10399* § 2 (*DMP* ## 42-45) (Sec. 3.2 b above), where the volume of a number of 'masts' with the upper diameter a and the lower diameter b are computed using the non-exact rule

$V = (1 - 4') \cdot$ sq. $d_m \cdot h$, where $d_m = (a + b)/2$.

However, in *P.Oxy. iii* 470, the volume is computed using another non-exact rule:

$V = \{3' \cdot (3\, d_m)\} \cdot \{4' \cdot (3\, d_m)\} \cdot h$, where again $d_m = (a + b)/2$.

The difference between the two approximate computation rules is that they use two different approximately correct rules for the area of a circle of diameter d, namely

a) $A = (1 - 4') \cdot$ sq. d,
b) $A = \{3' \cdot (3\, d)\} \cdot \{4' \cdot (3\, d)\} = 4' \cdot d \cdot a$, where a is the circumference of the circle.

Both rules, that is both a), the one used in the demotic **P.BM 10399 § 2**, and b), the one used in the Greek *P.Oxy. iii 470*, are also used in the demotic **P.Cairo § 9** (*DMP* ## 32-33) (see Sec. 3.1 h above).

4.3. *P.Vindobonensis G. 26740*. Five Illustrated Geometric Exercises

P.Vindobonensis G. 26740 (Bruins, Sijpesteijn, and Worp, *Janus* 61 (1974)) was probably originally inscribed on both sides with a demotic text. The demotic text on the obverse was later washed off and replaced by a Greek text, consisting of five geometric problems, a passage of Homer, and two metrological conversion problems. The five geometric problems will be discussed below. They are separately illustrated by drawings, schematically reproduced together in Fig. 4.3.1 below.

4.3 a. *P.Vindob. G. 26740* # 1. A segment of a circular band

4.3. P.Vindobonensis G. 26740. Five Illustrated Geometric Exercises

Let there be a crescent, of which the outer circumference is 10 schoinia, the inner one 12 sch., the basis 2 sch.
How many *arouras* is it? How one has to operate:
Add the two perimeters, result 22. Take one half of these, result 11.
Multiply these schoinia into the 2 of the basis, result 22.
So many *arouras* is the crescent, as asked for.

The exercise is accompanied by a drawing, which shows that the "crescent" (μηνίσκος) actually is a segment of a circular band, bounded by two concentric circular arcs and two straight line segments. The area A of the figure is computed as

A = the average length of the arcs times the width of the circular band.

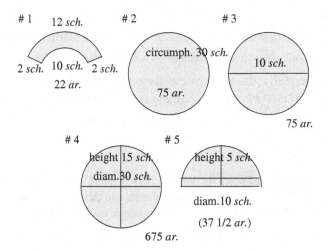

Fig. 4.3.1. *P.Vindob. G. 26740* ## 1-5. Drawings illustrating five geometric exercises.

A parallel text is **W 23291-x § 2** (Friberg, *et al, BaM* 21 (1990)), an isolated exercise in a Late Babylonian recombination text, illustrated by a drawing of a series of concentric circles. In that exercise, a circle of perimeter 1 00 ninda is given, and 4 inner concentric circles are formed by going inwards uniformly 4 times, 2 ninda each time. The perimeters of the four inner circles are then found to be 54, 42, 30, and 18 ninda. The original circle is in this way divided into four circular bands and a small central circle, with the areas of the five parts computed as follows:

(1 00 + 48)/2 · 2 = 1 48 sq. ninda, (48 + 36)/2 · 2 = 1 24 sq. ninda, *etc.*

Evidently, the rule used for the computation of the areas in W 23291-x § 2 is the same rule as the one used in *P.Vindob. G. 26740* # 1.

4.3 b. *P.Vindob. G. 26740* ## 2-4. A circle area found in three ways

P.Vindob. G. 26740 # 2

> Let there be a circle, of which the circumference is 30 schoinia
> How many *arouras* does it contain?
> How one has to operate:
> Multiply 30 sch. into themselves, result 900. Take one twelfth of those, result 75.
> So many *arouras* is the circle, as asked for.

In this exercise, the area A of a circle is computed as

$A = 1/12 \cdot$ sq. a, where a is the circumference of the circle.

This is the same rule as the one used in *Babylonian* mathematics, namely:

$A = ;05 \cdot$ sq. a, where a is the circumference of the circle, and $;05 =$ rec. 12.

It is also, essentially, the same rule as the one used in the *demotic* exercises ***P.Cairo* § 11** (*DMP* ## 36, 38), namely (see above, Sec. 3.1 j):

$A = (3' \cdot a) \cdot (4' \cdot a)$, where a is the circumference of the circle.

P.Vindob. G. 26740 # 3

> Let there be a circle, of which the diameter is 10 schoinia. How many *arouras* is it?
> How one has to operate: Multiply the 10 schoinia into themselves, result 100.
> Take of those 2' 4', result 75. So many *arouras* is the circle, as asked for.

In this exercise, the area A of a circle is computed *in a second way*, as

$A = 2' 4' \cdot$ sq. d, where d is the diameter of the circle.

This is, essentially, the same rule as the one applied in the *demotic* exercises ***P.Cairo* § 9** (*DMP* ## 32-33) (see above, Sec. 3.1 h), where the diameter d of a circle with given area A is computed as

$d =$ sqr. $(A + 3' \cdot A)$, which means that $d =$ sqr. $(1 3' \cdot A)$, so that, conversely, $A = 2' 4' \cdot$ sq. d.

P.Vindob. G. 26740 # 4

> Let there be a circle, of which the perpendicular is 15 schoinia, the diameter 30 sch.
> How many *arouras* is it?
> How one has to operate: Add the schoinia of the perpendicular <and the schoinia of

the diameter, result 45>.
Multiply with itself, result 2,025.
Take the 3' of these, making 675.
So many *arouras* is the circle, as asked for.

The area of a circle is computed in this exercise *in a third way*, as

$A = 3' \cdot$ sq. $(d + p)$,
where d is the diameter and p the "perpendicular" (κάθετος) of the circle.

Apparently here the "perpendicular" of the circle means the height of a semi-circle (half the given circle), and the area of the circle is equated with *twice the area of the semi-circle*. This is probably how the accompanying drawing must be interpreted. (See Fig. 4.3.1, # 4.) Indeed, if p is the height of the semi-circle, then

$3' \cdot$ sq. $(d + p) = p \cdot (d + p) = 2 \cdot p \cdot (d + p)/2 =$ twice the area of the semi-circle

(see below, *P.Vindob. G. 26740* # 5).

4.3 c. *P.Vindob. G. 26740* # 5. The area of a semicircle

> Let there be a semi-circle, of which the perpendicular is 5 schoinia
> and the diameter 10 sch. From these two, how many *arouras* is it?
> How one has to operate:
> Add the schoinia of the perpendicular and the schoinia of the diameter, result 15.
> Take 2' of these, result 7 2'.
> Multiply these schoinia into the 5 of the perpendicular, making 37 2'.
> So many *arouras* is the semi-circle, as asked for.

In this exercise, the area A of a semi-circle is computed as follows:

$A = p \cdot (d + p)/2$, where p is the height and d the diameter of the semi-circle.

In Bruins, *et al, Janus* 61 (1974), this area rule for a semi-circle is explained as a special case of the area rule for a segment of a circle attributed to 'the ancients' in Hero's ***Metrica* I: 30** (Heath, *HGM 2* (1921 (1981)), 330).) A more relevant reference is, perhaps, the "segment area rule" in the demotic *P.Cairo* §§ 11-12 (*DMP* ## 36-38), discussed in Sec. 3.1 j-k above. See, in particular, Fig. 3.1.10, right.

The application of the segment area rule in this exercise may be the reason why the associated drawing (Fig. 4.3.1, # 5) shows a semi-circle in which a chord cuts off a circle segment!

4.4. *P.Mich. 620.* Systems of Linear Equations. Tabular Arrays

P.Mich. 620 (Robbins, *CP* 24 (1929), Vogel, *CP* 25 (1930), Winter, *MP 3* (1936)) is a small Greek papyrus from the early part of the 2nd century CE. It is inscribed with three mathematical exercises, two of which are illustrated by preserved tabular arrays. Of the first exercise, there remains only the end of the solution procedure, the complete text of the verification (the "proof"), and a tabular array of numbers. However, in the verification the essential parts of the lost question are quoted, which makes it easy to reconstruct most of the question. Moreover, the tabular array repeats the essential steps of the lost solution procedure in an abbreviated form.

4.4 a. *P.Mich. 620* # 1. A system of linear equations: four unknowns

	$\bar{7}$		300 dr.		300 dr.		9,900 dr.
7 dr.'	8 dr.'	15 dr.'	300 dr.	30 dr.'	600 dr.		
						60 dr.'	900 dr.
						150	
1,050	1,200	2,550		5,100			

4.4.1. *P.Mich. 620* # 1. The tabular array.

(Below, italic style indicates reconstructed parts of the text of *P.Mich. 620*.)

9,900 drachmas, 4 ⋯ .
Let the second exceed the first by the 7th part.
Let the third exceed the two by 300 dr.
Let the fourth exceed the 3 by 300 dr.
Let the four numbers be found.
⋯ ⋯ ⋯ ⋯
⋯ again, multiply the 150 into the 30 numbers of the fourth: 4,500.
And the 600 dr. in its assigned value: 5,100. This is the fourth.
Then add together the four, 1,050 and 1,200 and 2,550 and 5,100: 9,900.
Proof.
Since he says, let the second exceed the first by the 7th part,
take the $\bar{7}$ of the first 1,050: 150.
Add this to the 1,050: 1,200, which is the second.
Again, since he says, let the third exceed the two by 300 dr., add the 1 and 2: 2,250,

4.4. P.Mich. 620. Systems of Linear Equations. Tabular Arrays

and add together the 300 dr. of the excess: 2,550, which is the third.
And since he says, let the fourth exceed the 3 by 300 dr., add together the three: 4,800, and the 300 dr. of the excess: 5,100, which is the fourth.
9,900 is the whole.

Robbins (*op. cit.*) explained this problem as *a system of four linear equations for four unknowns*:

1. $a + b + c + d = 9{,}900$ *drachmas*
2. $b = 1\overline{7} \cdot a$
3. $c = a + b + 300$ *drachmas*
4. $d = a + b + c + 300$ *drachmas*

Apparently, the one who made the tabular array had realized that the best way to solve this system of equations was to move equation 1 to the last place. That is why *the first line of the tabular array* mentions the constant 9,900 of equation 1 *after* the coefficient $\overline{7}$ in equation 2 and the constants 300 and 300 of equations 3 and 4.

The system of equations was solved by use of a simple application of the rule of false value. As shown by *the second line of the tabular array*, the initial step was to choose the false value 7 for the first unknown. Then equation 2 shows that the false value for the second unknown must be 8, equation 3 shows that the false value for the third unknown must be 15, plus 300 *drachmas*, and equation 4 shows that the false value for the fourth unknown must be 30, plus 600 *drachmas*.

Adding these results together, one finds that the false value for the sum of all four unknowns must be $7 + 8 + 15 + 30 = 60$, plus $300 + 600 = 900$ *drachmas*. These are the values recorded *in the third line of the tabular array*, to the far right, and two steps below the true sum 9,900, asked for in equation 1. The final result, then, is that 60 false units plus 900 *drachmas* = 9,900 *drachmas*. This is a single linear equation, trivially with the solution that the false unit (the "correction factor") must be 150 *drachmas*. The value 150 is recorded *in the fourth line of the tabular array*.

The answer to the problem is recorded *in the fifth line of the tabular array*. It is, without explicitly mentioning the *drachmas*,

$a = 150 \cdot 7 = 1{,}050$,
$b = 150 \cdot 8 = 1{,}200$,
$c = 150 \cdot 15 + 300 = 2{,}550$,
$d = 150 \cdot 30 + 600 = 5{,}100$.

4.4 b. *P.Mich. 620 # 2*. A system of linear equations: two unknowns

$\overline{6}$	12 mo.	$\hat{4}$	12 mo.
	1 dr.'	4 dr.'	12 mo.
		$\hat{3}''$	2 mo. / 3'' *14* mo.
3' 14 mo.			
1 42		168 12 / 180	

Fig. 4.4.2. *P.Mich. 620 # 2*. The tabular array

P.Mich. 620 # 2 can be interpreted as a system of *two linear equations for two unknowns*:

1. $a = \overline{6} \cdot b + 12$
2. $b = 4 \cdot a + 12$

This problem, too, is accompanied by a tabular array, summarizing the solution procedure:

In the first line of the tabular array are recorded the data for the problem. *In the second line* are recorded the initial false value 1 for *a* and (in view of equation 2) the resulting false value 4, plus 12 units, for *b*. *In the third line* are recorded first $\overline{6} \cdot b$ = the false value 3'', plus 2 units, then $\overline{6} \cdot b + 12 = 3''$, plus 14 units, which, according to equation 1, must be equal to the initial false value 1. Therefore, $1 - 3'' = 3'$ times this false unit must be equal to 14 units. This is the meaning of the notation *in the fourth line of the tabular array*. Now, if 3' times this false unit = 14 units, then 1 times the false unit must be 42 units. Then also the true value of the 1st unknown = 42 (units). This is the meaning of the notation *to the left in the fifth line of the tabular array*. The notation *to the right* means that the true value of the 2nd unknown is $4 \cdot 42 + 12 = 168 + 12 = 180$ (units).

4.4 c. *P.Mich. 620 # 3*. A system of linear equations: three unknowns

The tabular array which probably accompanied *P.Mich. 620 # 3* is lost. However, in this exercise a large part of the question is preserved, together with the beginning of the solution procedure:

> 3 numbers.
> The 3 are 5,300, and let *the first and the second be* the third multiplied by 24,

and let the second and the third be the first multiplied by 5.
Let the three numbers thus be found.
Since the first and the second are the third multiplied by 24,
the three together are 25 times the third.
Divide 5,300 by 25: 212¦. This is the third ⋯

Thus, it is clear that *P.Mich. 620 # 3* can be interpreted as a system of three linear equations for 3 unknowns, in the text called 'three numbers':

1. $a + b + c = 5,300$
2. $a + b = 24 \cdot c$
3. $b + c = 5 \cdot a$

The solution algorithm begins by noting that, in view of equation 2,

$a + b + c = 24 \cdot c + c = 25 \cdot c$.

On the other hand, in view of equation 1 the sum of the 3 numbers is 5,300. Therefore,

$25 \cdot c = 5,300$, so that $c = 5,300/25 = 4 \cdot 53 = 212$.

The remainder of the solution procedure is lost, but it is likely that it proceeded as follows:

Let a have the false value 1.

Then, in view of equation 3,

$b + c = 5$ times the false unit,

and in view of equation 2

6 times the false unit = 5,300.

Consequently,

a = the false unit = 5,300/6 = 883 3'.

Finally, in view of equation 1 again,

$b = 5,300 - 212 - 833\ 3' = 5,088 - 833\ 3' = 4,204\ 3"$.

A parallel to *P.Mich. 620 # 1* in a *demotic* mathematical papyrus is the system of four linear equations for four unknowns in **P.Carlsberg 30 # 2**, according to the proposed reconstruction in Sec. 3.5 b above. Unfortunately, it is impossible to know how similar the two exercises once were, because most of the solution procedure in *P.Carlsberg 30 # 2* is lost. Note, however, that in *P.Carlsberg 30 # 2* the four unknowns are referred to as 'silver, gold, copper, and lead', while in *P.Mich. 620* they are called 'the first', 'the second', *etc.*

Linear equations in hieratic mathematical papyri and in cuneiform mathematical texts were discussed above, in Sec. 2.1 b. A particularly interesting example is the Late Babylonian exercise on the fragment **BM 34800** (Fig. 2.1.2). According to the proposed reconstruction, the linear equation in that exercise can be expressed, in modern notations, as a *division problem*:

$a \cdot (1 - 1/5) \cdot (1 - 1/3) \cdot (1 - 2/5) \cdot (1 - 1/2) \cdot (1 - 3/5) = 1\ 00\ \text{gur} = 5\ 00\ \text{sìla}.\quad a = ?$

A somewhat less anachronistic interpretation of this and other similar division problems is as *a system of linear equations*, for instance, in the case of BM 34800, the following *chain of linear equations*:

$a \cdot (1 - 1/5) = b,\quad b \cdot (1 - 1/3) = c,\quad c \cdot (1 - 2/5) = d,$
$d \cdot (1 - 1/2) = e,\quad e \cdot (1 - 3/5) = 1\ 00\ \text{gur} = 5\ 00\ \text{sìla}.$

The solution procedure in BM 34800 makes use of the rule of false value, just as the solution procedure in *P.Mich. 620* ## 1-2.

The style and general layout of a mathematical exercise in a Greek mathematical papyrus such as *P.Vindob. G. 26740* or *P.Mich. 620* is quite similar to the style and layout of exercises in demotic mathematical papyri, or in Babylonian cuneiform texts. In the case of *P.Mich. 620*, there are, in addition, a pair of special, and very interesting similarities. One is the repeated use in *P.Mich. 620* # 1 of the phrase 'since he says' (ἐπεί λέγει), referring back to conditions stated in the question. A Late Babylonian parallel is the use o f the term *šá pi* 'as instructed' in BM 34800.

Alos the solution procedures in some Old Babylonian mathematical texts contain similar references to conditions stated in the question. One example is **YBC 8588**, Neugebauer and Sachs, *MCT* (1945), 75, where the question is

> ki.lá 1 30 uš 30 sag /
> *i-na iš-te-en ka-la-ak-ki-im* / 9 *ka-la-ak-ku* /
> *šu-up-lum* en.nam /

An excavation(?). 1 30 the length, 30 the front.
From one day of digging(?), 9 days of digging(?).
The depth is what?

What this probably means (the terms are otherwise undocumented) is that a rectangular underground room is excavated by a team of 9 diggers.

4.4. P.Mich. 620. Systems of Linear Equations. Tabular Arrays

The length and width of the underground room are known, and also, of course, the work norm for diggers. The depth has to be calculated. The solution procedure begins as follows:

```
1 30 uš gar.ra 30 sag gar.ra /
i-na iš-te-en ka-la-ak-ki-im /
9 ka-la-ak-ku ša iq-bu-ú /
iš-te-en ka-la-ak-kum ša iq-bu éš.kàr /
15 éš.kàr gar.ra /
9 ka-la-ak-ku ša iq-bu érin.ḫá /
9 érin.ḫá gar.ra /
```

1 30, the length, set. 30, the front, set.
From one day of digging,
9 days of digging, that he said.
One day of digging, that he said, the work norm?
15, the work norm, set.
9 days of digging, that he said, the workers?
9 workers set.

What is happening here is that the conditions in the questions are recapitulated and explained, and the associated numbers are written down (gar.ra), *probably in some kind of tabular array*, which is not preserved in this text. After these preliminaries, the computations can begin:

```
15 éš.kàr a-na 9 érin.ḫá / íl
2 15 saḫar.ḫá /
1 30 uš a-na 30 sag íl 45 a.šà /
igi 45 a.šà pu-ṭur-ma 1 20 /
1 20 a-na 2 15 saḫar.ḫá / íl 3 šu-up-lum
```

15, the work norm, to 9 days of digging, carry,
2 15 the mud (= volume).
1 30, the length, to 30, the front, carry, 45, the field (= area).
The opposite (= reciprocal) of 45 resolve, 1 20.
1 20 to 2 15, the mud, carry, 3, the depth.

Since the number of man-days is $m = 9$, for 9 days of digging, and since the fixed work norm is $w = 15$ volume-shekels (= 1/4 volume-šar) per man-day, the volume V can be computed here as

$V = 9$ (man-days) · ;15 (volume-šar/man-day) = 2;15 (volume-šar).

Next, the bottom area A of the excavated space is computed as

$A = 1;30 \text{ n.} \cdot ;30 \text{ n.} = ;45 \text{ (sq. ninda)}.$

The depth h of the excavated space can then be computed as

$h = V/A = 2;15 \text{ (volume-šar)} / ;45 \text{ (sq. ninda)} = 3 \text{ (cubits)}.$

Another example of a similar kind can be found in the OB mathematical text **VAT 8390**, where the phrase

ma-la uš ugu sag *i-te-ru uš-ta-ki-il*
As much as the length exceeds the front I squared.

in the question (lines 4-5) is echoed in the solution procedure (lines 14-15) with the words

aš-šum ma-la uš ugu sag *i-te-ru uš-ta-ki-il iq-bu-ú*
Since 'as much as the length exceeds the front I squared' he said

More examples of a similar kind can be found in the post-OB (Kassite) mathematical text **MS 5112** (Friberg, *MCTSC* (2005), Sec. 11.2). Thus, for instance, in MS 5112 § 5, two of the conditions in the question are recapitulated as follows at the beginning of the solution procedure:

aš-šum 2' igi.5.gál *qà-bu-kum*
Since '1/2 of the 5th-part' it was said to you

and

1 ninda 2 kùš *ša qà-bu-kum*
'1 ninda 2 cubits' that was said to you

Another interesting similarity between *P.Mich. 620* and Babylonian mathematical texts is that the kind of tabular arrays with numbers that summarize the solution procedures in *P.Mich. 620* ## 1-2 (and 3?) are of the same form and function as a number of known tabular arrays on OB round or square mathematical hand tablets. A prominent example is ***UET 6/2 274 rev.*** (Fig. 2.3.2 above, and Friberg, *RA* 94 (2000) § 2e), with its tabular array for the numerical solution of a quadratic-linear system of equations. Other examples are round hand tablets such as ***UET 6/2 233 rev.*** (Friberg, (*op. cit.*) § 2h) with tabular arrays for the numerical solution of exercises concerning 'the cost in man-days and silver for digging a canal, and the daily progress', round hand tablets such as ***UET 6/2 290 rev.*** (Friberg (*op. cit.*) § 2 i) with tabular arrays for the numerical solution of exercises concerning the computation of capacity measures of cylindrical containers, *etc.* There are, in particular, quite a few known examples of

4.4. P.Mich. 620. Systems of Linear Equations. Tabular Arrays

hand tablets such as **MS 2832**, **MS 2268/19**, *etc.*, (Friberg, *MCTSC* (2005), Sec. 7.2. a), or **YBC 7353**, **YBC 11125**, *etc.* (Neugebauer and Sachs, *MCT* (1945), 17; Friberg (*op. cit.*)), with tabular arrays for the numerical solution of "combined market rate problems". There are also OB mathematical texts such as **YBC 7326** (Neugebauer and Sachs, (*op. cit.*), 130: sheep and lambs) and **AO 8862** (Neugebauer, *MKT 1* (1935), 108 ff: unequal shares) with both problem texts and tabular arrays.

There is, however, one important difference between such OB tabular arrays and the tabular arrays in *P.Mich. 620*, namely the (well known) use of algebraic symbols in the latter. Thus, in the tabular array in *P.Mich. 620* # 1 there are several instances of the use of a common sign for *drachma*, originally an abbreviation. It is here transliterated as 'dr.'. In the first line of the array, the numbers marked dr. are the given numbers in the linear equations. It is not clear if dr. here stands for a small monetary unit, or if it simple indicates known numbers in general. The circumstance that the sign dr. *does not appear* in the last line of the tabular array, where the computed values of the four unknowns are recorded, probably speaks in favor of the assumption that dr. is *a symbol indicating known numbers*, and not a monetary unit.

There is also (as pointed out by Vogel, *CP* 25 (1930) another sign in the tabular array in *P.Mich. 620* # 1, which looks like the sign for *drachma* with an added diagonal stroke. This sign is transliterated here as 'dr.''. It is obvious that dr.' is a symbol standing for the "false (or unknown) unit", which plays a central role in the application of the rule of false value. Thus, it corresponds loosely to the modern use of x for an unknown number.

The first preserved line of text of *P.Mich. 620* # 1 mentions 'the 30 numbers (ἀριθμούc) of the fourth'. This passage clearly refers to what is called 30 dr.' in the tabular array. Therefore, 'number' may be the correct reading of the symbol dr.' in *P.Mich. 620* # 1.

The tabular array accompanying *P.Mich. 620* # 2 seems to show that the two exercises *P.Mich. 620* ## 1-2 are copied from different source documents. Indeed, in # 2 the symbol indicating known numbers is not 'dr.' as in # 1 but instead 'mo.', an abbreviation for μονάc 'unit'. Moreover, although the symbol dr.' is used twice to indicate the unknown unit, it is twice replaced by a short arc above the numbers, and twice there is no such symbol at all. Note that, just as in # 1, the computed numbers in the final line of the tabular array are written without symbols.

It is deplorable that no OB examples of tabular arrays are known of the kind where the need would have arisen to distinguish between known and unknown units, tabular arrays illustrating solution procedures for systems of linear equations, such as the ones in *P.Mich. 620*, or tabular arrays illustrating solution procedures for mixed quadratic equations solved by use of the method of 'completing the square', such as the equation sq. $m + 2 \cdot m = 2$ in **MS 5112** § 1 (Friberg, *MCTSC* (2005), Sec. 11.2 a; Fig. 11.2.2). Since no such OB tabular arrays are known, we don't know how OB mathematicians would have handled the difficulty. All that is known is that they thought of the unknown unit as the unknown length of a reed, usually in the end found to be equal to ;30 ninda, the value of a 'reed' as a standard unit of length. Cf. gi *ša lā ti-du-ú* 'the reed that you do not know' in the "broken reed problem" **VAT 7532** (Sec. 3.1 f above).

4.5. *P.Akhmîm* (7th C. CE). Calculations with Fractions

P.Akhmîm (Baillet, *PMA* (1892)) is a bound codex of 6 papyrus leaves, each leaf inscribed on both sides with mathematical tables or problem texts. The table of contents below shows that *P.Akhmîm* is a *mathematical recombination text* of the same kind as, for instance, the Egyptian hieratic papyrus *P.Rhind* (see Sec. 2.1 a above), the OB clay tablet BM 85194 (Fig. 2.4.1), or the Egyptian demotic papyrus *P.Cairo* (Sec. 3.1 a).

P.Akhmîm: Contents.

—	Tables of fractions. (Multiplication tables for 3", 3', $\overline{4}$, $\overline{5}$, ⋯ $\overline{20}$.)	
§ 1	**Capacity measure in Palestine *kor*??**	
	of a truncated circular cone. Constant: $\overline{3}$.	# 1
§ 2	**Capacity measure in *artabs***	
	of a rectangular granary. Constant: 3 $\overline{4}$ $\overline{8}$ (27/8)	# 2
§ 3 a-b	**Unequal sharing** of a crop in given proportions:	
	a) 3 2', 2 2', 3 2' $\overline{4}$, 6 $\overline{4}$, 4	# 3
	b) $\overline{7}$, $\overline{8}$, $\overline{9}$	# 4
§ 4 a-d	**Subtraction of fractions:**	
	a) 2' 3' – $\overline{9}$ $\overline{11}$ = $\overline{99}$ of 62 2'	# 6
	b) 3"– $\overline{9}$ $\overline{11}$ = $\overline{99}$ of 46	# 7

4.5. P.Akhmîm (7th C. CE). Calculations with Fractions

	c) $3''-\overline{3}\,\overline{9}\,\overline{99} = \overline{11}$ of $2\,\overline{3} = \overline{6}\,\overline{33}\,\overline{66}$	# 8
	d) $3''-\overline{4}\,\overline{44} = \overline{11}$ of $4\,\overline{3} = 3\,\overline{22}\,\overline{66}$	# 9
§ 3 c	Selling parts of a house in given proportions: $\overline{3}, \overline{4}, \overline{5}$.	# 10
§ 3 d	Paying seeding tax(?) proportionally for 7, 8, 9 *artabs*	# 11
§ 4 e	$3''- \overline{10}\,\overline{11}\,\overline{20}\,\overline{22}\,\overline{30}\,\overline{33}\,\overline{40}\,\overline{44}\,\overline{50}\,\overline{55}\,\overline{60}\,\overline{66}\,\overline{70}\,\overline{77}\,\overline{88}\,\overline{90}\,\overline{99}\,\overline{100}\,\overline{110}$	
	$= \overline{110}$ of $13\,\overline{5} = \overline{10}\,\overline{50}$	# 12
§ 5 a	**System of linear equations:**	
	$a - \overline{13}$ of $a = b$, $b - \overline{17}$ of $b = 150$ *artabs*. $a = \overline{192}$ of $33{,}150 = \cdots$	# 13
§ 4 f-g	f) $1 - \overline{3}\,\overline{11}\,\overline{33} = \overline{11}$ of $6 = 2'\,\overline{22}$	# 14
	g) $1 - 3''\,\overline{22}\,\overline{66} = \overline{11}$ of $3 = \overline{4}\,\overline{44}$	# 15
§ 6 a	**Decomposition of fractions:**	
	Expand $\overline{22}$ of 1 (sic!) into 3 parts. Answer: $\overline{55}\,\overline{70}\,\overline{77}$	# 16
§ 5 b	System of linear equations:	
	$a - \overline{17}$ of $a = b$, $b - \overline{19}$ of $b = 200$ *artabs*. $a = \overline{288}$ of $64{,}600 = \cdots$	# 17
§ 7 a	**Multiplication of fractions:**	
	$\overline{187}$ of $6\,\overline{15}\,\overline{40} = \overline{187}$ of $\overline{120}$ of $731 = \overline{22{,}440}$ of $731 = \overline{88}\,\overline{90}\,\overline{99}$	# 18
§ 6 b	Expand $\overline{55}\,\overline{56}\,\overline{70}$ into 4 parts.	
	Answer: $\overline{155}$ of $3{,}080 = \overline{63}\,\overline{77}\,\overline{88}\,\overline{99}$	# 19
§ 6 c	Expand $\overline{323}$ of 75 into 8 parts.	
	Answer: $\overline{17}\,\overline{19}\,\overline{34}\,\overline{38}\,\overline{51}\,\overline{57}\,\overline{68}\,\overline{76}$	# 20
§ 7 b	$\overline{323}$ of $11\,2'\,\overline{3}\,\overline{10}\,\overline{60} = \overline{323}$ of $\overline{20}$ of $6{,}460$	
	$= \overline{6{,}460}$ of $239 = \overline{68}\,\overline{85}\,\overline{95}$	# 21
§ 7 c	$\overline{323}$ of $7\,2'\,\overline{10}\,\overline{20} = \overline{323}$ of $\overline{20}$ of 153	
	$= \overline{6{,}460}$ of $153 = \overline{76}\,\overline{95}$	# 22
§ 7 d	$\overline{5}$ of $\overline{4}\,\overline{28} = \overline{5}$ of $\overline{7}$ of $2 = \overline{35}$ of $2 = \overline{30}\,\overline{42}$	# 23
§ 4 h	$\overline{11}\,\overline{13} - \overline{9} = ?$ (corrupt)	# 24
§ 8	**Multiplication and subtraction of fractions:**	
	$1\,3''\,\overline{11}\,\overline{22}\,\overline{66} \cdot 1\,2'\,\overline{29}\,\overline{58} - \overline{63}\,\overline{84}$ of the same	
	$= (\overline{11}\text{ of }20) \cdot (\overline{29}\text{ of }45) - \overline{63}\,\overline{84}$ of the same	
	$= \overline{319}$ of $(900 - 25) = \overline{319}$ of $875 = 2\,3''\,\overline{29}\,\overline{33}\,\overline{87}$	# 25
§ 9 a-b	**Prices and market rates:**	
	a) $1\,3''$ for cleaning 100 *art.*, what for 195 *art.*?	# 26
	b) 8 for 110, how much for 15?	# 27
§ 10	**Division problem:**	
	Capital plus interest = 100 *art.* Interest = $\overline{4}\,\overline{28} = \overline{7}$ of 2 of capital.	# 28
§ 4 i	$2'\,\overline{3} - \overline{4}\,\overline{28} = \overline{6}$ of $5 - \overline{7}$ of $2 = \overline{42}$ of $(5 \cdot 7 - 2 \cdot 6)$	
	$= \overline{42}$ of $23 = 2'\,\overline{21}$	# 29
§ 4 j-l	j) $2'\,\overline{4} - \overline{4}\,\overline{44}$	# 30
	k) $2'\,\overline{3}\,\overline{42} - \overline{6}\,\overline{66}$; l) $1 - \overline{12}\,\overline{51}\,\overline{68}$	# 32
§ 9 c-d	c) 100 *art.* for $7\,\overline{7}$ gold *nomismatia*. How much for 1 *nom.*?	# 33
	d) 100 *art.* for $5\,3''\,\overline{21}$ *nom.*	# 34
§ 9 e	$15\,2'\,\overline{4}$ *art.* for 1 *nom.* 100 *art.* for what?	# 35

§ 9 f-g	f) 500 *art.* for 85 3" $\overline{21}$ *nom.*, 100 *art.* for ?	# 36
	g) 500 *art.* for 31 2' $\overline{19}$ $\overline{38}$ *nom.*, 100 *art.* for ?	# 37
§ 7 e-g	e) $\overline{55}$ of 1 2' = $\overline{110}$ of 3 = $\overline{70}$ $\overline{77}$	# 38
	f) $\overline{88}$ of 3 2' = $\overline{176}$ of 7 = $\overline{70}$ $\overline{77}$ $\overline{80}$	# 39
	g) $\overline{119}$ of 9 3" = $\overline{357}$ of 29 = ⋯	# 40
§ 9 h-i	h) I gave 3 and received 9 3". If I gave 28, what?	# 41
	i) I gave 5 and received 9 3". If I gave 30, what?	# 42
§ 9 j-l	j) 17 2' $\overline{3}$ *art.* for 1 *nom.* 1 *art.* for what?	# 44
	k) 12 3" *art.* for 1 *nom.*; l) 11 2' $\overline{4}$ *art.* for 1 *nom.*	# 46
§ 3 e	Taking 60 *art.* proportionally out of 3 granaries with 200, 300, 500 *art.*	# 47
§ 3 f-g	f) 720 *art.* proportionally out of 320, 400, 480 *art.*	# 48
	g) 550 *art.* prop. out of 720, 830, 950 *art.*	# 49
§ 6 d	Expand $\overline{12}$ into 6 parts. Answer: $\overline{55}$ $\overline{63}$ $\overline{70}$ $\overline{77}$ $\overline{84}$ $\overline{99}$	# 50

As the table of contents shows, the 50 exercises in the problem section of *P.Akhmîm* can be organized into 10 separate paragraphs with different themes. However, the exercises belonging to any given paragraph do not always appear together. Thus, for instance, the 12 exercises belonging to § 4 (subtraction of fractions) appear in 5 different places in the text. This is a typical behavior of a recombination text, as opposed to an original theme text.

4.5 a. *P.Akhmîm*. Ten tables of fractions

P.Akhmîm begins with a series of 10 *tables of fractions*, in which the fractions 3", 3', $\overline{4}$, $\overline{5}$, $\overline{6}$, $\overline{7}$, $\overline{8}$, $\overline{9}$, $\overline{10}$ are applied to 6,000, called 'the number' (ἀριθμός), and to the whole numbers 1, 2, 3, ⋯ (α, β, γ, ⋯), 10, 20, 30, ⋯ (ι, κ, λ, ⋯), 100, 200, 300, ⋯ (ρ, σ, τ, ⋯), 1,000, 2,000, 3,000, ⋯ (α′, β′, γ′, ⋯), up to 10,000 = 1 myriad. Then follow 10 abbreviated tables of fractions, in which the fractions $\bar{n}$ = $\overline{11}$, $\overline{12}$, ⋯ , $\overline{20}$ are applied to 6,000 and to the whole numbers from 1 to *n*. The results of the applications are given in the traditional Egyptian way as sums of parts, as in the following examples:

$\overline{7}$ to 'the number'	857 $\overline{7}$	$\overline{11}$ to 'the number'	545 3' $\overline{11}$ $\overline{33}$
of 1 the $\overline{7}$	$\overline{7}$	of 1 the $\overline{11}$	$\overline{11}$
of 2	4' $\overline{28}$	of 2	$\overline{6}$ $\overline{66}$
of 3	3' $\overline{14}$ $\overline{42}$	of 3	$\overline{4}$ $\overline{44}$
⋯ ⋯	⋯ ⋯	⋯ ⋯	⋯ ⋯
of 1 myriad	1,428 2' $\overline{14}$	of 11	1

The ancestors of Greek-Egyptian fraction tables like the one in *P.Akhmîm* seem to be on one hand multiplication tables of the Babylonian type, on the other hand the $\overline{10}$ table in ***P.Rhind*** (9 lines), and, of course, the $\overline{90}$ and $\overline{150}$ tables in ***P.BM 10794*** (10 lines each; see Sec. 3.4 above). For a survey of all known Egyptian fraction tables from various periods, see Fowler, *MPA* (1987 (1999)), Sec. 7.5(a), where Fowler, somewhat reluctantly, agrees to call such tables 'division tables'.

4.5 b. *P.Akhmîm* § 1. The capacity measure(?) of a truncated cone

In the single exercise in § 1 of *P.Akhmîm*, the object considered is an excavated store room in the form of a truncated circular cone with the upper circumference $m = 20$ cubits, the lower circumference $n = 12$ cubits, and the depth $d = 6\ 1/2$ cubits. Its capacity measure(?) is computed as

$C = $ sq. $\{(m + n)/2\} \cdot d\ /36$.

This equation can be compared with the following equation in the OB exercise **VAT 8522 # 1** (see Sec. 3.2 b above) for the capacity measure of a truncated circular cone (a 'log') with the lower diameter a, the upper diameter b, and the height h:

$C = $;05 $\cdot$ sq. $\{(a + b)/2 \cdot 3\} \cdot h \cdot $ c.

Here c is a "storing number", used to make the desired transition from volume measure to capacity measure. Now, since ;05 = 1/12, and since $a \cdot 3$ = the lower circumference, and $b \cdot 3$ = the upper circumference, it follows that the equation in VAT 8522 # 1 can be rewritten in the form

$C = $ sq. $\{(m + n)/2\} \cdot h \cdot $ c/12.

Therefore, if the computation in *P.Akhmîm* § 1 can be explained in the same way as the computation in the mentioned OB exercise, then the constant $\overline{36}$ must be explained as $\overline{12} \cdot \overline{3}$, where $\overline{3}$ is a constant used to make the transition from volume measure to capacity measure. Cf. Baillet, *PMA* (1892), 35, where it is suggested that the constant is equal to the number of Palestinian *kor* measures in a cubic cubit.

4.5 c. *P.Akhmîm* § 2. The capacity measure of a rectangular granary

In *P.Akhmîm* § 2, the capacity measure of a rectangular 'granary' with given length a, width b, and height h is computed as follows

$C = a \cdot b \cdot h \cdot c$, where $c = 3\ \overline{4}\ \overline{8}\ (= 27/8)$.

Baillet (*op. cit.*), 64, explains the constant $c = 3\ \overline{4}\ \overline{8}$ as the number of *artabs* in a cubic cubit. This means that the Greek-Egyptian capacity measure *artab* was equal to a *cubic foot*, a foot being equal to 2/3 of a cubit.[37]

Parallels to *P.Akhmîm* § 2 are the exercises in the hieratic ***P.Rhind*** §§ **13-14** (*DMP* ## 44-46) (Sec. 2.1 d above), where the volumes of both round and square granaries are converted to capacity measure by use of the rule that 1 *khar* = 2/3 cubic cubit. An OB parallel is the exercise **BM 96954+BM 102366+SÉ 93 § 1** (Sec. 2.2 d above; Sec. 4.8 e below), where the volume of a ridge pyramid ('a granary') is converted to capacity measure by use of the equation 1 volume-šar = 90 gur of 300 sila.

4.5 d. *P.Akhmîm* §§ 3, 5. Unequal sharing, and division exercises

The exercises in *P.Akhmîm* § 3, with the theme "unequal sharing", have a parallel in ***P.Rhind* § 11 b** (# 63), where 700 loaves are shared by 4 men in the proportions 3", 2', 3', 4' (see Sec. 2.1 a). An OB parallel is **AO 8862 § 2 b** (Neugebauer, *MKT 1* (1935), 112), where 4 men carrying bricks share the work in the proportions 7, 11, 13, 14.

Hieratic and OB direct parallels to *P.Akhmîm* § 5 are discussed in Sec. 2.1 b above (iterated division exercises). There is also an interesting Late Babylonian parallel text, the fragment BM 34800 (Fig. 2.1.2).

4.5 e. *P.Akhmîm* §§ 4-10. Examples of counting with fractions

The main theme of *P.Akhmîm* is computation with fractions. This is particularly evident in § 4 (subtraction of fractions), §§ 7-8 (multiplication of fractions), and §§ 9-10 (division of fractions). Fractions are nominally of the traditional Egyptian type, that is *sums of parts*. However, in nearly all operations with fractions in the text, the preferred method is to convert the given sums of parts into *binomial fractions* (see above, Sec. 3.1 c), operating with those binomial fractions using the methods systematically displayed in ***P.BM 10520* § 5** (see above, Sec. 3.3 e), and then converting the

---

[37]. This explanation of the constant $3\ \overline{4}\ \overline{8}$ was later definitely confirmed by Shelton, *ZPE* 42 (1981).

resulting binomial fractions back to sums of parts. Ingenious methods for the unnecessary last step, the conversion of given binomial fractions into sums of parts, are demonstrated in § 6 (decomposition of fractions). See Baillet, *PMA* (1892), 32-62, for a very thorough and enlightening analysis of the nature of the various methods used in *P.Akhmîm* for computations with fractions.

An interesting element of the terminology in *P.Akhmîm* is the following phrase, repeatedly used to describe the conversion of a given sum of parts into an equivalent *binomial fraction*:

ἐν ποίᾳ ψήφῳ ταῦτα τῶν *m* τὸ *n*.   In which table(?) is this? Of *m* the $\bar{n}$.

Here, the term used for a binomial fraction is, except for the word order, 'the *n*-th part of *m*' or simply '$\bar{n}$ of *m*'. Here *m* is either an integer or a "quasi-integer", the latter term meaning an integer plus one or several basic fractions or parts. The terminology reveals how the author of *P.Akhmîm* was thinking of binomial fractions. It is clear that '$\bar{n}$ of *m*' does *not* mean *m/n*, that is '*m* divided by *n*'. It is equally clear that the term does *not* mean '*m* times $\bar{n}$'. Apparently, there can only be *one* n-th part, but that n-th part can be taken of an arbitrary integer or quasi-integer. The surprising circumstance that in exercise # 16 the 22nd part is called 'of 1 the $\overline{22}$', instead of just $\overline{22}$, indicates that the scribe thought of fractions *primarily* as binomial fractions, not as parts or sums of parts!

The most impressive example of a calculation with fractions in *P.Akhmîm* is the subtraction exercise # 12 (§ 4 e), where the student is instructed how to compute the difference

3"– *a*, with   *a* = $\overline{10}$ $\overline{11}$ $\overline{20}$ $\overline{22}$ $\overline{30}$ $\overline{33}$ $\overline{40}$ $\overline{44}$ $\overline{50}$ $\overline{55}$ $\overline{60}$ $\overline{66}$ $\overline{70}$ $\overline{77}$ $\overline{88}$ $\overline{90}$ $\overline{99}$ $\overline{100}$ $\overline{110}$.

The first step is to convert *a* from a sum of parts into a binomial fraction. Only the result is recorded in the text, but a reasonable conjecture is that *a* was converted into a binomial fraction as follows:

   a · 770 = 77 + 70 + 38 2' + 35 + 25 3" + 23 3' + 19 4' + 17 2' + 15 3' $\overline{15}$ + 14
             + 12 6" + 11 3" + 11 + 10 + 8 2' 4' + 8 2' $\overline{18}$ + 7 3" $\overline{9}$ + 7 2' $\overline{5}$ + 7
         = 414 + 5 · 2' + 2 · 3' + 3 · 3" + 2 · 4' + $\overline{5}$ + 6" $\overline{9}$ $\overline{15}$ $\overline{18}$ = 420 6" $\overline{10}$.

Hence

   a = $\overline{770}$ of 420 6" $\overline{10}$ = $\overline{110}$ of 60 $\overline{10}$ $\overline{30}$.

After *a*, somehow, has been converted into a binomial fraction, the text continues like this:

3" of 110 = 73 3',   and   73 3' − 60 $\overline{10}$ $\overline{30}$ = 13 $\overline{5}$.   Hence   3"− a = $\overline{110}$ of 13 $\overline{5}$.

Next, the unwanted term $\overline{5}$ is removed through raising the terms (cf. the discussion of the demotic exercise *P.BM 10520* § 5 a in 3.3 d above):

$\overline{110}$ of 13 $\overline{5}$ = $\overline{550}$ of 66.

This binomial fraction could have been the answer. However, bound by the tradition, the author of the text converts the binomial fraction to a sum of parts:

$\overline{550}$ of 66 = $\overline{550}$ of (55 + 11) = $\overline{10}$ $\overline{50}$.

Remark: The elegance of the repetitive construction of a is spoiled by the absence of the term $\overline{80}$. If the term $\overline{80}$ had not been left out, the result of the computation would have been instead

$\overline{10}$ $\overline{50}$ − $\overline{80}$ = $\overline{10}$ $\overline{200}$ $\overline{400}$.

As a curiosity, it can be mentioned that a similar repetitive construction of the data for a problem can be found in the OB combined market rate problem **VAT 7530 § 6** (Friberg, *MCTSC* (2005), Sec. 7.2), where 10 given market rates are

1 ma.na, 1 ma.na 10 gín, 2 ma.na, 2 3' ma.na, 3 ma.na, 3 2' ma.na,
4 ma.na, 4 3" ma.na, 5 ma.na, 5 6" ma.na
= 1, 1;10, 2, 2;20, 3, 3;30, 4, 4;40, 5, 5;50 minas (per silver shekel).

4.5 f. *P.Akhmîm* § 9. Prices and market rates

P.Akhmîm § 9 contains three exercises dealing with *prices and market rates* (inverted prices). Thus, for instance, in # 33 (§ 9 c), it is stated that the *price for 100 artabs* of wheat or barley is 7 $\overline{7}$ gold *nomismatia*. Then it is asked what the corresponding market rate will be, that is, how many *artabs* one will get for 1 *nomismation*. The answer is given in an unfinished form, as '700 divided by 50'. It is not clear why the answer was not given explicitly as '14 artabs for 1 nomismation'.

In # 35 (§ 9 e), the inverted kind of problem is stated: Given that the market rate for wheat or barley is 15 2' $\overline{4}$ *artabs* for 1 *nomismation*, what is the price for 100 *artabs*? The answer is again given in an unfinished form, as '400 divided by 63', a result that could easily have been replaced by, for instance, 6 and $\overline{63}$ of 22.

4.6. *WT.Michael. 62* (7th? C. CE). Prices and Market Rates

Michael. 62 (Crawford, *Aeg.* 33 (1953)) is a wooden tablet inscribed on the obverse with tables of fractions and on the reverse with 8 mathematical exercises, all about prices and market rates. As in *P.Akhmîm*, the tables of fractions begin with fractions of 6,000 (the number of *drachmas* in a talent). Nevertheless, the exercises count in terms of *nomismatia* and *keratia* (1 *nom.* = 24 *ker.*). Here is, for instance, the text of *Michael. 62* # 2:

> 100 *artabs* of wheat were bought at 8 3' *art.* per *nomismation* minus 4 *keratia*.
> They were sold at 7 2' *art.* per full-weight *nom.*
> I want to know how many *nom.* (3 3') were made in profit,
> and how many *art.* (25) were made in profit.

There is no solution procedure, only the answers interpolated into the text of the question. However, since 8 3' = 3' · 25, and 1 *nom.* – 4 *ker.* = 20 *ker.* = 6" *nom.*, those answers may have been obtained by arguing as follows:

> The wheat was bought at a market rate of 3' of 25 *art.* for 6" *nom.*,
> which is 3' of 1 $\overline{5}$ of 25 = 10 *art.* for 1 *nom.*
> The corresponding buying price for 100 *art.* is 10 *nom.*
> The wheat was then sold at a market rate of 7 2' *art.* = 2' of 15 *art.* per *nom.*
> The corresponding selling price for 100 art. is 2 · 100 · $\overline{15}$ = 13 3' *nom.*
> Hence, if all the 100 *art.* were sold, the profit would be (13 3' – 10) = 3 3' *nom.*
> However, if only enough was sold to recover the money spent,
> the profit would be (100 – 75) = 25 *art.*

Note how the price difference gives the profit in *nomismatia*, while the market rate difference gives the profit in *artabs*.

An interesting OB parallel is offered by some of the problems on the small clay tablet **YBC 4698** (Fig. 2.1.17) with its 17 mixed problems, all, like the problems on *Michael. 62*, about prices and market rates. Here is the text of 6 related problems:

YBC 4698 § 4 (## 6-11).

# 6	30 še gur /
	1_g še gur.ta šàm-*ma* / 4_{bg} še.ta búr.ra /
	kù.diri en.nam /
	7 2' gìn kù.babbar diri
# 7	sag kù.bi en.nam

# 8	1$_g$ gur ì.giš / 1$_{bn}$.ta šàm-*ma* / 8 sìla.ta búr.ra / kù.diri en.nam / 7 2' gìn kù.babbar diri
# 9	1$_g$ gur ì.giš / *i-na* šàm 1 gìn / 2 sìla sì / 7 2' gìn kù.babbar diri / en.nam šàm-*ma* / en.nam búr.ra 1$_{bn}$ šàm-*ma* / 8 sìla búr.ra
# 10	1$_g$ gur ì.giš / 9 sìla.ta šàm-*ma* / 7 2' sìla búr.ra / kù.diri en.nam / 6 3" gìn kù.diri
# 11	1$_g$ gur ì.giš / *i-na* šàm 1 gìn / 1 2' sìla sì / 6 3" gìn kù.diri / en.nam šàm-*ma* en.nam búr.ra / 9 sìla šàm-*ma* 7 2' sìla búr.ra

6   30 gur of barley.
       At 1 gur each I bought, then at 4 barig each I sold.
       The excess of silver was what?
       7 1/2 shekels of silver is the excess.

7   The head (initial amount) of the silver was what?

8   1 gur of oil.
       At 1(bán) each I bought, then at 8 sìla each I sold.
       The silver was what?
       7 1/2 shekels of silver was the excess.

9   1 gur of oil.
       From the buying (rate) for 1 shekel 2 sìla was given.
       7 1/2 shekels of silver was the excess.
       What did I buy (at), then what did I sell (at)?
       (At) 1(bán) I bought, then (at) 8 sìla I sold.

10   1 gur of oil.
       (At) 9 sìla I bought, then (at) 7 1/2 sìla I sold.
       What was the silver?
       6 2/3 shekels was the excess.

#11 1 gur of oil.
From the buying (rate) for 1 shekel 1 1/2 sila was given.
6 2/3 shekels of silver was the excess.
What did I buy (at), then what did I sell (at)?
(At) 9 sila I bought, (at) 7 1/2 sila I sold.

When YBC 4698 was first published, in Neugebauer, *MKT 3* (1937), pl. 5, the problem type and its terminology were both unknown. The first correct explanation was offered in Friberg, *Survey* (1982), 57. The solution procedures, which are missing in the text, are reconstructed below.

In **YBC 4698 # 6**, barley of the capacity measure $C = 30$ gur $(= 2\ 30\ 00$ sila$)$ is bought at the standard market rate $m_1 = 1$ gur $(5\ 00$ sila$)$ per shekel (gin) and then sold at the lower market rate $m_2 = 4$ barig $(4\ 00$ sila$)$ per shekel. The corresponding unit prices are, respectively, $p_1 =;00\ 12$ and $p_2 = ;00\ 15$ shekels per sila. Therefore, the *profit* made is

$P = C \cdot (p_2 - p_1) = 2\ 30\ 00$ sila $\cdot\ ;00\ 03$ shekel/sila $= 7;30$ shekels (of silver).

In **YBC 4698 # 7**, the question is what the initially invested silver was. The answer is, of course,

$S_1 = C \cdot p_1 = 2\ 30\ 00$ sila $\cdot\ ;00\ 12$ shekel/sila $= 30$ shekels $= 1/2$ mina.

In **YBC 4698 # 8**, oil of the capacity measure $C = 1$ gur $(= 5\ 00$ sila$)$ is bought at the normal market rate $m_1 = 1$ bán $(10$ sila$)$ per shekel and then sold at the lower market rate $m_2 = 8$ sila per shekel. The corresponding unit prices are, respectively, $p_1 = ;06$ and $p_2 = ;07\ 30$ shekels per sila. Therefore, the *profit* made is

$P = C \cdot (p_2 - p_1) = 5\ 00$ sila $\cdot\ ;01\ 30$ shekel/sila $= 7;30$ shekels (of silver).

YBC 4698 # 9 is an inverted variant of the problem in # 8. This time, it is known that oil of the capacity measure $C = 1$ gur $(= 5\ 00$ sila$)$ is bought at a high market rate and sold at a lower market rate. The difference between the two market rates and the profit are given as $m_1 - m_2 = 2$ sila per shekel, and $P = 7\ 1/2$ shekels of silver, respectively. The unknown values m_1 and m_2 of the buying and selling market rates have to be computed as the solutions to the following system of equations:

$(m_1 - m_2) = 2$ sila/shekel,
$C \cdot (p_2 - p_1) = P = 7;30$ shekels, where p_1 and p_2 are the reciprocals of m_1 and m_2.

The way in which the problem was solved is indicated by more complete

OB texts on the same theme (see below). Thus, the solution procedure is based on the observation that

$P \cdot m_1 \cdot m_2 = C \cdot (p_2 - p_1) \cdot m_1 \cdot m_2 = C \cdot (m_1 - m_2)$.

Therefore, the values of the market rates m_1 and m_2 can be found as the solutions to the following linear-rectangular system of equations of an OB standard type:

$m_1 \cdot m_2 = C \cdot (m_1 - m_2)/P = 5\ 00$ sìla · 2 sìla/sh. / 7;30 sh. = 1 20 sq. (sìla/sh.),
$(m_1 - m_2) = 2$ sìla/sh.

Without much trouble, the solution can be shown to be

$m_1 = 9 + 1 = 10$ sìla/shekel, $m_2 = 9 - 1 = 8$ sìla/shekel.

The exercises **YBC 4698 ## 10-11** are identical with ## 8-9, except for slightly different data.

A closely related OB text is the Susa text ***TMS* 13** (Bruins and Rutten, *TMS* (1961); Høyrup, *LWS* (2002), 206). It is a perfectly preserved text, with an explicit solution procedure. However, only the question is reproduced below.

TMS 13, lines 1-4.

> 2_g gur 2_{bg} barig 5 bán ì.giš šám/
> *i-na* šám 1 gín kù.babbar /
> 4 sìla.ta.àm ì.giš *ak-ší-it-ma* /
> 3" ma.na kù.babbar *ne-me-la a-mu-ur*
> *ki ma-ṣí* / *aš-šà-am ù ki ma-ṣí ap-šu-úr*

2 gur 2 barig 5 bán of oil I bought.
From the buying (rate) for 1 shekel of silver
4 sìla each of oil I cut away, then
2/3 mina of silver as profit I saw.
(At) how much did I buy, and (at) how much did I sell?

Here, with the same notations as above,

$C = 2$ gur 2 barig 5 bán = 12 50 sìla,
$(m_1 - m_2) = 4$ sìla/shekel,
$P = 2/3$ ma.na = 40 shekels.

Therefore, the equations for the unknown market rates are, as above,[38] but with the new data,

$m_1 \cdot m_2 = C \cdot (m_1 - m_2)/P = 12\ 50$ sìla $\cdot$ 4 sìla/sh. / 40 sh. = 1 17 sq. (sìla/sh.), $(m_1 - m_2) = 4$ sìla/sh.

In lines 5-17 of *TMS* 13, the solution is shown to be

$m_1 = 9 + 2 = 11$ sìla/shekel, $m_2 = 9 - 2 = 7$ sìla/shekel.

A third related, but not quite parallel, OB text is **MLC 1842** (Neugebauer/Sachs, *MCT* (1945), 106). This is a fragment of a "northern" single problem text, in which the question is perfectly preserved, but only the beginning of the solution procedure. Only the question is reproduced below.

MLC 1842, lines 1-7.

> ganba.e *i-li-i-ma* 30 še gur *a-ša-am* /
> ganba *iš-pí-il-ma* 30 še gur *a-ša-am* /
> *ma-ḫi-ri-ia ak-mu-ur-ma* 9 /
> kù.babbar *ma-ḫi-ri-ia ak-mu-ur-ma* /
> 1 ma.na 7 2' gín /
> ganba *a-ša-am ù ki-ia / ap-šu-ur*

The market rate increased, 30 gur of barley I bought,
The market rate decreased, 30 gur of barley I bought.
The market rates I added together, 9.
The silver of the market rates I added together, then
1 mina 7 1/2 shekels.
The market rate I bought (at), and how much I sold (at)?

In this exercise, two different lots of barley are purchased, both of the same size, but at different market rates. Instead of the difference of the two market rates as in the examples above, their sum is given, and instead of the difference between the silver invested and the silver returned, the sum of the amounts invested is given. Thus, with the same notations as above, the given values are

$C = 30$ gur $= 2\ 30\ 00$ sìla, $(m_1 + m_2) = 9\ 00$ sìla/sh., $S = C \cdot (p_1 + p_2) = 1\ 07;30$ sh.

This time, the solution procedure is based on the observation that

$S \cdot m_1 \cdot m_2 = C \cdot (p_2 + p_1) \cdot m_1 \cdot m_1 = C \cdot (m_1 + m_2)$.

Therefore, with the values of C, $m_1 + m_2$, and S, given, m_1 and m_2 can be found as the solutions to the following linear-rectangular system of equa-

---

38. Høyrup, (*op. cit.*), 206 ff, proposed a different explanation for the solution procedure.

tions of an OB standard type:

$m_1 \cdot m_2 = C \cdot (m_1 + m_2)/S$ = 2 30 00 sìla · 9 00 sìla/sh. / 1 07;30 sh.s
= 20 00 00 sq. (sìla/sh.),
$(m_1 + m_2)$ = 9 00 sìla/sh.

Without much trouble, the solution can be shown to be[39]

m_1 = 4 30 + 30 = 5 00 sìla/shekel = 1 gur/shekel,
m_2 = 4 30 – 30 = 4 00 sìla/shekel = 4 barig/shekel.

Note that these are the same market rates for barley as in YBC 4698 # 6.

4.7. Problems for Right Triangles and Quadrilaterals

4.7 a. *P.Genève 259*. Problems for the sides of a right triangle

P.Genève 259 (2nd c. CE; Rudhart, *MH* 35 (1978), Sesiano, *MH* 43 (1986), *MH* 56 (1999)) is a papyrus fragment with three geometric exercises. (A photo of the fragment was published in Sesiano *MH* 56 (1999).) Preserved drawings of right triangles illustrate two of the exercises.

In ***P.Gen. inv. 259* # 1**, a right triangle has the perpendicular a = 3 (feet) and the hypotenuse c = 5 (feet). A trivial application of the (*Babylonian*) *diagonal rule* gives the base b:

sq. b = sq. c – sq. a = 25 – 9 = 16, b = sqr. 16 = 4 (feet).

In ***P.Gen. inv. 259* # 2**, the sum $c + a$ = 8 (feet), and b = 4 (feet) are given. Hence, a and c can be computed as the solutions to the following quadratic-linear system of equations:

sq. c – sq. a = sq. b = 4, $c + a$ = 8.

The solution given in the text is, essentially, of the following form:

a = {$(c + a)$ – sq. $b /(c + a)$}/2 = (8 – 16/8)/2 = 3 (feet),
c = $(c + a) - a$ = 8 – 3 = 5 (feet).

In ***P.Gen. inv. 259* # 3**, finally, the sum $a + b$ = 17 (feet), and c = [13] (feet) are given. Hence, a and b can be computed as the solutions to the following quadratic-linear system of equations:

sq. a + sq. b = (sq. c =) 169, $a + b$ = 17.

---

39. The explanation proposed in Neugebauer and Sachs, *MCT* (1945), 106, is incorrect and does not lead to this solution.

Only a small part of the text is preserved. Nevertheless, as shown in Sesiano *MH* 56 (1999), it is likely that the lost solution was of the following form (in modern notations):

sq. $(b - a) = 2 \cdot$ (sq. a + sq. b) – sq. $(a + b) = 338 - [289] = [49]$,
$b - a = [7]$

(it is silently assumed that b is greater than a), and

$a = \{(a + b) - (b - a)\}/2 = (17 - 7)/2 = [5]$ (feet),
$b = (b - a) - a = 17 - 5 = [12]$ (feet).

An updated survey of the many known Old and Late Babylonian applications of the diagonal rule can be found in Friberg, *MCTSC* (2005), App. 7, section A7 f. A particularly important example is the Seleucid theme text **BM 34568** (Neugebauer, *MKT 3* (1937), 14 ff; Høyrup, *LWS* (2002), 391-399), with 18 mixed problems for the sides and the diagonal of a rectangle (or a right triangle). Two of these (## 4 and 10) are parallels to the interesting exercises *P.Gen. inv. 259* ## 2-3. However, it is interesting to note that the solution procedure for BM 34568 # 4 is slightly different from the solution procedure for *P.Gen. inv. 259* # 2, and that the solution procedure for BM 34568 # 10 is completely different from the solution procedure for *P.Gen. inv. 259* # 3.

In the demotic ***P.Cairo*** § 8, three variants of the "pole-against-a-wall" problem are also closely related to the three exercises in *P.Gen. inv. 259*. As mentioned in Sec. 3.1 b above, the Seleucid exercise BM 34568 # 12 is of the same type as *P.Cairo* § 8 g, while the OB exercise BM 85169 # 9 is essentially identical with *P.Cairo* § 8 d. For some reason, however, no other OB parallels are known to any one of the exercises in the Seleucid BM 34568 or to the exercises in the Greek-Egyptian *P.Gen. inv. 259*.

4.7 b. *P.Chicago litt. 3*. Three types of non-symmetric trapezoids

P.Chicago litt. 3 = *P.Ayer* (1st c. CE?; Goodspeed, *AJP* 19 (1898)) is a large papyrus fragment with parts of four geometric exercises, illustrated by three preserved drawings. According to Goodspeed,

> "it (is) not impossible that we have in this fragment one of those early mathematical works of whose materials Heron later became the organizer and compiler; in other words, the work of which this papyrus was a copy, if not itself one of Heron's sources, may fairly represent the character of the sources he had and used."

P.Chic. litt. 3 # 1 is mostly destroyed. Only the last two lines of the text and a drawing of a *symmetric trapezoid* are preserved. Apparently the set task was to find the area of the trapezoid when the lengths of its sides are given, with the parallel sides equal to 2 and 14 (*schoinia*), and with the sloping sides both equal to 10. The height was then found to be 8, and the area was computed as the sum of the areas of two right triangles, each equal to 24 (*arouras*), and the area of a central rectangle, equal to 16 (*ar.*). Thus, it is clear that the trapezoid was thought of as two right triangles with the sides 2 · (5, 4, 3) joined to a central rectangle with the sides 2 and 8.

P.Chic. litt. 3 # 2 is a well preserved exercise, in which the sides of a *non-symmetric trapezoid* are given, with 2 and 16 for the parallel sides, 13 and 15 for the sloping sides. The associated drawing is lost, but can be assumed to have looked more or less like the drawing in Fig. 4.7.1, left.

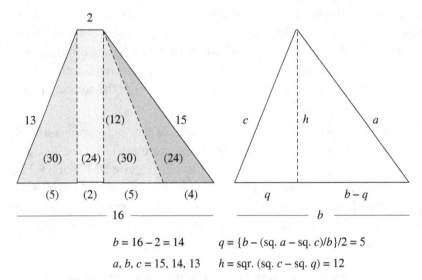

$b = 16 - 2 = 14$ $q = \{b - (\text{sq. } a - \text{sq. } c)/b\}/2 = 5$

$a, b, c = 15, 14, 13$ $h = \text{sqr. } (\text{sq. } c - \text{sq. } q) = 12$

Fig. 4.7.1. *P.Chic. litt. 3 # 2*. The area of a non-symmetric trapezoid.

The solution procedure begins by computing the length of the segment q (Fig. 4.7.1, right) as follows:

sq. 15 – sq. 13 = 56, 16 – 2 = 14, 56/14 = 4, 14 – 4 = 10, 10/2 = 5 = q.

After that, the height $h = 12$ is computed through a straightforward application of the diagonal rule. Finally, the area of the trapezoid is computed as the sum of the areas of the central rectangle, two flanking and equal

4.7. Problems for Right Triangles and Quadrilaterals

right triangles, and an 'obtuse-angled' triangle.

Apparently, the rule for the computation of the segment q was derived as follows. First q was shown to be the solution to a *quadratic equation* of the following type:

sq. $(b - q)$ – sq. q = sq. a – sq. c, where $a = 15$, $b = 16 - 2 = 14$, $c = 13$.

This quadratic equation was then reduced, by use of the conjugate rule,[40] to a *linear equation* for the unknown q:

$b \cdot (b - 2q)$ = sq. a – sq. c.

The solution to this linear equation was then easily computed. It is, of course,

$q = \{b - (\text{sq. } a - \text{sq. } c)/b\}/2$,

which is precisely the form in which q is computed in *P.Chic. litt. 3* # 2. (Cf. the discussion in Sec. 3.7 c above of the OB mathematical text VAT 7531 # 4, in particular Fig. 3.7.10.)[41]

P.Chic. litt. 3 # 3 is also well preserved, including a well preserved but

---

40. There is a simple *geometric* derivation of the conjugate rule. The difference sq. m – sq. n can be interpreted, for instance, as the area of the "square corner" between a square of side m and a smaller square of side n, situated in one corner of the larger square. It is easy to check that the square corner has the thickness $m - n$ and the average length $m + n$. It is equally simple to check that the square corner has the area $(m + n) \cdot (m - n)$. Since the area is also equal to sq. m – sq. n, it follows that sq. m – sq. $n = (m + n) \cdot (m - n)$.

41. Cf. the discussion in Heath, *HGM 2* (1921 (1981)), 320 and in Høyrup, *BSSM* 17 (1997) and *ANW* 7 (1997) of Hero's *Metrica* I: 5-6. In particular, the latter paper, which is very interesting, contains a detailed discussion of *three different methods* used in Hero's *Metrica* as well as in Arabic and other medieval mathematical treatises in order to compute the area of unsymmetric triangles or trapezoids with given sides. Høyrup, apparently unaware of the appearance of unsymmetric trapezoids in the Old Babylonian VAT 7531 and in the Greek-Egyptian mathematical papyri *P.Chic. litt. 3* = *P.Ayer* and *P.Cornell inv.69* (see below), assumed that all such methods used to compute the heights of unsymmetric triangles or trapezoids were based on Euclid's *Elements* II.6, II. 8, and II. 13. Høyrup's third method, which he calls "the algebraic alternative" is, essentially, the method used in *P.Chic. litt. 3* and *P.Cornell inv.69*, although Høyrup assumes that it was obtained by use of *Elements* II.6 rather than by use of the conjugate rule as proposed here. Moreover, in the examples considered by Høyrup, that third method is used only in the case of acute-angled triangles. From this observation, Høyrup draws the conclusion (possibly incorrect) that "the practical tradition knew the principle in its 'algebraic' form already before the Greeks, but that it had only applied it to internal heights".

quite inaccurate drawing. It is concerned with the computation of the area of what is called a 'parallelogram', actually an *oblique trapezoid*, with given sides, 10 and 6 for the two parallel sides, 13 and 15 for the two sloping sides (see Fig. 4.7.2, left). The solution procedure in this exercise begins, as in the preceding example, with the computation of q (Fig. 4.7.2, right), a segment of the base which in *P.Chic. litt. 3 # 3* is called 'the base of the right-angled triangle':

sq. 15 − sq. 13 = 56,   10 − 6 = 4,   56/4 = 14,   14 − 4 = 10,   10/2 = 5 = q.

The height $h = 12$ is then computed through a straightforward application of the diagonal rule, and the area of the trapezoid is computed as the sum of the areas of two right triangles and a narrow central rectangle. The computed values are explicitly recorded in the drawing of the trapezoid.

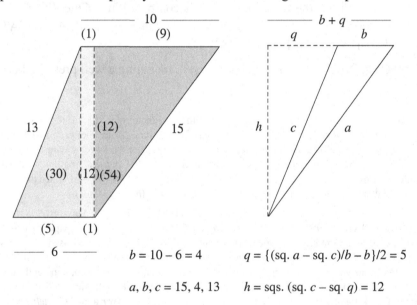

$b = 10 − 6 = 4$

$q = \{(\text{sq. } a − \text{sq. } c)/b − b\}/2 = 5$

$a, b, c = 15, 4, 13$

$h = \text{sqs. }(\text{sq. } c − \text{sq. } q) = 12$

Fig. 4.7.2. *P.Chic. litt. 3 # 3*. The area of a non-symmetric oblique trapezoid.

Apparently, the rule for the computation of the segment q was derived as follows. First q was shown to be the solution to a *quadratic equation* of the following type:

sq. $(b + q)$ − sq. q = sq. a − sq. c,   where $a = 15, b = 10 − 6 = 4, c = 13$.

This quadratic equation was then reduced, by use of the conjugate rule, to

4.7. Problems for Right Triangles and Quadrilaterals 225

a *linear equation*:

$b \cdot (b + 2q) = \text{sq. } a - \text{sq. } c.$

The solution to this linear equation is easily computed. It is, of course,

$q = \{(\text{sq. } a - \text{sq. } c)/b - b\}/2,$

which is precisely the form in which q is computed in *P.Chic. litt. 3 # 3*.

Note that the right and oblique trapezoids in *P.Chic. litt. 3 ## 2-3* are closely related. One is constructed as a central rectangle of height 12, to which is joined a right triangle with the sides 13, 12, 5 on the left and a right triangle with the sides 15, 12, 9 = 3 · (5, 4, 3) on the right. The other is constructed as another central rectangle of height 12, to which is joined again a right triangle with the sides 13, 12, 5 on the left, but an *upside-down* right triangle with the sides 15, 12, 9 on the right. Therefore, the oblique trapezoid in *P.Chic. litt. 3 # 3* is a *simple variant* of the right trapezoid in *P.Chic. litt. 3 # 2*. Thus, even if there are known OB precursors of the right trapezoid in *P.Chic. litt. 3 # 2* but not of the oblique trapezoid in *P.Chic. litt. 3 # 3*, that is not a fact of great significance. Note, by the way, that the symmetric trapezoid in *P.Chic. litt. 3 # 1* has a Late Babylonian precursor in **VAT 7848 § 3** (Friberg, *BaM* 28 (1997), Fig. 6.4).

P.Chic. litt. 3 # 4, finally, is concerned with a rhomb, constructed by joining four right triangles together, all four with the sides 10, 8, 6 = 2 · (5, 4, 3). Since both the sides of the rhombus and one of its diagonals are given, it is a trivial matter to compute the other diameter and the area.

In the present discussion of possible relations between Egyptian and Babylonian mathematical texts, *P.Chic. litt. 3* is particularly interesting, for the following reason: Although the topic of the text is close to the topics of known Old and Late Babylonian mathematical texts, and although the solution algorithms are of the same general character as typical Babylonian solution algorithms, the statements of the problems have some surprising features. Here is the text of the three preserved statements:

2: "If there is given a scalene trapezoid such as the one drawn below."
3: "If there is given a parallelogram such as the one drawn below."
4: "If there is given a rhomb such as the one drawn below."

Thus, the data for the exercises are not given explicitly but are to be obtained from inspection of the associated drawings. This, on the other hand, is not as simple as it may sound, since the drawings, just like many drawings illustrating Old and Late Babylonian mathematical exercises, show not just the given numbers but also the numbers computed in the course of the solution procedure.

The terms 'scalene trapezoid' (τραπέζιον σκαληνόν), 'parallelogram' (παραλληλόγραμμον), and 'rhomb' ('ρόμβοc) used in these statements, and also the terms 'right-angled (ὀρθογώνιοc), 'obtuse-angled' (ἀμβλυγώνιοc), and 'oblong, of different lengths = rectangular' (ἑτερόμηκεc), have no known Babylonian counterparts.[42] The presence of these terms may possibly indicate that the author of this text was in some way acquainted with high level Greek mathematics (cf. the discussion of *Elements* I, Definitions 20-22, in Heath, *TBE 1* (1956), 154,187-190), although the erroneous use of the word 'parallelogram' in *P.Chic. litt. 3* # 3 may indicate that this acquaintance was not profound.

It is interesting that, in spite of various similarities between *P.Chic. litt. 3* and *P.Gen. inv. 259* (Sec. 4.7 a), the former expresses lengths (implicitly) in *schoinia*, while the latter expresses lengths in 'feet'. The reason for the difference is not known to the present author, but it is interesting that the *schoinion* of 100 cubits may have had a counterpart in Late Babylonian texts. This is shown, for instance, by the mathematical exercise W 23291 § 2 b.2 (Friberg, *BaM* 28 (1997), 277), where the "seed measure" for a square with the side 100 cubits is found to be 1 barig. Note, by the way, that measuring lengths in *schoinia* and areas in *arouras* is suitable for actual fields, while measuring lengths in feet (and areas in square feet) is suitable for smaller figures.

4.7 c. *P.Cornell 69*. Non-symmetric trapezoids, and a birectangle

P.Cornell 69 (2nd c. CE; Bülow-Jacobsen and Taisbak, *FPR* (2003)) is a papyrus fragment with parts preserved of three geometric exercises, illustrated by two preserved drawings and written in two columns. Of the first column, only the rightmost third is preserved, of the second column the left half. The lower part of the fragment is lost, as well. See Fig. 4.7.6.

Because the fragment is so extensively damaged, it is impossible to

---

42. Remember that the concept of angles played no role in Babylonian mathematics.

reconstruct much of the original text. On the other hand, the two preserved drawings in the second column make it fairly clear what exercises ## 2-3 were about, and the preserved numbers in what remains of # 1 make it quite clear what that exercise was about, too.

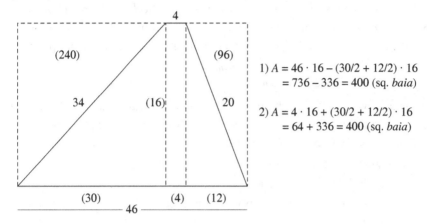

Fig. 4.7.3. *P.Cornell* 69 # 1. Computation of the area of a right unsymmetric trapezoid.

***P.Cornell* 69 # 1** deals with an unsymmetric trapezoid of the same kind as *P.Chic. litt. 3* # 2 (Fig. 4.7.1 above). The two parallel sides are 4 and 46 (*baia*), the left sloping side (called 'south') is 20, and the right sloping side ('north') is 34. One of the segments of the base, p, is computed as follows:

[46] − 4 = [42], [sq. 34 − sq. 20] = 1156 − [400] = 756 (sq. *baia*),
[756/42] = 18, [18 + 42 = 60], [60/2 = 30 = p].

Next the height h is computed by use of the diagonal rule,

[sq. 34 − sq. 30] = 256, sqr. 256 = 16 = h.

Although the continuation is much less clear, apparently the area A of the trapezoid is computed in two different ways, first as

$A = [46 \cdot 16 − (30/2 + 12/2) \cdot 16] = 736 − 336 = 400$,

and then as

$A = 4 \cdot 16 + (30/2 + 12/2) \cdot 16 = 64 + 336 = 400$.

The length units used in this text are multiples of the *baion* 'palm leave'. According to Shelton, *ZPE* 42 (1981), a *baion* was 1/2 *hamma*, and a *hamma* was 1/8 *schoinion*. Since a 'land measuring' *aroura* contained 96 cubits (see Sec. 4.1 above, fn. 2), it follows that a *hamma* contained 12 cubits and a *baion* 6 cubits. This is the same relation as between three common OB units of length. Indeed, 1 ninda contained 12 cubits and 1 'reed' contained 6 cubits.[43] (Interestingly, another name for the *baion* was *kalamos* 'reed'. See Bell, *GP 5* (1917), 161.) The mentioned numerical relations between the *schoinion*, the *hamma*, and the *baion* were proved by Shelton (*op. cit.*) through an analysis of the computations of the areas of some fields in the text ***VBP IV 92***. Here is one of Shelton's examples:

> South 41 *baia*, north 49 *baia*, east 72 *baia*, west 69 *baia*.
> Area 12 *arouras* and 25 $\overline{8}$ *hammata*.

If the area was computed by use of the *quadrilateral area rule* (Sec. 3.3 f above), and if the mentioned numerical relations are correct, then the result can have been be obtained as follows:

> $(41 + 49)/2$ *b.* · $(72 + 69)/2$ *b.* = 45 · 70 2' sq. *b.* = 22 2' *h.* · 35 4' *h.*
> = 7,938 $\overline{8}$ sq. *h.* = 12 *a.* 25 $\overline{8}$ sq. *h.*

(Note that in the text the same notation is used for a square *hamma* as for a linear *hamma*.)

P.Cornell 69 # 2 contains the computation of the area of an oblique unsymmetric trapezoid (Fig. 4.7.4, left). The lengths of the two sloping sides are 15 *baia* (south), 13 *baia* (north), and the lengths of the two parallel sides are 8 *baia* (west) and 4 *baia* (east). The height of the trapezoid is computed by essentially the same method as the height h of the trapezoid in *P.Chic. litt. 3 # 3* (Fig. 4.7.2 above), although in *P.Cornell 69 # 2* the computation of h is based on the intermediate computation of the segment $p = q + b$, not the segment q as in *P.Chic. litt. 3 # 3*. Another difference is that the area in *P.Chic. litt. 3 # 3* is computed as the sum of the areas of the central rectangle and the areas of the two flanking right triangles, while the area in *P.Cornell 69 # 2* is computed as the sum of the areas of two non-right triangles. The reason for this difference is that the trapezoid in *P.Cornell 69 # 2* *cannot* be interpreted as a central rectangle extended on two sides through the addition of right triangles, one of them upside down. Instead, it must be interpreted as a rectangle with the sides 4 and 12, to the left side of which is *added* a triangle with the sides 15, 12, 9, and from the

---

43. For some unknown reason, a reed usually contains 7 cubits in Late Babylonian texts.

4.7. Problems for Right Triangles and Quadrilaterals 229

right side of which is *subtracted* a right triangle with the sides 13, 12, 5!

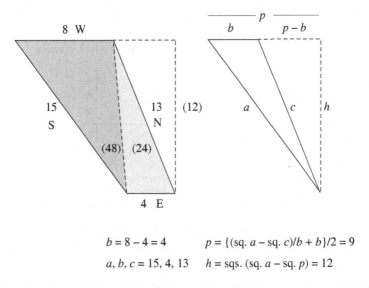

$b = 8 - 4 = 4$ $\quad p = \{(\text{sq. } a - \text{sq. } c)/b + b\}/2 = 9$

$a, b, c = 15, 4, 13$ $\quad h = \text{sqs. }(\text{sq. } a - \text{sq. } p) = 12$

Fig. 4.7.4. *P.Cornell 69 # 2*. Computation of the area of an oblique unsymmetric trapezoid.

P.Cornell 69 # 3 probably begins in col. *ii*, line 22. The associated drawing is perfectly preserved, but the text of the exercise is almost completely destroyed. All that remains of the question is a schematic drawing of the same kind as in ***P.BM 10520 § 7 b*** (Fig. 3.3.3), and in the land survey ostraca ***O.Bodl. ii 1847*** (Sec. 4.1) and ***Theban O. D 12*** (Fig. 3.3.4), plus a few uninformative words. The schematic drawing shows 15 south (left), 5 north (right), 5 east (below), and 15 west (above). What remains of the solution procedure is, essentially, only the phrases 'the 5 in the west upon itself', '[the 15] in the south upon itself ⋯ 225'. In other words, the squares of 5 'in the west' and 15 'in the south' were computed at some stage of the solution procedure.

Strangely enough, the drawing shows a quadrilateral with two sides of length 15 above and to the right, and two sides of length 5 below and to the left. Apparently, the original drawing were replaced by its mirror image at some time when a copy was made of the original text. Maybe the reason is that the original text was a demotic mathematical exercise, written from right to left, and that the direction of the drawing was changed when the direction of writing was changed, as the text was translated into Greek?

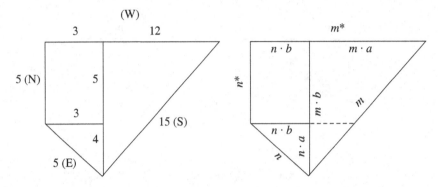

Fig. 4.7.5. P.Cornell 69 # 3. Computation of the area of a birectangle.

It is likely that it was silently understood that the object considered in this exercise is what may be called a "birectangle", a quadrilateral with two opposite right angles, but not with parallel sides as in a rectangle. It is also likely that the sides of the birectangle were given, and that the set task was to find first the height against one of the long sides, then the area. For that purpose, two lines were drawn inside the birectangle, a vertical line from the lower vertex to the upper (west) side, and a horizontal line from the left vertex to that vertical line. In this way the given birectangle was divided into a rectangle and two right triangles. It is not difficult to see that the two right triangles are *similar triangles*. For an OB mathematician, for instance, this would have been obvious because he would intuitively "know" that a line parallel to one of the short sides of a right triangle, like the dotted line in Fig. 4.7.5, right, cuts off a small triangle that is of the same shape as the original triangle. He would also intuitively "know" that the height against the diagonal of a right triangle cuts the right triangle into two small triangles of the same shape as the original triangle. Now, if the birectangle is divided in this way into a rectangle and two right triangles of the same shape, then both triangles must be multiples of some right triangle, say a right triangle *normalized* so that the length of its diagonal is 1. If the sides of this normalized right triangle are called 1, b, a, and if the given diagonals of the two triangles in the birectangle are called m and n, then it is clear that the two small triangles must have the sides m, $m \cdot b$,

4.7. Problems for Right Triangles and Quadrilaterals

$m \cdot a$, and n, $n \cdot b$, $n \cdot a$, respectively (Fig. 4.7.5, right).

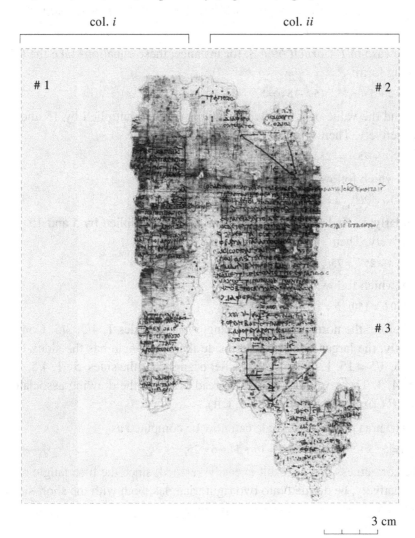

Fig. 4.7.6. *P.Cornell 69, obv.* (The likely original size of the papyrus is shown in grey.)

Let the sides opposite to m and n be called n^* and m^*. Then it is clear that if all the four sides of the birectangle are known, the sides a and b of the normalized right triangle 1, b, a must be, in modern notations, the solutions to the following *system of two linear equations for two unknowns*:

$m \cdot a + n \cdot b = m^*, \quad m \cdot b - n \cdot a = n^*.$

In the case of *P.Cornell 69 # 3*, for instance, these equations take the following form:

$15\, a + 5\, b = 15, \quad 15\, b - 5\, a = 5.$

To find the value of a, the two equations can be multiplied by 15 and 5, respectively. Then

$225\, a + 75\, b = 225, \quad 75\, b - 25\, a = 25,$

from which follows that

$250\, a = 200, \quad a = 4/5.$

Similarly, to find b, the two equations can be multiplied by 5 and 15, respectively, Then

$75\, a + 25\, b = 75, \quad 225\, b - 75\, a = 75,$

from which follows that

$250\, b = 150, \quad b = 3/5.$

Therefore, the normalized right triangle has the sides 1, 4/5, 3/5. Consequently, the larger right triangle inside the birectangle has the sides $15 \cdot$ 1, 4/5, 3/5 = 15, 12, 9, and the smaller triangle has the sides $5 \cdot 1$, 4/5, 3/5 = 5, 4, 3. These values can also be read off from the drawing associated with *P.Cornell 69 # 3* (Fig. 4.7.5, left).

The area of the birectangle can now be computed as

$A = 5 \cdot 3 + 9 \cdot 12/2 + 4 \cdot 3/2 = 15 + 54 + 6 = 75.$

The correctness of this result is easily verified, since the birectangle can, alternatively, be divided into two right triangles, both with the short sides 15 and 5. Therefore, A can be computed as

$A = 2 \cdot 15 \cdot 5/2 = 75.$

In the general case, the solutions to the system of equations $m \cdot a + n \cdot b = m^*$, $m \cdot b - n \cdot a = n^*$ can be computed in a similar way. Thus, to find the value of a, one proceeds as follows:

$m \cdot m^* = \text{sq.}\, m \cdot a + m \cdot n \cdot b, \quad \text{and} \quad n \cdot n^* = m \cdot n \cdot b - \text{sq.}\, n \cdot a.$

Therefore,

$(sq.\ m + sq.\ n) \cdot a = m \cdot m^* - n \cdot n^*$, so that $a = (m \cdot m^* - n \cdot n^*)/(sq.\ m + sq.\ n)$.

Analogously,

$n \cdot m^* = m \cdot n \cdot a + sq.\ n \cdot b$, and $m \cdot n^* = sq.\ m \cdot b - m \cdot n \cdot a$.

Therefore,

$(sq.\ m + sq.\ n) \cdot b = n \cdot m^* + m \cdot n^*$,

so that

$b = (n \cdot m^* + m \cdot n^*)/(sq.\ m + sq.\ n)$.

Obviously, the situation in *P.Cornell 69* # 3 is quite special, with $m = m^* = 15$ and $n = n^* = 5$.

Much more can be said about the interesting properties of birectangles, but for now the interested reader is referred to the discussion in Friberg, *BaM* 28 (1997) § 8.c, in particular to the reference given there to the OB general method for the construction of "confluent equipartitioned trapezoids" by use of birectangles, or some similar device.

4.8. *P.Vindobonensis G. 19996* (1st C. CE?). Stereometric Exercises

P.Vindob. G. 19996 is a fairly well preserved papyrus roll inscribed on one side with 38 stereometric exercises (Gerstinger and Vogel, *GLP 1* (1932)), on the other side with an addition table (Harrauer and Sijpesteijn, *NTAU* (1985), 151.) The complexity of the addition table is caused by the Greek use of letters for numbers, which in no way reflects the structure of the decimal number system.

The side of the papyrus with the 38 stereometric exercises is a typical recombination text with a mixed bag of exercises sharing a common topic, but in apparent disarray. The text begins with some metrological notes, for instance the remark that volumes will be measured in (cubic) feet, and with a number of trivial computations of volumes of cubes or various kinds of rectangular slabs. The rest of the text, beginning with # 10, is devoted to correct computations of volumes of whole or truncated pyramids, cones, and prisms. Below is presented a table of contents for *P.Vindob. G. 19996*, essentially the same as the one in Gerstinger, *et al.* (*op. cit.*), 74-75, but in condensed form:

4.8 a. *P.Vindob. G. 19996*: Contents

...	General remarks about length and volume measures (feet, fingers, ...)	## 1, 4, 7
§ 1	Cubes: $s =$ a) 10, b) 5, c) 16, d) 4, e) 6 and 3 3/4	## 2-3, 5-6
§ 2	Rectangular prisms: $a, b, c =$ a) 24, 6, 4, b) [...], c) 4, 3, [...]	## 4, 9, 36
§ 3	Triangular prisms: $a, b, c =$ a) 20, 3, 4, b) no numbers	## 16-17
§ 4	Trapezoidal prisms: a), b) no numbers	## 33, 38
§ 5	Circular cylinders: $a, h =$ a) 30, 10, b) 30, 20, c) 4, 10, d) 6, 20	## 20-23
	e), f) no numbers	## 34, 37
§ 6	Triangular pyramids: $a, s =$ a) 12, 8, b) 12, 7	## 10, 12
	c) $s, h_1, h_2 =$ 12, 4, 6	# 15
§ 7	Square pyramids: $a, s =$ a) 12, 9, b) 20, 20, c) no numbers	## 18, 29, 30
	d) $a, b, h =$ 6, 3, 14	# 32
§ 8	Cone: $a, s =$ 6, [?]	# 35
§ 9	Truncated triangular pyramids: $a, b, s =$ a) 14, 2, 13	# 11
	b) 18, 3, 10, c) 12, 2, 10, d) 12, 4, 13	## 13-14, 27
§ 10	Truncated square pyramids: $a, b, s =$ a) 14, 2, 9, b) 12, 9, [?]	## 19, 26
	c) 10, 2, 6, d) no numbers	## 28, 31
§ 11	Truncated cones: $a, b, s =$ a) 24, 6, 5, b) 10, 2, 5	## 24-25

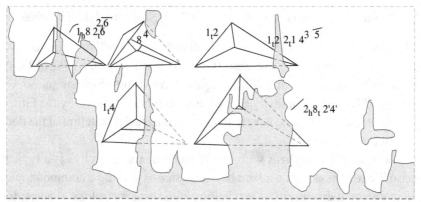

Fig. 4.8.1. *P.Vindob. G. 19996*, col. *vi*, bottom. Whole or truncated triangular pyramids.

A number of drawings are inserted more or less at random in the text. See, for instance, Fig. 4.8.1 above, which shows five drawings below exercise # 13 in col. *vi* of the papyrus. (Cf. the photo of col. *vi* in Fowler, *MPA* (1987 (1999)), pl. 8. Two other photos of the text are published in Gerstinger and Vogel, *GLP 1* (1932), pl. 1: col. *x*, ## 24-25, and in Weitzmann, *ABI* (1959), pl. 1: col. *xiii*, ## 28-32.) Just like drawings of three-dimensional objects in OB mathematical texts, some of these drawings are somewhat awkwardly designed and correspondingly hard to understand. The numbers associated with the drawings are often more or less obliterated and difficult to read. For this and other reasons, it is in several cases difficult to see to which exercises the drawings belong, if any.

4.8 b. *P.Vindob. G. 19996* # 10. A pyramid with a triangular base

In *P.Vindob. G. 19996* # 10 is computed the volume of *a right pyramid with an equilateral triangular base.* Given are the side of the triangular base, $a = 12$ (feet), and each one of the sloping edges, $s = 8$ (feet). The solution procedure begins with the computation of the height h of the pyramid by use of the *diagonal rule*, applied to a right triangle with the sides h, s, r, where r *is* the radius of the circumscribed circle for the triangular base. Thus, the height is computed as follows:

sq. a = sq. 12 = 144,   sq. r = (sq. a)/3 = 48,   sq. s = 64,
sq. h = sq. s – sq. r = 64 – 48 = 16, $h = 4$.

The indicated value of sq. r can be obtained, in several ways, by use of the diagonal rule. It is, for instance, easy to see that $r = 2/3$ of the height of the equilateral triangle (see Fig. 4.8.2). Therefore,

sq. $(3/2 \cdot r)$ = sq. a – sq. $(a/2)$ = 3/4 · sq. a,

so that

sq. r = 4/9· 3/4 · sq. a = 1/3 · sq. a.

After the height $h = 4$ (feet) has been computed in *P.Vindob. G. 19996* # 10, the computation of the volume of the pyramid continues as follows:

> The area of the equilateral triangular base of side 12 is 62 2 $\overline{5}$.
> This multiplied with the height 4 of the pyramid is 249 3 $\overline{5}$.
> 3' of that is 83 $\overline{5}$. So many (cubic) feet is the pyramid.

It is not disclosed how the area was computed. It was probably done this way:

A = $\overline{3}$ $\overline{10}$ · sq. s = $\overline{3}$ $\overline{10}$ · 144 = 48 + 14 2/5 = 62 2/5.

This is *the square of the side of the equilateral triangle times the constant*

c = $\overline{3}$ $\overline{10}$ = 26/60 (= ;26).

The fact that the radius of the circumscribed circle for an equilateral triangle is 2/3 of the height of the triangle was, of course, well known in Egypt at the time when *P.Vindob. G. 19996* was written. See the demotic *P.Cairo*

§ 11 (Sec.3.1 j above), where the 'diameter' (meaning the height) of a circular segment cut off from the circumscribed circle by one of the sides of the triangle is computed as 1/3 of the height of the triangle (line 23), and where the diameter of the circumscribed circle (called 'the diameter of the triangle rounded') is then computed as the height of the triangle plus the 'diameter' of the segment (line 33). See also the discussion in Friberg, *et al, BaM* 21 (1990) § 1, of exercise § 1 in the Late Babylonian recombination text **W 23291-x**, in particular Fig. 2 a, right.

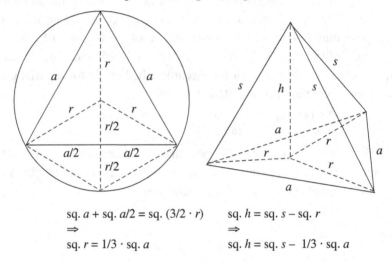

sq. a + sq. $a/2$ = sq. $(3/2 \cdot r)$ sq. h = sq. s − sq. r
⇒   ⇒
sq. r = 1/3 · sq. a sq. h = sq. s − 1/3 · sq. a

Fig. 4.8.2. *P.Vindob. G. 19996* # 10. Computation of the height of a pyramid with an equilateral triangular base.

In the demotic *P.Cairo* § 11, the area of the equilateral triangle is computed directly as the height of the triangle times half the side, *not* as the square of the side times some constant as in *P.Vindob. G. 19996* # 10. On the other hand the use of constants in similar situation is known from a number of Babylonian mathematical texts. A particularly interesting example can be found in the Kassite (post-OB) text **MS 3876** (Friberg, *MCTSC* (2005), Sec. 11.3). What is computed there, in MS 3876 # 3 is, apparently, the area and the weight of the outer shell of an icosahedron (called a 'horn figure' gán.si), with that outer shell consisting of 20 finger-thick copper sheets informed as equilateral triangles (called 'gaming-piece figures' gán.za.na), each with a side of 3 cubits.

4.8. P.Vindobonensis G. 19996 (1st C. CE?). Stereometric Exercises

Here is an excerpt from MS 3876 # 3, showing how the area of an equilateral triangle of side 12 cubits is computed there:

MS 3876 # 3, lines 5-7.

5	*šum-ma* 3 kùš.ta.àm za.na *im-ta-aḫ-ru saḫar! mi-nu ba-ma-at* 15 sag *ḫe-pe-ma* 7 30 /
6	7 30 a.rá 15 sag *ša-ni-tim* 1 52 30 *ba-am-ta* 14 03 45 *sa-am-na-ti-šu* zi-*ma* /
7	1 38 26 15 *qá-qá-ar* gán.za.na *iš-te-en ša ta-mar*

If 3 cubits each (the sides of) a gaming-piece are equal,
the volume (is) what?
Half (of) 15, the front, break, then 7 30.
7 30 steps of 15, the second front, 1 52 30, the halved.
14 03 45, its eighth tear off, then
1 38 26 15 (is) the ground (of) one gaming-piece-field that you see.

Thus, the area of the equilateral triangle of side 12 cubits is computed in MS 3876 as follows:

a = 3 cubits = 3 · ;05 (ninda) = ;15 (ninda),
1/2 of ;15 (ninda) = ;07 30 (ninda),
(sq. a)/2 = ;07 30 (ninda) · ;15 (ninda) = ;01 52 30 (sq. ninda),
(sq. a)/2 – the eighth of (sq. a)/2 = ;01 52 30 – ;00 14 03 45
 = ;01 38 26 15 (sq. ninda) = the area.

This rule for the computation of the area of an equilateral triangle can be interpreted as follows:

$A = (1 - 1/8) \cdot$ (sq. a)/2,

or

$A = h \cdot a/2$, with $h = (1 - 1/8) \cdot a$ = the height of the equilateral triangle.

The same rule for the computation of the area of an equilateral triangle resurfaces in § 4 b of the Late Babylonian recombination text **W 23291** (Friberg, *BaM* 28 (1997). In the first part of that text, an equilateral triangle of unspecified side length is presented as an equilateral 'peg-head' (triangle) of the kind that has 'its 8th torn off'.

W 23291 § 4 b.

1 gán.sag.kak ur.a *ša* 5 *ša* 8-*šú na-as-ḫu* /
mi-ḫi-il-tú / a.rá ki.2 *ù* a.rá 2[6 1]5 rá

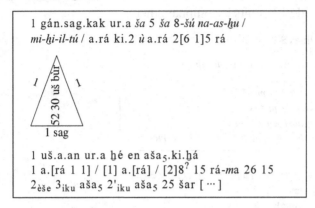

1 uš.a.an ur.a ḫé en aša₅.ki.ḫá
1 a.[rá 1 1] / [1] a.[rá] / [2]8⁷ 15 rá-*ma* 26 15
2èše 3iku aša₅ 2'iku aša₅ 25 šar [⋯]

1 peghead-field, equilateral, <<that with 5>> that with its 8th torn off.
Line steps of ditto, and steps of 26 15 go.

1 front

1 uš each way, equilateral. What shall the field be?
1 ste*ps of 1 is 1. 1* ste*ps of 26*¹ 15 go, it is 26 15.
2(èše) 3(iku) 1/2(iku) 25 šar ⋯ .

Presumably this means that, as in MS 3876 # 3, the *height* of an equilateral triangle is computed as *the side minus 1/8 of the side*. That this is really the case is shown by the continuation in line 2, where the curious phrase 'line (side) steps of (times) ditto, and steps of (times) 26 15 go' means that the area shall be computed as

$A = {;26\ 15} \cdot \text{sq. } a,$

where the constant ;26 15 can be explained as follows:

;26 15 = ;52 30 · 1/2 = (1 − ;07 30) · 1/2 = (1 − 1/8) · 1/2.

A numerical example of the general rule is given in lines 3-4:

$a = 1$ uš = 1 00 ninda,
$A = 1\ (00) \cdot 1\ (00) \cdot {;26\ 15} = 26\ 15$ sq. ninda = 2(èše) 3 1/2(iku) 25 šar.

It is likely that W 23291 § 4 b was meant to show how a traditional

4.8. P.Vindobonensis G. 19996 (1st C. CE?). Stereometric Exercises

method to compute the area of an equilateral triangle worked, a method inherited from the *Old Babylonian* predecessors. The proper *Late Babylonian* method with a better constant is demonstrated in W 23291 § 4 c (cf. the discussion of *P.BM 10520* § 6 b in Sec. 3.3 e above).

W 23291 § 4 c.

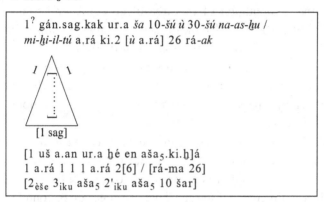

1? gán.sag.kak ur.a *ša* 10-*šú ù* 30-*šú na-as-ḫu* /
mi-ḫi-il-tú a.rá ki.2 [*ù* a.rá] 26 rá-*ak*

[1 sag]
[1 uš a.an ur.a ḫé en aša₅.ki.ḫ]á
1 a.rá 1 1 1 a.rá 2[6] / [rá-ma 26]
[2èše 3iku aša₅ 2'iku aša₅ 10 šar]

1 peghead-field, equilateral, that with its 10th and its 30th torn off.
Line steps of ditto, *and steps of* 26 you go.

1, the front

1 uš each way, equilateral. What shall the field be?
1 steps of 1 is 1. 1 steps of 26 go, it is 26.
2(èše) 3(iku) 1/2(iku) 10 šar ⋯

In this text is presented an equilateral triangle of the kind that has 'its 10th and its 30th torn off'. This means that the *height* of an equilateral triangle shall be computed as *the side minus 1/10 1/30 of the side*. That this is a correct interpretation is shown by the continuation in line 2, where the phrase 'line steps of ditto, and steps of 26 go' means that the area shall be computed as

$A = ;26 \cdot$ sq. *a*,

where the constant ;26 can be explained as follows:

$;26 = ;52 \cdot 1/2 = (1 - ;06 - ;02) \cdot 1/2 = \{1 - (1/10\ 1/30)\} \cdot 1/2.$

A numerical example of the general rule is given in lines 3-4 (partially reconstructed here):

a = 1 uš = 1 00 ninda,
A = 1 (00) · 1 (00) · ;26 = 26 (00) sq. ninda = 2(èše) 3 1/2(iku) 10 šar.

Note that the constant ;26 = 26/60 = 1/3 1/10 is identical with the constant used for the computation of the area of an equilateral triangle in *P.Vindob. G. 19996* # 10. Note also the quite remarkable circumstance that in line 1 of W 23291 § 4 c the expression 10-*šú ù* 30-*šú* 'its 10th and its 30th' is the only known example of the use of a *sum of parts* in a Babylonian mathematical text! More precisely, this is a *sexagesimally adapted* sum of parts, just like many of the sums of parts in the demotic text *P.Cairo* (Sec. 3.1 above).

In *P.Vindob. G. 19996* # 10, the diagonal rule is used in a 3-dimensional setting, to compute the height of the triangular pyramid. It is interesting that there is now also a Babylonian example of an application of the diagonal rule in a 3-dimensional setting. That example can be found in MS 3049, a fragment of an OB (or Kassite) mathematical recombination text, in which § 5 (Friberg, *MCTSC* (2005), Sec. 11.1 d) contains the computation of the šà.bar 'inner diagonal' of a gate in a wall. The gate has the height h = 5 cubits 10 fingers = ;26 40 ninda, the width w = ;08 53 20 ninda, and the depth t = ;06 40 ninda = the thickness of the wall. (Note that this means that h, w, t = ;02 13 20 ninda · 12, 4, 3.) The inner diagonal d is computed as follows:

sq. d = sq. h + sq. w + sq. t = ;13 54 34 14 26 40,
d = ;28 53 20 (= ;02 13 20 ninda · 13).

One last remark concerning *P.Vindob. G. 19996* # 10: The computed bottom area of the triangular pyramid is expressed as 62 $2^{\overline{5}}$, clearly meaning 62 2/5. Similarly, this bottom area multiplied by the height of the pyramid is expressed as 249 $3^{\overline{5}}$, which must mean 249 3/5. Thus, it is likely that $2^{\overline{5}}$, for instance, is an abbreviated expression for τῶν β το ε' 'of 2 the 5th part', the kind of expression for binomial fractions that occurs repeatedly in *P.Akhmîm* (Sec. 4.5 above).[44]

---

44. For a conflicting explanation, see Fowler, *MPA* (1987 (1999)), Sec. 7.3(d).

4.8 c. *P. Vindob. G. 19996* # 13. A truncated triangular pyramid

In *P. Vindob. G. 19996* # 13 is computed the volume of *a truncated right pyramid with an equilateral triangular base*. Given are the side of the triangular base, $a = 18$ (feet), the side of the triangular top, $b = 3$ (feet), and each one of the sloping edges, $s = 10$ (feet). The solution procedure begins with the computation of the height h of the truncated pyramid by use of the diagonal rule, applied to a right triangle with the sides h, s, r, where r is the radius of the circumscribed circle for an equilateral triangle with the side $a - b = 12$ (feet). Thus, the height h is computed as follows:

sq. h = sq. s − sq. r = sq. s − {sq. $(a-b)$}/3 = 100 − 75 = 25, $h = 5$.

After that, the volume V is computed as the volume of right prism of height h and with a base in the form of an equilateral triangle with the side $(a + b)/2$, plus a correction term:

$V = (c \cdot$ sq. $\{(a+b)/2\} + c \cdot$ sq. $\{(a-b)/2\} \cdot \overline{3}) \cdot h$,   where $c = \overline{3}\,\overline{10}$ (= 26/60).

4.8 d. *P. Vindob. G. 19996* # 18. A square pyramid

P. Vindob. G. 19996 # 18 is a computation of the volume of a *right pyramid with a square base*. Given are the side of the square base, $a = 12$, and the sloping edges of the pyramid, $s = 9$. The solution procedure begins with the simple computation of sq. r, the square of the radius of the circumscribed circle for the square base, which is then used to compute the height of the pyramid, $h = 3$. The volume V is computed as

$V = c \cdot$ sq. $a \cdot h$,   where $c = \overline{3}$.

4.8 e. *P. Vindob. G. 19996* # 19. A truncated square pyramid

The volume of a *truncated right pyramid with a square base* is computed in *P. Vindob. G. 19996* # 19. Given are the side of the square base, $a = 14$, the side of the square top, $b = 2$, and the length of the sloping edge, $s = 9$. The solution procedure does not explicitly mention the computation of the height h of the truncated pyramid. It was probably found by use of the diagonal rule, applied to a right triangle with the sides h, s, r, where r is the radius of the circumscribed circle for a square with the side $a - b = 12$ (feet). Thus, the height h can have been computed as follows:

sq. h = sq. s − sq. r = sq. s − {sq. $(a-b)$}/2 = 81 − 72 = 9, $h = 3$.

The volume is correctly computed as

$$V = \text{sq.}\{(a+b)/2\} + \text{sq.}\{(a-b)/2\} \cdot \overline{3}) \cdot h.$$

4.8 f. *P.Vindob. G. 19996 # 24.* **A truncated circular cone**

There is no well preserved computation of the volume of a cone in *P.Vindob. G. 19996*. There is, however, a correct computation of the volume of *a truncated right cone with a circular base* in *P.Vindob. G. 19996 # 24*. Given are the circumference $p = 24$ (feet) of the base, the circumference $q = 6$ of the top, and the length $s = 5$ of a sloping edge. The first step of the solution procedure is to compute the diameters of the base and the top as

$$a = p/3 = 8, \; b = q/3 = 2.$$

Then the volume is computed as

$$V = (c \cdot \text{sq.}\{(a+b)/2\} + c \cdot \text{sq.}\{(a-b)/2\} \cdot \overline{3}) \cdot h, \text{ where } c = 1 - 4' \; (= 3/4).$$

Essentially, then, the purpose of *P.Vindob. G. 19996* was to exhibit rules for the computation of volumes of pyramids and truncated pyramids, cones and truncated cones. In this connection, it may be of interest to make the following observation. Assuming that the rule for the computation of the volume of a whole pyramid with a square base had been found, in some way, it was not difficult to find also the rule for the computation of the volume of a truncated pyramid with a square base, namely by cutting up the truncated cone in various pieces, a number of prisms and small pyramids in the corners. The corresponding volumes of whole or truncated pyramids with an equilateral triangle for base could then be found by multiplication with the constant $\overline{3} \; \overline{10}$. Naively, this is obvious, because a triangular pyramid can be thought of as composed of a large number of thin slices, and the volume of each slice is $\overline{3} \; \overline{10}$ of the volume of a corresponding slice from a square pyramid with the same height and with the same side of the base. The idea works also in the case of whole and truncated circular cones, with the only difference that the constant in these cases is $1 - 1/4$.[45]

---

45. In Gerstinger and Vogel, *GLP 1* (1932), 40, the generalization from the cases of truncated triangular and square pyramids (the drawings to the left and in the middle) to the case of a truncated cone (the drawing to the right) makes no sense at all!

4.8 g. Pyramids and cones in Old Babylonian mathematical texts

The volume of a truncated square pyramid

A survey of OB texts concerned with both correct and incorrect computations of volumes of whole or truncated square pyramids was published in Friberg, *PCHM* 6 (1996) § 1.5. A particularly interesting example is exercise # 28 in the OB mathematical recombination text BM 85194 (Thureau-Dangin, *TMB* (1938), text 72).

BM 85194 # 28.

ḫi-ri-tum
10.ta.an *mu-ḫu* 18 sukud *i-na* 1 kùš 1 šà.gal /
za.zum saḫar.ḫi.a za.e
5 *ù* 5 ul.gar 10 *ta-mar* / [10] *a-na* 18 sukud *i-ši* 3 *ta-mar*
3 *i-na* 10 ba.zi 7 / [*ta-mar*] za.zum
nigín.na
za.zum *ù mu-ḫa* ul.gar 17 *ta-mar* / [2' 17 *ḫe-pé*] 8 30 *ta-mar*
nigin 1 12 15 *ta-mar* / 1 12 [15 gar].ra
igi.2?.gál 3 dirig *ša mu-ḫu* ugu / za.zum nig[in] 45
a-na 1 12 15 daḫ.ḫa-*ma* / 1 13 *ta-mar*
18 *a-na* 1 13 *i-ši* 22 30 *ta-mar* /
2$_{eše}$ 1$_{iku}$ 2'$_{iku}$ aša$_5$ saḫar.ḫi.a
ki-<a-am> ne-pé-šum

An excavation.
10 (ninda) each way is the top, 18 (cubits) the height, in 1 cubit 1 is the feed. The base and the mud (volume)?
You:
5 and 5 gather, 10 you see. *10* to 18 the height raise, 3 you see.
3 from 10 tear off, 7 *you see*, the base.
Turn around.
The base and the top gather, 17 you see. *1/2 of 17 break*, 8 30 you see.
Square (it), 1 12 15 you see. 1 12 *15* set.
The 12th^l-part of 3, the excess of the top over the base, squared, 45.
To 1 12 15 add (it) on, then 1 13 you see.
18 to 1 13 raise, 22 30 you see,
2(eše) 1 1/2(iku) is the mud.
Such is the doing.

The exercise deals with an upside-down *truncated pyramid,* uninformatively called an 'excavation'. The text of the exercise is only slightly damaged but the damaged part is situated in the middle of the most crucial passage. It has long been suspected that a correct expression for the volume of a truncated cone was used in this problem, but the evidence has been deemed to be inconclusive. However, a new reading of the crucial passage (see below) confirms the suspicion.

Given are the side a (= 10 ninda) of the square top, and the height, or rather the depth, of the excavation, h (= 18 cubits). Given is also the inclination of the side walls of the excavation, expressed through the customary phrase *i-na* 1 kùš 1 šà.gal 'in 1 cubit 1 is the feed'. What this means is that the inclination is 1 cubit/cubit, so that the width of the excavation decreases with 2 · 1 cubit = ;10 ninda for each descent of 1 cubit.

The solution procedure begins with the computation of the side b of the square bottom of the excavation. Since the 'feed' is 1 cubit/cubit, and since all side walls of the excavation slope inwards, the width of the excavation has decreased with 2 · ;05 · 18 ninda = 3 ninda after a descent of 18 cubits. Therefore $b = (10 - 3)$ ninda = 7 ninda. Next is computed the area of the excavation halfway between the top and the bottom, which is

sq. $\{(a + b)/2\}$ = sq. 8;30 = 1 12;15 square ninda.

This result is recorded somewhere, and then the solution procedure continues with the computation of

1/12 · sq. $(a - b)$ = 1/12 · 9 = ;45 square ninda.

Finally, the scribe added this "correction term" to the initial approximation sq. $\{(a + b)/2\}$ = 1 12;15 square ninda, obtaining the correct result 1 13 square ninda. This area, multiplied by the height, h = 18 cubits, would then have given the correct volume 21 54 volume-šar, but the scribe misread an entry in his multiplication table and got instead the incorrect result 1 5 square ninda · 18 cubits = 22 30 volume-šar, which he then converted correctly to an area number. Anyway, the expression used by the author of BM 85 194 § 28 for the volume of a truncated square pyramid seems to have been, correctly,

$V = [\text{sq. } \{(a + b)/2\} + 1/12 \cdot \text{sq. } (a - b)] \cdot h$.

This is, essentially, the same equation for V as the one used in *P.Vindob. G. 19996* # 19.

The volume and grain measure of a ridge pyramid

TMS 14 (Bruins and Rutten, *TMB* (1961), Friberg, *PCHM* 6 (1996)) is an OB text from Susa, dealing with a certain three-dimensional solid, here called a "ridge pyramid". The Sumerian/Babylonian term for a ridge pyramid is not known. The term gur_7 mentioned at the beginning of the exercise has the general meaning 'granary', that is a large building used to store grain. Another possibility is that the term means 'grain measure', since the content of the ridge pyramid is expressed not in terms of its *volume V*, but in terms of its (storage) *capacity* $C = c \cdot V$, where c is a fixed storing number,

c = 8 (00 00) sìla/šar, where 1 (volume-)šar = 1 sq. ninda · 1 cubit.

It is known (Friberg, *BaM* 28 (1997), Fig. 7.1) that a "granary sìla" with the storing number 8 (00 00) can be understood as the content of a cylindrical measuring vessel with the diameter 6 fingers and the height 5 fingers (appr. 0.63 liters).

TMS 14.

```
gur₇ a.na 14 24 saḫar 3 gi me-lu-[um] /
a.na 14 24 saḫar uš sag ù qà-aq-qa-da / mi-na gar
za.e
igi 12 šu-up-li pu-ṭú-úr / 5 ta-mar
5 a-na 14 24 saḫar i-ší-ma / 1 12 ta-mar
3 gi me-la-a-am nigin 9 ta-mar / 9 a-na 3 me-le-e te-er-ma 27 ta-mar /
i-na 1 a.rá ka-a-a-ma-ni <saḫar> 20 <<saḫar>>
šà-lu-uš-ti / šà ka-a-a-ma-ni saḫar tu-uṣ-ṣa-bu zi / 40 ta-mar
40 aš-šum 2 sag gur₇ a-na 2 tab.ba / 2 20 ta-mar
1 20 a-na 27 i-ší-ma 36 ta-mar / 36 i-na [1 12] zi 36 [ta-mar]
[tu]-úr-ma /
3 me-la-a-am nigin 9 ta-[mar igi 9 pu-ṭ]ú-úr / 6 40 ta-mar
6 40 a-na 36 i-š[í-ma] / 4 ta-mar 4 qà-aq-qa-du
3 me-la-[a-am] / [aš]-šum i-na am-ma-at am-ma-at [šà.gal]? /
[a-na 2] tab.ba 6 ta-mar 6 sag
6 [a-na 4] / [qà-aq-qa]-di daḫ 10 ta-mar 10 [uš] / [ ⋯  ⋯  ⋯ ] /
[12 a-na] 3 me-la-am [i-ší-ma] / 36 ta-mar
[36 a-na 24 i-ší-ma] / [14] 24 ta-mar saḫar
14 2[4] [saḫar] / a-na 8 na-aš-pa-ak gur₇ [i-ší-ma] / 1 55 12 ta-mar
23 [gur₇]? / ù 2+šu 24_g gur še-[um]
[ki-a]-am ne-[pé-šum]
```

A granary, as much as 14 24 the mud (volume), 3, reeds, the height.
(If) as much as 14 24 the mud, the length, the front and the top what do I set?
You:
The opposite of 12 of the depth, release, 5 you see,
5 to 14 24 the mud raise, then 1 12 you see.
3, reeds, the height, square, 9 you see. 9 to 3, the height, again, 27 you see.
From 1 the regular step of mud 20,
a third of what <as> the regular mud you added, tear off, 40 you see.
40, since (there are) 2 fronts of the granary, to 2 repeat, then 1^1 20 you see.
1 20 to 27 raise, then 36 you see. 36 from *1 12* tear off, 36 *you see*.
Return.
3, the height, square, 9 *you see. The opposite of* 9 break off, 6 40 you see.
6 40 to 36 raise, then 4 you see, 4 the top.
3, the height, since in a cubit a cubit is *the feed*,
to 2 double, 6 you see. 6 (is) the front.
6 *to 4, the top*, heap, 10 you see, 10 *the length*.
... ...
12 to 3 the height *raise*, then 36 you see.
36 to 24 raise, then 14 24 you see, the mud.
14 *24*, the mud, to 8, the storing number of the granary, *raise, then* 1 55 12 you see.
23 *gur*₇ 2 24 gur of barley.
Such is the *doing*.

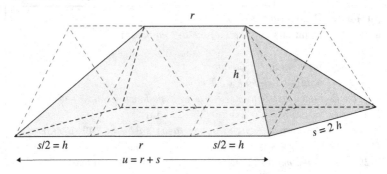

$f = 1$ c./c. $\Rightarrow$ $s = 2h$, and $V = r \cdot$ sq. $h +$ sq. $(2h) \cdot h/3$

Fig. 4.8.3. The ridge pyramid in *TMS* 14.

The nature of the ridge pyramid in *TMS* 14 can be deduced from the computations in the text which mention the 'length' (or long side) u and the 'front' (or short side) s at the base, the 'ridge' or 'top' r, and the 'height' h. As shown in Fig. 4.8.3 above, the ridge pyramid must have the form of a roof, sloping uniformly from the ridge all the way to the ground in four directions.

The statement of the problem is clear: What are the length u, the front s, and the ridge r (the 'top') of the ridge pyramid, when it is known that its volume V = 14 24 šar, and that its height h = 3 ninda? In order to eliminate the risk of misunderstanding, 3 ninda is written as 3 gi '3, reeds', meant to emphasize that the order of magnitude of the given length number is that of a small number of reeds, 1 reed being 1/2 ninda = 6 cubits. The 'feed' of the ridge pyramid is assumed to be 1 cubit/cubit, just as in the case of the 'excavation' in BM 85194 # 28. By mistake this assumption is not mentioned in the question, but in line 15 the mistake is corrected with the phrase 'since in 1 cubit 1 cubit is the feed ⋯ '.

The very explicit solution procedure in *TMS* 14 is interesting. It begins by expressing the volume in cubic ninda instead of in volume-šar:

V = 14 24 šar = 14 24 sq. n. · c. = 14 24 sq. n. · ;05 n. = 1 12 sq. n. · n.

Next the cube of the height h = 3 ninda is computed as follows:

sq. $h \cdot h$ = sq. (3 n.) · 3 n. = 9 sq. n. · 3 n. = 27 sq. n. · n.

The awkwardly worded passage of the text which then follows,

> From 1 the regular step of mud 20,
> a third of what <as> the regular mud you added, tear off, 40 you see.
> 40, since (there are) 2 fronts of the granary, to 2 repeat, then 1! 20 you see.
> 1 20 to 27 raise, then 36 you see.

can be interpreted as describing the computation of 2 · 2/3 of the cube of h:

2 · (1 – 1/3) · sq. $h \cdot h$ = 2 · ;40 · 27 sq. n. · n. = 1;20 · 27 sq. n. · n. = 36 sq. n. · n.

The exact meaning of the word *kayyamānu* 'regular, usual' as a technical term in a mathematical cuneiform text is elusive, and there are also other difficulties in the text. Anyway, the mathematical meaning of the passage as a whole is clear, namely that the volumes of the two rectangular pyramids at the ends of the ridge pyramid are computed as *two thirds of the volumes of the two wedges (triangular prisms) containing them*. Indeed, since the feed of the ridge pyramid is 1 c./c., each end pyramid has the height h, while its sides at the base are s = 2 h and 1/2 ($u - r$) = h. Therefore the volume of each containing wedge is 1/2 · 2 $h \cdot h \cdot h$ = 1 sq. $h \cdot h$, and the combined volume of the two end pyramids is 2 · (1 - 1/3) · 1 sq. $h \cdot h$.

In the next step of the procedure, the volume V_c of the central wedge is computed as the given volume of the whole ridge pyramid, diminished by the combined volume of the two end pyramids, that is

$V_c = V - 2 \cdot (1 - 1/3) \cdot$ sq. $h \cdot h = 1\ 12$ sq. n. $\cdot$ n. $- 36$ sq. n. $\cdot$ n. $= 36$ sq. n. $\cdot$ n.

On the other hand, since the feed f is 1 c./c. so that $s = 2\ h$, it is also true that

$V_c = 1/2 \cdot r \cdot s \cdot h = 1/2 \cdot r \cdot 2\ h \cdot h = r \cdot$ sq. $h = r \cdot 9$ sq. n.

Consequently, the ridge r can be computed as follows:

$r \cdot 9$ sq. n. $= 36$ sq. n. $\cdot$ n. $\Rightarrow r = 36$ sq. n. $\cdot$ n./(9 sq. n.) $= 4$ n.

Next it is shown, explicitly, that

$h = 3$ n. and $f = 1$ c./c. ($= 1$ n./n.) $\Rightarrow s = 2 \cdot f \cdot h = 2 \cdot 1 \cdot 3$ n. $= 6$ n.

Finally, the length u is computed as

$u = r + s = 4$ n. $+ 6$ n. $= 10$ n.

This is correct because all sides of the ridge pyramid have the same feed, so that $f = s/h = (u - r)/h$ and therefore $s = u - r$.

Summing up, one can conclude that the computation of the ridge r of the ridge pyramid in *TMS* 14 is based on the following correct observation: If V is the volume of a ridge pyramid with the feed 1 c./c. everywhere, if V_c is the volume of the of the central wedge, and if V_p is the combined volume of the two end pyramids, then

$V = V_c + V_p = r \cdot$ sq. $h +$ sq. $(2\ h) \cdot h/3$.

Now, consider the second part of the solution procedure in *TMS* 14, the *verification* of the result of the computation in the first part. The beginning of this verification is lost, but apparently it starts with the computation of

$r \cdot h + 4\ h \cdot h/3 = 24$ sq. n.

Then the volume V is obtained through multiplication by the height h, and the result is, as it should be, that

$V = 24$ sq. n. $\cdot 36$ c. $= 14\ 24$ sq. n. $\cdot$ c. $= 14\ 24$ šar.

After this successful verification, the computation is carried one step further, in that the *grain measure* of the granary is computed. (The grain measure *was not* mentioned in the statement of the problem in lines 1-3.) With the grain constant c = '8' = 8 00 00 sìla/šar, and with 1 gur = 5 00 sìla, the computation proceeds as follows:

$C = c \cdot V = 14\ 24 \cdot 8\ 00\ 00$ sìla $= 1\ 55\ 12\ 00\ 00$ sìla $= 23\ 02\ 24$ gur.

In the final step of the computation, this sexagesimal multiple of the large capacity unit gur is expressed in *traditional* number notation as

$C = 23\ [\cdots]$ and 2 sixties 24 gur of barley.

4.8. P.Vindobonensis G. 19996 (1st C. CE?). Stereometric Exercises

BM 96954 + 102366 + SÉ 93 (Friberg, *PCHM* 6 (1996), Robson, *MMTC* (1999), App. 3) is a text composed of three fragments of a large clay tablet. As shown by the outline below of the clay tablet and the table of contents, it is a recombination text which, just like *P.Vindob. G. 19996*, has "whole or truncated pyramids and cones" as its dominating topic.

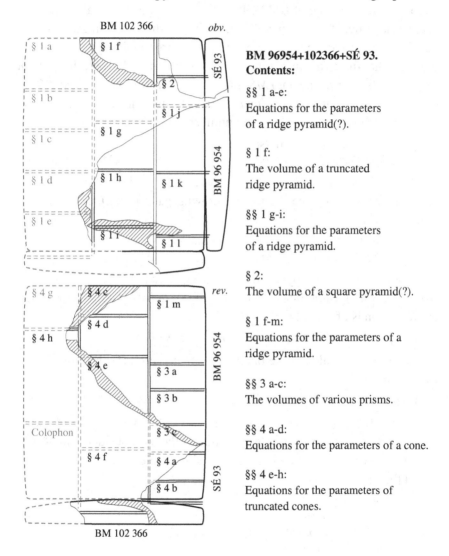

**BM 96954+102366+SÉ 93.
Contents:**

§§ 1 a-e:
Equations for the parameters
of a ridge pyramid(?).

§ 1 f:
The volume of a truncated
ridge pyramid.

§§ 1 g-i:
Equations for the parameters
of a ridge pyramid.

§ 2:
The volume of a square pyramid(?).

§ 1 f-m:
Equations for the parameters of a
ridge pyramid.

§§ 3 a-c:
The volumes of various prisms.

§§ 4 a-d:
Equations for the parameters of a cone.

§§ 4 e-h:
Equations for the parameters of
truncated cones.

Fig. 4.8.4. BM 96954+. An OB recombination text with the theme pyramids and cones.

§ 1 of the recombination text may have consisted originally of 13 mathematical exercises, all dealing with a certain *ridge pyramid*, similar to the one treated in *TMS* 14 (Fig. 4.8.3 above). The Sumerian or Babylonian term for a ridge pyramid is not known. The term gur$_7$, mentioned in *TMS* 14 as well as at the beginning of all preserved exercises in § 1 of BM 96954+, refers in § 3 b of BM 96954+ to a solid (a trapezoidal prism) that is *not* a ridge pyramid. As in *TMS* 14, it is likely that the term gur$_7$ simply has the general meaning 'granary', that is a large building used to store grain, or that the term simply means 'grain measure'. Indeed, the content of each solid appearing in *TMS* 14 and in BM 96954+ §§ 1 and 3, is expressed not in terms of its *volume* V, but in terms of its (storage) *capacity* $C = c \cdot V$, where c is a fixed storing number:

c = 1 36 gur/šar = 8 00 00 sìla/šar^V in *TMS* 14
c = 1 30 gur/šar = 7 30 00 sìla/šar^V in BM 96954+ §§ 1, 3

The eight preserved exercises BM 96954+ §§ 1 f-1m, all deal with the same ridge pyramid, one with the following parameters (see again Fig. 4.8.3 above):

u, s, r = 10, 6, 4 ninda, h = 48 cubits (= 4 ninda),
V = 19 12 šar, C = 28 48 00 gur.

The volume of the ridge pyramid is, again, equal to the volume V_c of *a central wedge* plus the combined volume V_p of *two halves of a square pyramid* at the two ends of the ridge pyramid:

$V = V_c + V_p = r \cdot s \cdot h/2 + $ sq. $s \cdot h/3 = (r/2 + s/3) \cdot s \cdot h$.

Since $s = u - r$, an alternative but equivalent equation for the volume of a ridge pyramid is

$V = (u + r/2) \cdot s \cdot h/3$.

It is likely that the lost exercises in the first column of BM 96954+ were all concerned with the same ridge pyramid as the one considered in §§ 1 f–1 m. In that case, the following series of simple questions may have been asked there:

§ 1 a: u, s, r, h given C = ?
§ 1 b: s, r, h, C given u = ?
§ 1 c: u, r, h, C given s = ?
§ 1 d: u, s, h, C given r = ?
§ 1 e: u, s, r, C given h = ?

4.8. P.Vindobonensis G. 19996 (1st C. CE?). Stereometric Exercises 251

Similar series of questions can be found elsewhere in the corpus of OB mathematical texts. A good example is **YBC 4663** (Neugebauer and Sachs, *MCT* (1945), text H), a text with a series of 8 exercises. In the first exercise, the linear parameters of a rectangular 'excavation' are specified: the length u and the front s of the base, the height h, and two constants, namely the 'work norm' w = ;10 volume-šar per man-day, and the 'pay rate' (the wages) p = 6 barley-corns (= ;02 shekels) of silver per man-day. The base area A, the volume V, the man-days M needed for the excavation and the resulting expense E *in silver* are then computed as follows:

$A = u \cdot s, \quad V = A \cdot h, \quad M = V \cdot 1/w, \quad E = M \cdot p,$

so that

$E = u \cdot s \cdot h \cdot 1/w \cdot p.$

In the next 5 exercises, the parameters u, s, h, and the constants w and p are computed, one at a time, by use of the equation above for E and simple arithmetic. In the last two exercises, the situation is complicated in an artificial way, typical of OB mathematics, when it is assumed that E, h, w, and p are known, in addition to the sum $u + s$ or the difference $u - s$. There are then two unknowns, u and s, which are computed as solutions to a *rectangular-linear system of equations*:

$u \cdot s = E \cdot w \cdot 1/(h \cdot p), \quad u + s = m \quad (\text{or } u - s = n).$

The conjectured analogy between BM 96954 § 1 and YBC 4663 suggests that after the initial series of exercises, §§ 1 a-e, the next exercise, § 1 f, ought to be one of the artificial type leading to a rectangular-linear system of equations. This is not the case, however, possibly because, being a recombination text, BM 96954 is somewhat chaotically organized. Instead, § 1 f deals with a *truncated* ridge pyramid, and the expected continuation of the series of exercises for the ridge pyramid of §§ 1 a-1 e is moved up one step, to § 1g.

The grain measure of a ridge pyramid truncated at mid-height

Thus, in BM 96954+ § 1f, which is the first preserved exercise of § 1, the ridge pyramid common to all the exercises in § 1 is *truncated at mid-height* (Fig. 4.8.5 below), and is thus comparable in both size and form to the well known *mastaba* graves from the Old Kingdom in Egypt.

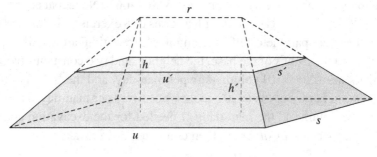

$u = 10$ n., $s = 6$ n., $r = 4$ n. $u' = 7$ n., $s' = 3$ n.
$h = 48$ c., $h - h' = 24$ c. $h' = 24$ c.
$V = 19\ 12$ šar $V' = 15\ 36$ šar
$C = 28\ 48\ 00$ gur $C' = 23\ 24\ 00$ gur

Fig. 4.8.5. BM 96954 § 1 f. A ridge pyramid truncated at mid-height.

Here is the text of § 1 f, of which only the first part is preserved:

BM 96954 + BM 102366 + SÉ 93 § 1 f.

```
gur₇
10 uš 6 sag 4 sagšu 28 48 <še.gur> /
48 sukud 24 ur-dam
dal ù še-um en.nam /
[za].e
igi 48 sukud du₈.a 1 15 ta-mar
1 15 a-na / [6 dirig?] ša? uš ugu sagšu? i-ši 7 30 ta-mar /
[7 30 a-na] 24 i-ši 3 ta-mar
3 i-na 10 uš / [ba.zi 7 ta-mar 7] dal
3 i-na 6 sag ba.zi? / [3 ta-mar ··· ]
[ ··· ··· ] [x x] i-ši 1 ta-mar / ··· ··· ··· ···
```

A granary.
10 the length, 6 the front, 4 the ridge (head-piece), 28 48 <gur of barley>,
48 the height, 24 I went down.
The transversal(s) and the barley are what?
*Yo*u:
The opposite of 48 the height release, 1 15 you see.
1 15 to *6, the excess* that the length is beyond the ridge, raise, 7 30 you see.
7 30 to 24 raise, 3 you see.

4.8. *P.Vindobonensis G. 19996 (1st C. CE?). Stereometric Exercises*

3 from 10 the length *tear off, 7 you see, 7* the transversal.
3 from 6 the front tear off, *3 you see, the* ⋯ .
⋯ ⋯ ⋯ ⋯ raise, 1 you see. ⋯ ⋯ ⋯ ⋯ ⋯

Although only the first part of the text of § 1 f is preserved, it is clear that the equation for the volume of the truncated ridge pyramid must have been expressed in terms of its linear parameters, the length u and front s at the base, the "upper length" $u´$ and "the upper front" $s´$ at mid-height, and the lower height $h´$. (Unless, of course, the volume of the truncated ridge pyramid was computed as the volume of the whole ridge pyramid minus the volume of the small upper ridge pyramid.) The volume $V´$ of the truncated ridge pyramid is easily seen to be equal to the volume $V´_c$ of *a central trapezoidal prism* plus the combined volume $V´_p$ of *two halves of a ridge pyramid* at the two ends of the truncated ridge pyramid:

$$V = V_c + V_p = u´ \cdot (s + s´) \cdot h´/2 + (u - u´) \cdot (s + s´/2) \cdot h´/3.$$

An equivalent, more elegant, form of this equation for V is

$$V = \{(u \cdot s + u´ \cdot s´) + (u \cdot s´ + u´ \cdot s)/2\} \cdot h´/3.$$

However, before this equation can be used to compute the volume of the truncated ridge pyramid, the values of $u´$ and $s´$ must be known.

The preserved first part of BM 96954 § 1 f is devoted to the computation of these values. The first step of the computation is to find the *combined feed* for the two ends of the ridge pyramid:

$$2 \cdot f = (u - r)/h = (10 - 4) \text{ ninda} \cdot 1/(48 \text{ cubits}) = 6 \cdot ;01\ 15 \text{ n./c.} = ;07\ 30 \text{ n./c.}$$

The double feed is multiplied by the height of the truncated ridge pyramid:

$$2 \cdot f \cdot h´ = ;07\ 30 \text{ n./c.} \cdot 24 \text{ c.} = 3 \text{ n.}$$

This is how much smaller the upper length and the upper front of the truncated ridge pyramid are than the lower length and the lower front, respectively. Thus,

$u´ = u - 3 \text{ n.} = 10 \text{ n.} - 3 \text{ n.} = 7 \text{ n.},$
$s´ = s - 3 \text{ n.} = 6 \text{ n.} - 3 \text{ n.} = 3 \text{ n.}$

Inserting these computed values into the equation for the volume of the truncated ridge pyramid, one obtains the following result (unfortunately not present in the preserved part of the text):

$V´ = \{(10 \cdot 6 + 7 \cdot 3) + (10 \cdot 3 + 7 \cdot 6)/2\} \text{ sq. n.} \cdot 24 \text{ c./3}$
$ = 1\ 57 \text{ sq. n.} \cdot 8 \text{ c.} = 15\ 36 \text{ šar},$
$C´ = c \cdot V´ = 1\ 30 \text{ gur/šar} \cdot 15\ 36 \text{ šar} = 23\ 24\ 00 \text{ gur.}$

Note that the first step in the computation of the volume V' would be to compute the product of u and s as $10 \cdot 6 = 1 \ (00)$. This proposed first step of the computation agrees well with the only preserved part of the calculation of V', which is '[⋯] times [⋯] = 1' in the last preserved line of § 1 f.

Systems of linear equations for the length u and the ridge r

BM 96954 + BM 102366 + SÉ 93 § 1 g.

[gur₇]
[6 sag 48 sukud 28 48 še gur] uš *ù* / ⋯   ⋯
[uš sagšu en].nam
za.[e]
igi 48 / [du₈.a 1 15 *ta-mar*] 1 15 *a-na* 28 48 še / [*i-ši* 36] *ta-mar*
igi 1 30 igi.gub.ba du₈.a / [40 *a-n*]*a* 36 *i-ši* 24 *ta-mar*
igi 6 sag / [du₈.a] 10 *ta-mar* 24 *a-na* 10 *i-ši* 4 *ta-mar* /
[4] *i-na* ul.<gar> ba.zi 10 *ta-mar* 10 uš 4ᵥ sagšu /
[*ki-a*]-*am ne-pé-šum*

A granary.
6 the front, 48 the height 28 48 g ur the barley. The length and ⋯   ⋯ .
The length and the ridge are what?
Y*ou:*
The opposite of 48 *release, 1 15 you* see. 1 15 to 28 48 the barley *raise*, 36 you see.
The opposite of 1 30, the constant, release, *40 to* the 36 raise, 24 you see.
The opposite of 6 the front re*lease*, 10 you see. 24 to 10 raise, 4 you see.
4 from the gath<ering>, tear off, 10 you see. 10 the length, 4 the ridge.
Such is the procedure.

The beginning of this exercise, including the question, is almost completely lost. Nevertheless, it is not difficult to reconstruct both the problem statement and the missing part of the solution procedure. Indeed, enough of the text is preserved so that it is clear that the solution procedure begins with the computation of the following quantities:

$1/h \cdot C = 1/48$ c. $\cdot \ 28 \ 48 \ 00$ gur $= 36 \ 00$ gur/c.,
$1/h \cdot C \cdot 1/c = 36 \ 00$ gur/c. $\cdot \ 1/(1 \ 30 \ \text{gur/šar}) = 24$ šar/c. $= 24$ sq. n.,
$1/h \cdot C \cdot 1/c \cdot 1/s = 24$ sq. n. $\cdot \ 1/ 6$ n. $= 4$ n.

In view of the known equation for the grain measure of the ridge pyramid, the last of these equations can be interpreted as saying that

$C/(c \cdot s \cdot h) = \{c \cdot (r/2 + s/3) \cdot s \cdot h\}/(c \cdot s \cdot h) = (r/2 + s/3) = 4$ n.

Then this quantity is subtracted from a certain 'gathering', that is from *a certain sum S*, and it is stated that the remainder is equal to the length u of the ridge pyramid. Counting backwards, one finds that

$S - (r/2 + s/3) = u \Rightarrow S = u + r/2 + s/3 = 14$ n. (given).

What this means is that the stated problem must have been (in modern symbolic notations) the following *system of linear equations* for the parameters of the ridge pyramid:

$s = 6$ n., $h = 48$ c.,
$C = c \cdot (u + r/2)/3 \cdot s \cdot h = 28\ 48\ 00$ gur (where c = 1 30 gur/šar),
$u + r/2 + s/3 = S = 14$ n.

Through simple arithmetic, this complicated system can be reduced to its essential part, namely the following *pair of linear equations* for the two unknowns:

$r/2 + s/3 = C/(c \cdot s \cdot h) = 4$ n.
$u + r/2 + s/3 = S = 14$ n.

The form of the given sum S was chosen in such a way that this pair of linear equations can be solved *in a trivial way*, simply by subtraction.

A similar problem in **BM 96954+ § 1 h** has to be solved in a less trivial way. In § 1 h, the values of s, h, and C are once again given. In addition, there is the following equation:

[2' sa]gšu ki-ma igi.5.gál uš
1/2 the ridge is like the 5th-part of the length.

Therefore, the length u and the ridge r have to be found as the solutions to the following *system of linear equations*:

$(u + r/2)/3 = C/(c \cdot s \cdot h) = 4$ n., $1/2\ r = 1/5\ u$.

This system of linear of equations is solved by use of the rule of false value. The first step of the solution algorithm is to assume the false values u^* of u and r^* of r to be

5 'bal uš', 2 'bal sagšu'
(the exact meaning of the term bal is not clear)

Next follows the computation of

$(u + r/2)/6 = 1/2 \cdot C/(c \cdot s \cdot h) = 4$ n./2 = 2 n.

The necessary correction factor k is computed as follows (although the text is unclear here):

$(u^* + r^*/2)/6 = (5 + 1/2 \cdot 2)/6 = 1, \quad k = 2 \text{ n.}/1 = 2 \text{ n.}$

The last step of the computation is to compute

$u = u^* \cdot k = 5 \cdot 2 \text{ n.} = 10 \text{ n.}, \quad r = r^* \cdot k = 2 \cdot 2 \text{ n.} = 4 \text{ n.}$

Rectangular-linear systems of equations

In **BM 96954+ § 1 i**, the length, the height, and the grain measure of the ridge pyramid are given, $u = 10$ n., $h = 48$ c., $C = 28\ 48$ gur. In addition, there is a given linear relation between the front s and the ridge r, namely $2/3 \cdot s = r$. As a result, the front and the ridge can be determined as the solutions to the following *rectangular-linear system of equations*:

$(10 \text{ n.} + r/2)/3 \cdot s = C/(c \cdot h) = 24$ sq. n.,
$2/3\ s = r.$

The solution procedure in the text of § 1 i is brief and omits several crucial steps. Nevertheless, it is possible to figure out what was in the mind of the author of the text. With the missing steps reinstated, the solution procedure gets along as follows. The first step is to replace the rectangular-linear system of equations for s and r with a single *quadratic equation* for s:

$(10 \text{ n.} + 1/2 \cdot 2/3 \cdot s)/3 \cdot s = 24$ sq. n.

Next, s is replaced by $s^* = 1/2 \cdot 2/3 \cdot 1/3 \cdot s$. The resulting quadratic equation for s^* is then

$(3;20 \text{ n.} + s^*) \cdot s^* = 1/2 \cdot 2/3 \cdot 1/3 \cdot 24$ sq. n. $= 2;40$ sq. n.

By use of a Babylonian standard procedure, the solution to this quadratic equation is found to be $s^* = ;40$ n. On the other hand, $1/2 \cdot 2/3 \cdot 1/3 = ;06\ 40$. Therefore,

$s = s^* \cdot 1/;06\ 40 = ;40 \text{ n.} \cdot 9 = 6 \text{ n.}, \quad \text{and} \quad r = 2/3\ s = 4 \text{ n.}$

BM 96954+ §§ 1 j, 1 k, 1 m, are three further examples of problems leading to rectangular-linear systems of equations for u and s, or for r and s. The remaining paragraph of the first section, **BM 96954+ § 1 l**, is inserted somewhat out of place between two problems leading to rectangular-linear systems of equations. In § 1 i, the ridge r and the height h are given (as well as the grain measure C, which is not needed in this exercise). In addition, the *inclination f* of the sides of the ridge pyramid, the feed, is given through the phrase

ša 1 kùš 7 30 kùš šà.gal

of 1 cubit 7 30, cubits, is the feed.

What this obscure phrase means is that for each cubit of descent from the top, the side goes out ;07 30 ninda = 1 1/2 cubits. It is, therefore, a trivial matter to compute the length:

$u = r + f \cdot h = 4$ n. $+;07\ 30$ n./c. $\cdot 48$ c. $= 4$ n. $+ 6$ n. $= 10$ n.

Problems for a circular cone and its 'feed'

With **BM 96954 + § 4** begins a new part of the text, differing in several ways from the first part, consisting of § 1 and § 3. The solids considered in § 4 are whole or truncated circular cones (Figs. 4.8.6-7).[46] The height of the whole cone is '1', not 48 (cubits) as in the case of all the solids in §§ 1 and 3. Most significantly, the content of the cone is expressed in terms of its *volume*, not its grain measure.

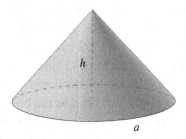

Fig. 4.8.6. BM 96954+ §§ 4 a-d: A circular cone.

About half of the text of this exercise is destroyed. However, since §§ 4 a and 4 b are closely related, the reconstruction below is fairly certain and will be taken for granted here. The object of the exercise is to find the 'feed' of a circular cone with the height h = '1', almost certainly meaning 1 00 cubits = 5 n. (30 meters), and with the 'arc' a, that is the circumference of the circular base, correspondingly equal to or 30 n. The diameter of the circular base is then 10 n. (if the diameter is assumed to be 1/3 of the circumference, as usual). (Cf. BM 96954 § 4 e below, where a cone with the height '1' is truncated 2 1/2 ninda = 30 cubits below the top.)

---

46. What is here called § 4 of BM 96954+ is called problems xx-xxviii in Robson, *MMTC* (1999), App. 3. Robson's attempted comments in her footnotes 40-49 show that she has absolutely no clue what is going on in these exercises.

BM 96954+ § 4 a.

> [... ...]
> [30 gúr 1 sukud] *i-na* 1 kùš / [šà.gal en.nam]
> [za].e
> igi 1 sukud du₈.a 1 *ta-mar* / [1 *a-na* 30 *i*]-*ši* 30 *ta-mar* /
> 2' 30 *ḫe-pé* / [15 *ta-mar*] 15 *i-na* 1 [kùš] šà.gal /
> [*ki-a-a*]*m ne-pé-šum*

A ... ...
30 the arc, 1 the height. In 1 cubit, the feed is what?
You:
The opposite of 1, the height, release, 1 you see. *1 to 30 rai*se, 30 you see.
1/2 of the 30 break, 15 you see. 15 in 1 *cubit* is the feed.
Such is the doing.

In the case of the ridge pyramid considered in BM 96954+ § 1, the term šà.gal 'feed' was a measure for the inclination of the sides of the pyramid. More precisely, in §§ 1 g and 1 l the feed f of the ridge pyramid was equal to $s/h = (u - r)/h =$;07 30 n./c. (= 1 1/2 c./c.). This is in accordance with what is known about the use of the term šà.gal 'feed' and the related term kú 'food' in several other OB mathematical texts. See, for instance, Neugebauer and Sachs, *MCT* (1945), text K, and the brief survey (*ibid.*), p. 81, fn. 191. The meaning of both terms is clearly "the inclination of the sides" (of a canal, an excavation, or a construction), measured in terms of the horizontal increase (or decrease) of the width (in ninda, cubits, or fingers) for each cubit of vertical descent.

Interestingly, this interpretation of šà.gal 'feed' is no longer adequate in the case of the present text, where the term appears in §§ 4 a, 4 b, and 4 d. What the correct definition is here is not immediately clear. Note, however, the following rather explicit question and answer mentioning the feed in the statement of the problem in § 4 a:

> *i-na* 1 kùš / [šà.gal en.nam] 15 *i-na* 1 [kùš] šà.gal
> 'In 1 cubit, the feed is what? 15 in 1 cubit is the feed'.

The interpretation offered here is that the *intended* definition of the 'feed' in § 4 of BM 96954 + is a "quadratic inclination", equal to the *square* of ratio of the 'arc' of a horizontal section of the cone to the distance (in cubits) of that arc from the top of the cone. More precisely, this "quadratic feed" F is computed in the following way:

4.8. P.Vindobonensis G. 19996 (1st C. CE?). Stereometric Exercises

$F = $ sq. $(a/h) = a/h \cdot a/h$.

This means that in § 4a, where $a = 30$ n. and $h = 1$ 00c., this feed F ought to be equal to

$F = $ sq. (30 n. /1 00 c.) = ;15 sq. (n./c.)

However, what actually happens in § 4 a is that F is computed as follows, in floating place value notation, and without explicit mention of the names of the units of length:

$F = 1/2 \cdot a/h = 1/2 \cdot 30/1 = 15$.

Numerically, and in the given situation, the result is correct. However, the scribe who wrote the exercise seems to have been carried away by his success, continuing to apply the incorrect relation $F = 1/2 \cdot a/h$ also in § 4 b.

BM 96954 + § 4 b

```
[ ··· ]x-ru-tum
30 gúr i-na 1 kùš 15 šà.gal / sukud en.nam
za.e
15 šà.gal tab.ba 30 ta-mar / igi 30 du₈ 2 ta-mar
30 gúr! a-na 2 i-ši / 1 ta-mar sukud
ne-pé-šum
```

A ··· ··· ··· .
30 the arc, in 1 cubit 15 is the feed. The height is what?
You:
15 the feed repeat, 30 you see. The opposite of 30 release, 2 you see.
30 the arc to 2 raise, 1 you see, the height.
The doing.

Apparently, the text of all the exercises BM 96954 §§ 4 a-h begins by mentioning the name of a 'cone' or a 'truncated cone'. In § 4 b the name for a cone is [···]x-ru-tum. In § 4 d, it is [··· ···]-tum. Everywhere else, the name is lost. Jursa and Radner, who published the piece SÉ 93 (*AfO* 42/ 43 (1995/96)), suggested that the name could be [b]é-ru-tum 'mound'. Another possibility is that the name should be read [ga]m-ru-tum 'the whole', and that it is the name for a whole (cone), as opposed to a truncated (cone).

In § 4b, which is completely preserved, the arc $a = $ '30' and the feed $F = $ '15' are given, but the height h is unknown. The correct way of computing the height would have been as follows:

$F = $ sq. (a/h) ⇒ sq. $h = $ (sq. a)$/F$, $h = a/($sqr. $F)$, or, in relative numbers,

$F = 30/(\text{sqr. } 15) = 30/30 = 1$

Instead, the height h is computed, incorrectly, in the following way:

$F = 1/2 \cdot a/h$ (as in § 4 a) $\Rightarrow$ $h = a/(2\,F)$,

or

$2\,F = 2 \cdot 15 = 30$, $1/(2\,F) = 1/30 = 2$, $h = a/(2\,F) = 30 \cdot 2 = 1$.

Here again, the result is numerically correct, in the given situation.

The next exercise, **BM 96954+ § 4 c**, is a brief exercise in four lines. The text of the exercise is heavily damaged, only the lower right corner is preserved. It is clear, anyway, that the arc $a = 30$ n. is mentioned, after 30 has been computed as the square root of something, obviously 15.

A reconstruction of the damaged text of § 4 c is suggested below. Note that, according to the reconstruction, the quadratic feed plays no role in this exercise.

BM 96954+ § 4 c.

[⋯ ⋯] [25 saḫar.ḫi.a 1 sukud]
[za].e /
[3 *a-na* 25 *i-ši* 1 15 *ta-mar*] gar.ra /
[igi 5 du₈.a 12 *a-na* 1]15 *i-ši* 15 *ta-mar* /
[en.nam íb.si₈ 30 í]b.si₈ 30 ninda gúr /
[*ki-a-am n*]*e-pé-šum*

A ⋯ ⋯ ⋯ 25 the mud (volume), 1 the height.
You:
3 to 1 15 raise, 1 15 you see. Set it.
The opposite of 5 resolve. To 1 15 raise (it), 15 you see.
What is it equalsided? 30 it is equalsided. 30 n. the arc.
*Such is the do*ing.

Here the given parameters are the volume V and the height h, and the arc a is computed in the following way:

$V = A \cdot h/3$ $\Rightarrow$ $A = 3 \cdot V/h = 3 \cdot 25\ 00$ šar/$1\ 00$ c. $= 1\ 15$ sq. n.,
$A = ;05 \cdot$ sq. a $\Rightarrow$ sq. $a = 1/;05 \cdot 1\ 15$ sq. n. $= 12 \cdot 1\ 15$ sq. n.
$= 15\ 00$ sq. n., $a = $ sqr. $15\ 00$ sq. n. $= 30$ n.

(Remember that the Old Babylonian author of the text computed with *relative* sexagesimal numbers. The zeros, *etc.* are inserted here, as elsewhere in this account, only for the readers' convenience.)

The full text of **BM 96954+ § 4d** is preserved, with the exception of the name of the solid considered, presumably as in §§ 4 a-c a 'mound', meaning 'cone'. The fact that the name is damaged here, as well as in all the other exercises of this paragraph, is unfortunate. Note that § 4 d contains a couple of other syllabically written, previously unknown terms, the Akkadian names for the arc and the height of a circular cone. The syllabically written name u for the arc is *ma-ṣa-rum*, a term derived from *maṣārum* 'to go around'. Cf. the phrase 10 š u . s i *am-ṣú-ur* '10, fingers, I went around', meaning 'I made the circumference (of a cylindrical barig-measure) equal to ;10 n. = 1 00 fingers', in the Sippar text BM 85194 § 35 (Sec. 2.3 a above). The syllabically written name for the height is *di-ik-šum*, derived from *dakāšum*, a verb with an elusive meaning, apparently something like 'stick through, stick out'. See the discussion in Friberg, *RlA 7* (1990) Sec. 5.4 1. Other noteworthy examples of the appearance of this word can be found in the OB mathematical catalog text BM 80209, also a Sippar text (Friberg, *JCS* 33 (1981)), and in the Late Babylonian mathematical recombination text W 23 291-x (Friberg, *et al.*, *BaM* 21 (1990) § 1).

BM 96954 + § 4 d.

```
[ ··· ]-tum
25 saḫar.ḫi.a <15 šà.gal> ma-ṣa-rum ù di-ik-šum / en.nam
za.e
25 šu-ul-li-iš 1 15 ta-mar / igi 5 igi.gub dug.a 12 ta-mar
12 a-na 1 15 i-ši / 15 ta-mar gar.ra nígin.na
igi 15 šà.gal dug.a 4 ta-mar / 15 a-na 4 i-ši 1 ta-mar
1 en.nam íb.sig 1 íb.sig / sukud
15 šà.gal tab.ba 30 ta-mar 30 a-na 1 sukud i-ši /
[30] ta-mar ma-ṣa-rum
ne-pé-šum
```

A ··· ··· ···
25 the mud, <15 the feed>. The go-around and the stick-out are what?
You:
25 triple, 1 15 you see. The opposite of 5, the constant, release, 12 you see.
12 to 1 15 raise, 15 you see. Set it! Turn around!
The opposite of 15, the feed, release, 4 you see. 15 to 4 raise, 1 you see.
1, what (is it) equalsided? 1 (it is) equalsided, the height.

15, the feed, double (sic!), 30 you see. 30 to 1, the height, raise,
30 you see, the go-around.
The procedure.

In the statement of the problem in § 4d, the only numerical parameter given is the volume V = '25', probably meaning 25 00 šar. It is silently understood that the feed is the same as in §§ 4a-b, F = ;15 sq. (n./c.). The computation of the height h is straightforward, beginning with

$h \cdot A$ = 3 · V = 3 · 25 00 šar = 1 15 00 šar,
$h \cdot$ sq. a = 1/;05 · $h \cdot A$ = 15 šar = 15 00 00 sq. n. · c.

The next step of the computation makes use of the reciprocal value of the quadratic feed F:

$h \cdot$ sq. $h = h \cdot$ sq. $a \cdot 1/F$ = (15 00 00 sq. n. · c.) · 1/(;15 sq. (n./c.))
= 1 00 00 00 sq. c. · c.

After that, the value of h is found as a cube root:

h = cube root of 1 sq. c. · c. = 1 c.

Finally, the correct value of the arc a is found by use of the incorrect computation

F = 1/2 · a/h (as in § 4 a) ⇒ $a = 2 \cdot F \cdot h = 2 \cdot 15 \cdot 1 = 30$.

Problems for truncated circular cones

Only the upper right corner of the text of **BM 96954+** § 4 e, is preserved. See Fig. 4.8.4 above. (Missing parts of the text are italicized in the translation below and put within square brackets in the transliteration.) Fortunately, in spite of the extensive damage to § 4 e, it is possible to reconstruct some of the most important missing parts of the text. In particular, the whole question, except for the name of the solid considered and the crucial term [*ur-dam*] 'I went down', is still there. (Cf. the occurrence of the term *ur-dam* in § 1 f, the exercise dealing with a truncated ridge pyramid.) Thus, it is clear that the object of the exercise is to compute the volume of a truncated cone.

The truncated cone in § 4 e is what remains after a truncation 2 1/2 n. (= 30 c.) below the top, that is at mid-height, of a cone with the given volume V = 25 00 šar, the height h = 1 00 cubits = 5 ninda, and the arc a at the base = 30 ninda. (See Fig. 4.8.7 below, left.)

4.8. P.Vindobonensis G. 19996 (1st C. CE?). Stereometric Exercises

BM 96954 + § 4 e.

```
... ...
30 ma-ṣa-rum
1 sukud 25 saḫar.ḫi.a 2 2' ninda / [ur-dam]
saḫar.ḫi.a en.nam
za.e
igi 1 sukud du₈.a / [1 ta-mar]
[a-na] 30 ma-ṣa-rum i-ši 30 ta-mar
30 a-na / [30 ⋯ ⋯ i]-ši 15 ta-mar
15 i-na 30 ma-ṣa-rum / [ba.zi 15 ta-mar]
[saḫar].ḫi.a ki.ta¹ <en.nam>
30 ma-ṣa-rum ki.ta / [nigin 15 ta-mar]
[a-na 5 ig]i.gub.ba i-ši 1 15 ta-mar /
[15 ma-ṣa-rum an.ta nigin 3 4]5 ta-mar
3 45 a-na / [5 igi.gub.ba i-ši 18 45 ta-mar]
[1 1]5 ú 18 45 / [ul.gar 1 33 45 ta-mar]
[2' 1 33 45 ḫe-pé 46 5]2 30 ta-mar /
[30 ù 45 ul.gar 45 ta-mar 2'] 45 ḫe-pé / [22 30 ta-mar]
[22 30 nigin 8 26 15 ta-mar]
[8 2]6 15 / [a-na 5 igi.gub.ba i-ši 42 11 15 ta-mar] ⋯ ⋯ ⋯ ⋯
```

... ...
30, the go-around,
1 the height, 25 the mud. 2 1/2 ninda *I went down.*
The <lower> mud is what?
You:
The opposite of 1, the height, release, *1 you see.*
To 30, the go-around, raise (it), 30 you see.
30 to *30 ⋯ ⋯ r*aise, 15 you see.
15 from 30, the go-around, *tear off, 15 you see.*
The lower *mud* <is what>?
30, the lower go-around, *square, 15 you see.*
*To 5, the co*nstant, raise (it), 1 15 you see.
15, the upper go-around, square, 3 45 you see.
3 45 to *5 the constant raise, 18 45 you see.*
1 15 and 18 45 *gather, 1 33 45 you see.*
1/2 of 1 33 45 break, 46 52 30 *you see.*
30 and 45 gather, 45 you see. 1/2 of 45 break, 22 30 you see.
22 30 square, 8 26 15 you see.
8 26 15 *to 5, the constant,* raise, 42 11 15 you see. ⋯ ⋯ ⋯ ⋯

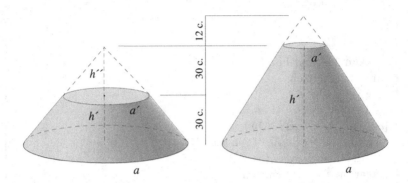

Fig. 4.8.7. BM 96954+ §§ 4 e-h: Two truncated cones.

In the first step of the solution procedure, the upper arc $a´$ of the truncated cone (that is, the circumference of the circular top of the truncated cone) is computed, by use of what is in effect a similarity argument, namely as

$a´ = a – a/h · h´$ = 30 n. – (30 n./1 00 c.) · 30 c. = 30 n. - 15 n. = 15 n.

Here $h´$ is the height of the truncated cone.

Next, the lower area A at the base of the truncated cone and the upper area $A´$ at its top are computed as

A = ;05 · sq. a = ;05 · sq. (30 n.) = ;05 · 15 00 sq. n. = 1 15 sq. n.,
$A´$ = ;05 · sq. $a´$ = ;05 · sq. (15 n.) = ;05 · 3 45 sq. n. = 18;45 sq. n.

The arithmetic mean of A and $A´$, the "average area" A_a, is computed as

A_a = ;05 · (sq. a + sq. $a´$)/2 = (A + $A´$)/2 = 1/2 · 1 33;45 sq. n. = 46;52 30 sq. n.

In the next lines (of which not much remains), the "middle area" A_m (the area at mid-height of the truncated cone) seems to be computed in the following way:

sq. {(a + $a´$)/2} = sq. (22;30 n.) = 8 26;15 sq. n.,
A_m = ;05 · sq. {(a + $a´$)/2} = ;05 · 8 26;15 sq. n. = 42;11 15 sq. n.265

Note that here (and always) the average area A_a is larger than the middle area A_m.

The rest of the solution procedure is not preserved. A couple of formally slightly different reconstructions are possible. The first steps of both are the following: Let the "average volume" V_a and the "middle volume"

V_m be the result when A_a and A_m, respectively, are multiplied by the height h' of the truncated cone:

$V_a = A_a \cdot h' = 46;52\ 30$ sq. n. $\cdot\ 30$ c. $= 23\ 26;15$ sq. n. $\cdot$ c. $= 23\ 26;15$ šar,

$V_m = A_m \cdot h' = 42;11\ 15$ sq. n. $\cdot\ 30$ c. $= 21\ 05;37\ 30$ sq. n. $\cdot$ c $= 21\ 05;37\ 30$ šar.

Obviously, V_a and V_m are two different *approximations* to the correct volume V'' of the truncated cone. (Actually, V_a is an approximation from above, V_m an approximation from below.) The difference between the two approximations is

$V_a - V_m = 23\ 26;15$ šar $- 21\ 05;37\ 30$ šar $= 2\ 20;37\ 30$ šar.

Continuing from here, one possibility is that the volume V'' was computed as (cf. Friberg, *PCHM* 6 (1996) § 4)

$V' = V_m + (V_a - V_m)/3 = 21\ 05;37\ 30$ šar $+ 46;52\ 30$ šar $= 21\ 52;30$ šar.

A mathematically equivalent alternative is that V' was computed as

$V' = 1/3 \cdot V_a + 2/3 \cdot V_m = 7;48\ 45$ šar $+ 14;03\ 45$ šar $= 21\ 52;30$ šar.

In both cases, the result is the same.

This result can be checked in the following way. The difference between the volume V of the whole cone and the volume V'' of the truncated cone ought to be equal to the volume V''' of the small cone *above* the truncation plane. Since the cone is truncated at mid-height, and since the small cone and the whole cone are *similar solids*, V''' ought to be equal to 1/8 of V. In other words, another way of computing the volume of the truncated cone is to proceed as follows: Since

$V = A \cdot h/3 = 1\ 15$ sq. n. $\cdot\ 20$ c. $= 25\ 00$ sq. n. $\cdot$ c. $= 25$ šar

(as stated in the text), it follows that

$V' = (1 - 1/8) \cdot V = 25\ 00$ šar $- 3;07\ 30$ šar $= 21\ 52;30$ šar.

The result is, as it should be, the same as before. (A similar method would work in the case of the computation of the volume of the truncated ridge pyramid in § 1 f. In that case, however, the whole ridge pyramid and the smaller ridge pyramid above the truncation plane *are not similar solids*, so it would be incorrect to draw the conclusion that the volume of the smaller ridge pyramid is simply 1/8 of the volume of the whole ridge pyramid.)

The text of **BM 96954 § 4 f** is even more damaged than the text of § 4 e. Only parts of the last few lines are preserved on the edge of the clay tablet.

See again Fig. 4.8.4 above. Luckily, that is all that is needed for the understanding of what happens in this exercise.

BM 96954 + § 4 f.

```
[ ···   ···   ···   ···   ···   ···   ··· ta-m]ar /
[ ···   ···   ···   ···   ···   ···   ··· ] i-ši /
[ ···   ···   ···   ···   ···   ···   ··· ] 29 51 40 [ta-mar ··· ]
[29 51 40] / a-na 1 sukud i-ši 29 51 40 ta-mar saḫar.[ḫá]
```

··· ··· ··· ··· ··· ··· ··· *you see.*
··· ··· ··· ··· ··· ··· ··· raise,
··· ··· ··· ··· ··· ··· ··· 29 51 40 *you see,* [··· .]
The *29 51 40* to 1 the height raise, 29 51 40 you see, the mud.

The last preserved line of § 4 f contains the *last operation* of the solution procedure and the *answer* to the problem. Those two pieces of information show that the object considered in § 4 f is a solid with the height '1' and the volume '29 51 40'. It is reasonable to assume that here again '1' means 1 00 cubits, as, presumably, in all the preceding exercises of § 4, and that, correspondingly, the volume is $V = 29\ 51;40$ šar. It is also reasonable to assume that the solid considered here is a *truncated cone*, as in the preceding exercise § 4 e.

Now, in the comment above to § 4 e, it was shown, through an argument based on similarity, that the volume of a circular cone truncated at mid-height is $(1 - 1/8)$ of the volume of the whole cone. The same type of argument can be used to show that if a circular cone with the height h is truncated at a distance of $1/n \cdot h$ from the top, where n is an integer, then the volume of the truncated cone is $(1 - 1/n^3)$ of the volume of the whole cone. Consider the sexagesimal number 29 51;40. It is close to the *round number* 30 00. The difference between 30 00 and 29 51;40 is only 8;20, which in its turn is *a simple fraction* of 30 00. Indeed,

$6 \cdot 8;20 = 50, \quad 6 \cdot 50 = 5\ 00, \text{ and } 6 \cdot 5\ 00 = 30\ 00.$

Therefore, the volume mentioned in the last line of § 4 f is

$V = 29\ 51;40 \text{ šar} = (1 - 1/6^3) \cdot 30\ 00 \text{ šar}.$

This result implies that the solid in § 4 f is what remains when a circular cone is truncated by a plane a distance 1/6 of the height h of the cone below the top of the cone. On the other hand, the height h' of the truncated cone is 1 cubit. Taken together, these facts demonstrate that

$h' = 1\ 00$ c. $= (1 - 1/6) \cdot h$,

so that the height of the whole cone must be

$h = (1 + 1/5) \cdot h' = 1\ 12$ cubits $= 13$ ninda.

Taking one's cue from § 4 c, one can now compute the area A and the arc a (that is, the circumference) of the base of the cone as follows:

$A = 3 \cdot V/h = 3 \cdot 30\ 00$ sq. n. $\cdot$ c./1 12 c. = 1 15 sq. n.,
sq. $a = 1/;05 \cdot 1\ 15$ sq. n. $= 15\ 00$ sq. n.,
$a =$ sqr. 15 00 n. $= 30$ n.

The arc a' at the top of the truncated cone in § 4 f must then be

$a' = 1/6 \cdot a = 5$ n.

See Fig. 4.8.7 above, right.

According to the suggested reconstructed outline above of BM 96954+ and its division into paragraphs, there seems to be space left at the top of col. *iii* on the reverse of the clay tablet for a brief paragraph, here called **BM 96954+ § 4 g**. However, the space available is so restricted that it is possible that the first few lines of col. *iii* were just a continuation of § 4 f. (Note that there is no end phrase *ne-pé-šum* at the bottom of col. *ii.*)

BM 96954 + § 4 h.

```
[ ···   ···   ···   ···   ···   ···   ··· ] 29 51 40 saḫar.ḫi.<a> |
[ ···   ···   ···   ···   ···   ···   ···   ··· ] x2 30 |
[ ···   ···   ···   ···   ···   ···   ···   ··· ] x
```

··· ··· ··· ··· ··· ··· ··· 29 51 40 the mud.
··· ··· ··· ··· ··· ··· ··· ··· ··· x 2 30.
··· ··· ··· ··· ··· ··· ··· ··· ···

Of the text of this exercise, only a few signs are preserved at the ends of the first couple of lines. This time, by a lucky coincidence, a part of the question is preserved, prescribing that the volume of the solid considered should be '29 51 40', just as in § 4 f. Therefore it is clear that § 4 h is a problem about *the same truncated cone* as the one in § 4 f. The only difference is that, apparently, in § 4 f the three linear parameters $a = 30$ n., $a' = 5$ n., and $h' = 1\ 00$ c. were given and the volume V'' unknown, whereas in § 4 h the volume and two of the linear parameters are given while the third linear parameter is unknown.

4.9. Conclusion

Relations between Greek-Egyptian and Babylonian mathematics

The discussion above has clearly demonstrated, as promised in the Introduction to Chapter 4, that there is no discernible difference between the form and content of demotic and (non-Euclidean) Greek-Egyptian mathematical papyri. In addition, there is little difference in form and content between these two groups of Egyptian mathematical documents and Late or Old Babylonian mathematical cuneiform texts. As a matter of fact, the main difference between the mathematics of these (non-Euclidean) Ptolemaic/Roman Egyptian mathematical papyri and cuneiform mathematical texts seems to have been not only the counting with sums of parts but also the counting with binomial fractions in demotic mathematical texts such as *P.Cairo* § 1 (Sec. 3.1.c), *P.Carlsberg 30 #* 2 (Sec. 3.5 b), *P.BM 10520* § 5 (Sec. 3.3 e), and in Greek mathematical papyri such as *P.Akhmîm* §§ 4-10 (Sec. 4.5 f). As mentioned in Sec. 3.1 c, this counting with binomial fractions may have had its roots in the counting with red auxiliaries in hieratic mathematical texts, as in the example in Fig. 3.1.3. Another seemingly important difference in comparison with Babylonian mathematical texts is the new(?) idea of using a symbol for the unknown quantity in the tabular arrays illustrating solution procedures for systems of linear equations in the Greek mathematical papyrus *P.Mich. 620* (Sec. 4.4).

New Thoughts About the History of Ancient Mathematics

The aim of the discussion in this book has been to try to spread some light over the difficult question of differences and similarities between form and content of mathematics in the corpus of mathematical papyri from Egypt on one hand and the corpus of mathematical cuneiform texts from Mesopotamia on the other. The result of the discussion should be a better understanding of the development of what may be called the ancient mathematical tradition. Here is a brief summary of what the situation seems to be like, in the opinion of the present writer.

Conceivably, the initial development of mathematical ideas started at a very early date with the invention of words for sexagesimal or decimal numbers in various ancient languages, and with the widespread use of number tokens in the Middle East. A major step forward was then the invention of an integrated family of number and measure systems, in connection with the invention of writing in Mesopotamia and neighboring areas of Iran in the late fourth millennium BCE. There must have been a similar development in Egypt, about which not much is known at present. A small number of known examples of proto-Sumerian, Old Sumerian, Old Akkadian, and Eblaite mathematical exercises and table texts are witnesses of the continuing important role played by education in mathematics in the scribe schools of Mesopotamia throughout the third millennium.

Then there is a strange gap in the documentation, with almost no mathematical texts known from the Ur III period in Mesopotamia towards the end of the third millennium BCE. Nevertheless, at some time in the Ur III period a new major step in the development of mathematics was taken with the invention of sexagesimal place value notation. To a large part as a result of that invention, mathematics flourished in the Old Babylonian scribe schools in Mesopotamia. Simultaneously, mathematics may have reached a comparable level in Egypt, and, in spite of the fundamentally different ways of counting in the two regions, there was clearly some communication of mathematical ideas between Egypt and Mesopotamia.

A few late Kassite mathematical texts seem to indicate that the Old Babylonian mathematical tradition was still operative, although reduced to a small trickle, in the second half of the second millennium BCE.

Then follows a new strange gap in the documentation. When mathematics flourished again in Mesopotamia in the Late Babylonian and Seleucid periods in the second half of the first millennium BCE, possibly in connection with the rise of mathematical astronomy, a great part of the Old Babylonian corpus of mathematical knowledge had been taken over relatively intact. However, for some reason, the transmission of knowledge cannot have been direct, which is shown by an almost complete transformation of the mathematical vocabulary.

Similarly in Egypt, after a comparable gap in the documentation, there was a new flourishing of mathematics, documented by demotic and Greek mathematical papyri and ostraca from the Ptolemaic and Roman periods. Some of the Greek mathematical texts are associated with the Euclidean type of high-level mathematics. Except for those, the remainder of the demotic and Greek mathematical texts show clear signs of having been influenced both by Egyptian traditions, principally the counting with sums of parts, and by Babylonian traditions. An interesting new development was the experimentation with new kinds of representations of fractions, first sexagesimally adapted sums of parts, soon to be abandoned in favor of binomial fractions, the predecessors of our common fractions.

The observation that Greek ostraca and papyri with Euclidean style mathematics existed side by side with demotic and Greek papyri with Babylonian style mathematics is important for the reason that this surprising circumstance is an indication that when the Greeks themselves claimed that they got their mathematics from Egypt, they can really have meant that they got their mathematical inspiration from Egyptian texts with mathematics of the Babylonian type. To make this thought much more explicit would be a natural continuation of the present investigation.

Index of Texts

Text	Sec.	Fig.	Topic	Kind
AO 5449	1.2 d		growth of a heard of cows	Ur III
AO 6484 # 1	1.1 e		sum of 10 first powers of 2	Sel.
AO 6484 # 2	3.3 a		sum of 10 first square numbers	Sel.
AO 8862 § 2	4.4 c, 4.5 d		tabular array, unequal shares	OB
BM 13901 # 13	2.3 a		quadratic-linear system of equations	OB
BM 22706	1.1 b		table of powers	OB
BM 34568	4.7 a		exerc. for sides and diagonal of rect.	Sel.
BM 34568 # 12	3.1 b		pole against a wall	Sel.
BM 34601	3.3 c		explicit multiplication algorithm	LB
BM 34800	2.1 b, 4.4 c	2.1.2-3	repeated division problem	OB
BM 47437	3.3 g		field plan	LB
BM 67314	3.3 g		application of quadrilateral area rule	LB
BM 85194	2.4	2.4.1	outline, contents	OB
BM 85194 # 4	3.7 c		volume of a ring wall; work norm	OB
BM 85194 ## 22, 35	2.3 b		content of cylindrical vessels	OB
BM 85194 # 28	4.8 g		volume of a truncated pyramid	OB
BM 85194 # 29	3.1 k		circle segment (corrupt text)	OB
BM 85194 # 36	3.1 j		orientation of circular segments	OB
BM 85196 # 9	3.1 b		pole against a wall	OB
BM 85210	2.2 e		constants for semicircle	OB
BM 96954+	2.1 d, 2.2 d, 4.5 c		pyramids and cones	OB
BM 96954+	4.8 g	4.8.4	outline, table of contents	OB
Elements XII	2.2 d		volume of a pyramid	Greek
Haddad 104 # 2	3.2 b		capacity measure of a truncated cone	OB
IM 43996	2.1 d, 3.7 c	2.1.9	striped triangle	OB
IM 52301	3.3 g		the general quadrilateral area rule	OB
IM 53957	2.1 b		applied division problem	OB
IM 58045	3.7 c		equipartitioned trapezoid	OAkk
IM 67118 = Db_2-146	3.1 i		metric algebra, rectangle	OB
IM 73355	1.1 b	1.1.3	two tables of powers	OB
IM 121565	3.3 g		quadrilateral area rule	OB
IM 121613 # 1	2.2 c, 3.1 e	1.4.4	a rectangle, metric algebra	OB
Inventor of chess	1.2 d		64 powers of 2	Ind.

Ist. O 236	2.1 e	2.1.21	bread rations of various sizes	OB
Ist. O 3826	1.1 b		table of powers	OB
JZSS 1:35-36	3.1 k		segment area rule	Chin.
JZSS 5	2.2 d		volume of a pyramid	Chin.
JZSS 6:18	2.1 c		arithmetic progression	Chin.
Liber abaci, fol. 138 r	1.1 g		*Septem vetule vadunt romam*	Eur.
M. 7857	1.1 a	1.1.2	sum of geometric progression	OB
M. 8631	1.2	1.2.1	geometric progression., 30 terms	OB
Metrica I: 5-6	4.7 b		height of unsymmetric triangles	Greek
Metrica I: 30	3.1 k, 4.3 c		segment area rule	Greek
Michael 62	1.2 c, 4.6		prices and market rates	Greek
MLC 1354	2.2 e		problem for a semicircle	OB
MLC 1842	4.6		buying and selling exercise	OB
Mother Goose	1.1 g		sum of geometric progression	Eur.
MS 1844	1.1 d, 3.1 g	1.1.6	7 brothers, geometric progression	OB
MS 2049, *obv.*	3.1 j		orientation of circular segment	OB
MS 2221, *rev.*	2.2 b	2.2.1	a combined work norm	OB
MS 2242	1.1 b	1.1.4	descending table of powers	OB
MS 2268/19	4.4 c		tabular array, combined market rate	OB
MS 2830, *obv.*	3.1 g		4 brothers, geometric progression	OB
MS 2832	4.4 c		tabular array, combined market rate	OB
MS 3037	1.1 b		descending table of powers	OB
MS 3048	3.3 a	3.3.1	a quasi-cube table	OB
MS 3049	3.2 a		circle-and-chord problem	late OB?
MS 3050	3.1 k	3.1.12	square in a circle	OB
MS 3051	3.1 j	3.1.8	equilateral triangle in a circle	OB
MS 3052 § 1	3.2 a		growth rate of a mud wall	OB
MS 3866	3.3 b		smallest head number	OB
MS 3876	4.8 a		outer shell of an icosahedron	Kassite
MS 5112 § 1	4.4 c		completing the square	OB
MS 5112 §§ 1, 5	4.4 c		'since it was said to you'	Kassite?
MS 3971 § 2	3.1 i		metric algebra, rectangle	OB
MS 5112 § 9	3.1 f		reshaping a rectangle	OB
MSVO 4, 66	2.1 e	2.1.18-19	bread-and-beer-text	proto-cun.
NCBT 1913	2.1 d	2.1.10	almost round area number	OB
O.Bodl. ii 1847	4.1, 4.7 c		land survey, eight fields	Greek
Pack no. 2323	4, Intr.		*El.* XIII, 10, 16	Greek
P.Akhmîm	1.2 c, 4.5		table of contents	Greek
P.Berlin 6619 # 1	2.3 a	2.3.1	quadratic-linear system of equations	hier.
P.BM 10399 § 1	3.2 a	3.2.2	circle-and-chord problem?	dem.
P.BM 10399 § 2	3.2 b, 4.3		capacity measure of masts	dem.
P.BM 10399 § 3	3.2 c		the reciprocal of $1 + 1/n$	dem.
P.BM 10520 § 1	3.3 a		the iterated sum of 1 through 10	dem.
P.BM 10520 § 2	3.3 b		a multiplication table for 64	dem.
P.BM 10520 § 3	3.3 c		a new multiplication rule	dem.
P.BM 10520 § 4	3.3 d		2/35 expressed as a sum of parts	dem.

Index of Texts

P.BM 10520 § 5	3.3 e, 4.5 e		operations with fractions	dem.
P.BM 10520 § 6	3.1 d, 3.3 f		the square side rule	dem.
P.BM 10520 § 7 a	3.3 g	3.3.2	the quadrilateral area rule	dem.
P.BM 10520 § 7 b	3.3 g, 3.7 a, 4.7 c		schematic quadrilateral	dem.
P.BM 10794	3.4, 4.5 a		the $\overline{90}$ and $\overline{150}$ tables	dem.
P.Cairo	3.1 a		table of contents	dem.
P.Cairo § 1	3.1 c	3.1.2	two related division problems	dem
P.Cairo § 2	3.1 d		completion problems	dem
P.Cairo § 3	3.1 e		a rectangular sail	dem
P.Cairo § 4	3.1 f, 3.2 c	3.1.4	reshaping a rectangular cloth	dem
P.Cairo § 7	3.1 g		shares in a geometric progression	dem
P.Cairo § 8	3.1 b, 4.7 a	3.1.1	pole against a wall	dem.
P.Cairo § 9	4.2, 4.3 b		circle area rule	dem.
P.Cairo § 9 a	3.1 h	3.1.5	diameter of a circle w. given area	dem.
P.Cairo § 10	3.1 b, 3.1 i	3.1.6	rectangle w. given area and diagonal	dem.
P.Cairo § 11	4.3 b, 4.8 a		circle area rule	dem.
P.Cairo § 11 a	3.1 j, 3.3 f	3.1.7	equilateral triangle in a circle	dem.
P.Cairo § 12	3.1 k	3.1.9	square in a circle	dem.
P.Cairo §§ 13-14	3.1 l		pyramids with a square base	dem.
P.Cairo §§ 15-16	3.1 m	3.1.13	metric algebra	dem.
P.Carlsberg 30 # 1	3.1 b, 3.5 a	3.5.1	drawings of geometric figures	dem.
P.Carlsberg 30 # 2	3,5 b, 4.4 c		linear equations, four metals	dem.
P.Chicago litt. 3	4.7 b	4.7.1-2	non-symmetric trapezoids	Greek
P.Cornell 69	4.7 c	4.7.3-6	non-symm. trap. and birectangle	Greek
P.Fay. 9	4, Intr.		El. I, 39, 41	Greek
P.Genève 259	4.7 a		problems for right triangles	Greek
P.Griffith Inst. I. E. 7	3.6		linear equations	dem.
P.Heidelberg 663	3.1 j, 3.7		striped trapezoids	dem.
P.Herc. 1061	4, Intr.		essay on El. I	Greek
P.IFAO 88	1.2 c	1.2.3	geometric progression, 30 terms	Greek
Plato, Laws vii	3.5 b		counting w. gold, bronze, silver	Greek
Plimpton 322	3.3 f		semi-regular square sides	OB
P.Mich. 620	4.4		systems of linear equations	Greek
P.Mich. iii 143	4, Intr.		El. I, Definitions	Greek
P.Moscow	2.2 b		table of contents	hier.
P.Moscow # 10	2.2 e	2.2.6	area of semicircle (basket)	hier.
P.Moscow # 14	2.2 d	2.2.5	volume of truncated pyramid	hier.
P.Moscow # 17	3.1 e, 2.2 c	2.2.2	a triangle, metric algebra	hier.
P.Moscow # 23	2.2 a		a combined work norm	hier.
P.Oxy. i 29	4, Intr.		El. II exercises	Greek
P.Oxy. iii 470	4.2		truncated cone (water clock)	Greek
P.Rhind	2.1 a		table of contents	hier.
P.Rhind	3.3 d		the 2/n table	hier.
P.Rhind	4.5 a		the $\overline{10}$ table	hier.
P.Rhind # 23	3.1 c	3.1.3	completion problem	hier.
P.Rhind ## 28-29	2.1 b	2.1.1	incomplete division exercises	hier.

P.Rhind ## 35-38	2.1 b		applied division problem	hier.
P.Rhind ## 40, 64	2.1 c		sharing problems	hier.
P.Rhind ## 41-55	2.4	2.4.3	outline	hier.
P.Rhind ## 41-43	2.1 d, 4.5 c		cylindrical granaries	hier.
P.Rhind ## 44-46	2.1 d, 4.5 c		rectangular granaries	hier.
P.Rhind ## 48, 50	2.1 d	2.1.4	circle area rule, hieratic	hier.
P.Rhind # 51	3.3 g		the area of a triangle	hier.
P.Rhind # 53 a	2.1 d, 3.7 c	2.1.7-8	a striped triangle	hier.
P.Rhind # 53 b	2.1 d	2.1.7, 15	almost round area number	hier.
P.Rhind ## 54-55	2.1 d		subtracting pieces of land	hier.
P.Rhind ## 56-60	2.1 d	2.1.16	slope of pyramids and cone	hier.
P.Rhind # 62	2.1 f, 3.5 b		bag of gold, silver, lead	hier.
P.Rhind # 63	4.5 d		unequal sharing	hier.
P.Rhind # 64	2.1 c, 2.3 c		unequal sharing	hier.
P.Rhind ## 65-81	2.4	2.4.4	outline	hier.
P.Rhind ## 69-78	2.1 e		baking and brewing numbers	hier.
P.Rhind # 79	1.1.f	1.1.7	sum of powers of 7	hier.
P.UC 32160	2.4	2.4.2	outline	hier.
P.UC 32160 # 1	2.3 b	2.3.3	content of cylindrical granary	hier.
P.UC 32160 # 2	2.3 c	2.3.4	problem for arithm. progression	hier.
P.UC 32161	2.3 d	2.3.5	many-digit decimal numbers	hier.
P.Vindob. G 26740	3.1 k		segment area rule	Greek
P.Vindob. G. 19996	3.3 f, 4.8	4.8.1	pyramids and cones	Greek
P.Vindob. G. 26740	4.3	4.3.1	five illustrated geom. exercises	Greek
Str. 362 # 1	2.1 c		10 brothers, arithmetic progression	OB
Str. 364	2.1 d		striped triangle	OB
Theban O. D 12	3.3 g, 3.7 a	3.3.4	land survey, four fields	dem.
TMS 1	3.2 a	3.2.1	circle-and-chord problem	OB
TMS 3 (BR)	2.1 d, 3.1 k	3.1.11	constants for circle segments	OB
TMS 13	4.6		buying and selling exercise	OB
TMS 14	2.2 d, 4.8 g		volume of a ridge pyramid	OB
UET 6/2 222, rev.	3.3 f		semi-regular square sides	OB
UET 6/2 233, rev.	4.4 c		tabular array for cost of digging	OB
UET 6/2 274, rev.	2.3 a, 4.4 c	2.3.2	tabular array for quadratic equations	OB
UET 6/2 290, rev.	4.4 c		tabular array for capacity measures	OB
UM 29.13.21	1.2.b	1.2.2	doubling and halving algorithm	OB
VAT 7530 § 6	4.5 e		repetitive construction of data	OB
VAT 7531	3.7 c	3.7.8-10	trapezoid with vertical transversals	OB
VAT 7532	3.1 f, 3.2 c		unknown original length of a reed	OB
VAT 7535	3.1 f, 3.2 c		reciprocal of $1 - 1/5$	OB
VAT 7621 # 1	3.7 c		striped trapezoid	OB
VAT 7848 § 1	3.1 j		area of equilateral triangle	LB
VAT 7848 § 3	4.7 b		symmetric trapezoid	OB
VAT 8389	2.1 f, 3.5 b		system of linear equations	OB
VAT 8390	4.4 c		'since he said'	OB
VAT 8391	2.1 f, 3.5 b		system of linear equations	OB

Index of Texts 275

VAT 8522 # 1	3.2 b, 4.5 b		capacity measure of a truncated cone	OB
VAT 8522 # 2	2.1 c		arithmetic progression	OB
VPB IV 92	4.7 c		areas of fields	Greek
W 14148	2.1 d	2.1.12	almost round area number	proto-cun.
W 19408, 76	3.3 g		quadrilateral area rule	proto-cun.
W 20044, 28	2.1 d	2.1.12	almost round area number	proto-cun.
W 23021	1.1 c	1.1.5	trailing part algorithm	LB
W 23291 § 1 f	3.2 a		metric algebra, seed measure	LB
W 23291 §§ 4 b-c	3.1 j		drawings of heights in triangles	LB
W 23291 § 4 c	3.1 j		area of equilateral triangle	LB
W 23291-x § 1	4.8 a		circle segment	LB
W 23291-x § 2	4.3 a		four concentric circular bands	LB
W 23291-x § 11	3.3 g		tables for reed measure	LB
YBC 4652 # 9	2.1 b		repeated division problem	OB
YBC 4663	4.8 g		series of exercises	OB
YBC 4698	2.1 e	2.1.17	outline, table of contents	OB
YBC 4698 § 1	1.2 d		interest on capital	OB
YBC 4698 § 2 a	2.1 e	2.1.17	combined market rate	OB
YBC 4698 § 3 a	2.1 f, 3.5 b		linear equations: iron and gold	OB
YBC 4698 § 4	4.6		prices and market rates	OB
YBC 5022 (NSd)	2.1 d, 2.2 e		circle constants	OB
YBC 7243 (NSe)	2.1 d, 2.2 e		circle constants	OB
YBC 7290	2.1 d	2.1.11	almost round area number	OB
YBC 7326	4.4 c		tabular array, sheep and lambs	OB
YBC 7353	4.4 c		tabular array, combined market rate	OB
YBC 8588	4.4 c		'that he said'	OB
YBC 9856	2.1 c		arithmetic progression	OB
YBC 11125	4.4 c		tabular array, combined market rate	OB
YBC 11126	2.2 c, 3.1 e	2.2.3	metric algebra: trapezoid	OB

Index of Subjects

$2/n$ table	*P.Rhind*	29, 96, 144, 150
$\overline{10} \cdot n$ table	*P.Rhind*	29, 32, 211
$\overline{90}$ and $\overline{150}$ tables	*P.BM 10794*	165, 211
algorithm, doubling and halving	UM 29.13.21	18
algorithm, trailing part	W 23021	8
almost round area numbers	NCBT 1913	51
	P.Rhind # 53 a-b	56, 114
	W 14148	53
	W 20044, 28	53
	YBC 7290	52
basket, area of semicircle	*P.Moscow* # 10	78
bread rations of various sizes	Ist. O 236	66
	MSVO 4, 66	63
buying and selling exercise	*TMS* 13	218
	YBC 4698 # 9	217
circle area rule, Babylonian	*TMS* 3 (BR)	42
circle area rule, demotic	*P.BM 10399* § 2	142
	P.Cairo § 9 a	123
	P.Cairo § 11	127
	P.Cairo § 12	131
circle area rule, Greek	*P.Oxy. iii 470*	195
	P.Vindob. G.26740 ## 2-4	198
circle area rule, hieratic	*P.Rhind* # 48	40, 42, 44, 98
	P.UC 32160	85
circle constants	*TMS* 3 (BR)	42
	YBC 5022 (NSd)	42
	YBC 7243 (NSe)	42
circle segment (corrupt text)	BM 85194 # 29	133
circle segments, constants	*TMS* 3 (BR)	134
circle segment area rule	see *segment area rule*	
circle-and-chord problem	MS 3049	141
	P.BM 10399 § 1	138
	TMS 1	139
circle, given area, unknown diameter	*P.Cairo* § 9 a	122
circles and a segment of a circular band	*P.Vindob. G. 26740* ## 1-5	196

circular bands, concentric	W 23291-x § 2	197
combined baking number	*P.Rhind* # 76	60
combined market rate	MS 2832, YBC 7535, *etc.*	207
	VAT 7530 § 6	214
	YBC 4698 § 2 a	60, 103
combined work norm	MS 2221	70
	P.Moscow # 23	69
completing the square	MS 5112 § 1	208
completion problem	*P.Cairo* § 2	106, 112
	P.Rhind ## 21-23	27, 29, 96, 112
cone, truncated	BM 96954+ § 4 e	262
	P.Vindob. G. 19996	234, 242
——————— ("log")	Haddad 104 # 2	143
	VAT 8522 # 1	142, 211
——————— (mast)	*P.BM 10399* § 2	142
——————— (store room)	*P.Akhmîm* § 1	211
——————— (water clock)	*P.Oxy. iii 470*	195
corrupt text	BM 85194 # 29	133
	P.BM 10520 § 5 c	153
	P.Rhind ## 28-29	30
	P.Rhind ## 40, 43	37, 44
cylindrical vessels, content	BM 85194 ## 22, 35	86, 88
	UET 6/2 290 rev.	206
division problems	*P.Cairo* ## 2-3	109
	P.Rhind ## 24-38	28, 30
division problem, applied	IM 53957	25, 36
	P.Rhind ## 35-38	35
division problem, repeated	BM 34800	33, 204
	P.Rhind ## 28-29	30, 102
	YBC 4652 # 9	32
doubling and halving algorithm	UM 29.13.21	18, 22
drawing of basket (semicircle?)	*P.Moscow* # 10	79
——————— birectangle	*P.Cornell 69* # 3	230
——————— circle	*P.Rhind* # 50, *P.UC 32160*	42, 85, 98
——————— circle and square	*P.Rhind* # 48	41
——————— circles, *etc.*	*P.Vindob. G. 26740* ## 1-5	197
——————— equilateral triangle in circle	*P.Cairo* # 36	128
	MS 3051	130
——————— pyramid and cone	*P.Rhind* ## 56, 60	58
——————— striped trapezoids	*P.Heidelberg 663* ## 1-3	175 ff
——————— rectangles and trapezoid	*P.Carlsberg 30, obv.*	167
——————— rectangular field	*P.BM 10520* § 7 a	157
——————— rectangular field, schematic	*P.BM 10520* § 7 b	158
——————— rectangular granary	*P.Rhind* # 44	44
——————— square in circle	*P.Cairo* # 37	131
	MS 3050	135

Index of Subjects

———— striped triangle	P.Rhind # 53a	45 ff
———— striped triangle	IM 43996	49
———— trapezoid	YBC 7290, YBC 11126	52, 72
———— trapezoids, nonsymmetric	P Chic. litt. 3, P.Cornell 69 ## 1-2	221 ff
———— triangle in circle	TMS 1	139
———— triangular pyramids	P.Vindob. G. 19996, col. vi	234
———— truncated pyramid	P.Moscow # 14	76
———— upright triangles	W 23291 § 4 b-c	238
Elements I, essay on	P.Herc. 1061	193
Elements I, Definitions	P.Mich. iii 143	193
Elements I, 39, 41	P.Fay. 9	193
Elements II, exercises	P.Oxy. i 29	193
Elements XIII, 10, 16	Pack no. 2323	193
equilateral triangle, area	MS 3876 # 3	237
	VAT 7848 § 1	129
	W 23291 §§ 4 b-c	127, 156, 238
equilateral triangle in a circle	MS 3051	129
	P.Cairo # 36	126
equilateral triangular base of pyramid	P.Vindob. G. 19996 # 10	235
equipartitioned trapezoid	IM 58045	181
	VAT 7621 # 1	188
fractions, counting with	P.BM 10520 § 5	150 ff
fractions, operations with	P.Akhmîm §§ 4-10	212 ff
granary, cylindrical	P.UC 32160 # 1	85
————, rectangular	P.Akhmîm § 2	211
————, repeated division problem	BM 34800	33
————, square and circular	P.Rhind ## 44-46	44
height of unsymmetric triangle	Metrica I: 5-6	224
icosahedron, outer shell	MS 3876	236
iterated sum of 10 first integers	P.BM 10520 § 1	145
land survey, one field	BM 47437	162
————, four fields	Theban Ostracon D 12	159, 194
————, eight fields	O.Bodl. ii 1847	194
linear equations	P.Griffith Inst. I. E. 7	174
linear equations, four metals	P.Carlsberg 30 # 2	67, 168
linear equations, iron and gold	YBC 4698 # 4	67, 170
linear equations, system of	VAT 8389, VAT 8391	68, 170
linear equations, systems of	P.Mich. 620	173, 200 ff
many-digit decimal numbers	P.UC 32161	90
metric algebra, rectangle	BM 34568	221
	P.Cairo ## "32-333"	136
————, area and side ratio	IM 121613 # 1	73, 115
	P.Moscow # 6	75 fn.16
————, diagonal and side ratio	P.Berlin 6619 # 1	81
————, expanded rectangle	P.Rhind # 53 b	57
————, area and diagonal	P.Cairo ## 34-35	125

280   *Unexpected Links Between Egyptian and Babylonian Mathematics*

	IM 67118 = Db2-146	126
	MS 3971 § 2	126
————, striped triangle	P.Rhind # 53 a	45 f, 54
————, reshaping a rectangle	P.Cairo § 1	110
	P.Cairo § 3	115
	MS 5112 § 9	120
————, quadrilateral	IM 121565	160
————, trapezoid	YBC 11126	72, 115
————, triangle	P.Moscow # 17	71
multiplication algorithm, explicit	BM 34601 = *LBAT* 1644	149
multiplication rule	*P.BM 10520* § 3	149
multiplication table, times 64	*P.BM 10520* § 2 (# 54)	149
non-symmetric trapezoids	*P.Chicago litt. 3* (= *P.Ayer*)	221
non-symmetric trapezoids, birectangle	*P.Cornell 69*	224 fn. 40, 226
orientation of circular segment	BM 85194 § 11 (## 20-21)	128
	MS 2049, *obv.*	128
original length of a reed	VAT 7532	117, 144, 208
outline of text	BM 85194	95
	BM 96954+	249
	P.Rhind ## 41-55	100
	P.Rhind ## 65-81	101
	P.UC 32160	98
	YBC 4698	61
Plimpton 322		82, 93, 155
pole against a wall	BM 34568 # 12	108
	BM 85196 # 9	108
	P.Cairo § 8 (## 24-31)	107
powers of 2, 64	*Inventor of game of chess*	23
prices and market rates	*Michael 62*	215
	P.Akhmîm § 9	214
	YBC 4698	61, 68, 215 f
progression, arithmetic, 5 persons	*Jiu Zhang Suan Shu* 6:18	39
————, 5 brothers	YBC 9856	38
————, 10 terms	*P.UC 32160* # 2	89
————, 10 brothers	Str. 362 # 1	37
progression, geometric, sum of	M. 7857	2
	Mother Goose rhyme	12
	P.Rhind # 79	11
————, sum of, 7 brothers	MS 1844	9
————, 4 terms (*tgs*?)	*P.Cairo* § 7 (# 23)	121
————, 4 brothers	MS 2830, *obv.*	122
————, 30 terms	M. 8631	14
	P.IFAO 88	21
progression, quasi-arithmetic, 5 brothers	VAT 8522 # 2	38
pyramid and cone? *seked*	*P.Rhind* §§ 21-22 (## 56, 60)	58
pyramids and cones	BM 96954+	76, 249

	P.Vindob. G. 19996	233 ff
quadratic-linear system of equations	BM 13901 # 13	83
	P.Berlin 6619 # 1	81
	UET 6/2 274 rev.	84, 206
quadrilateral, schematic	*P.BM 10520* § 7 b	158, 194
quadrilateral area rule	BM 67314	165
	IM 121565	160
	P.BM 10520 § 7 a	157
	W 19408, 76	159
—————, general	IM 52301	161
quasi-cube table	MS 3048	148
reciprocal of $1 - 1/5$	VAT 7535	117
————— $1 - 1/6$	VAT 7532	117
————— $1 + 1/n$	*P.BM 10399* ## 46-51	143
reshaping a rectangle	MS 5112 § 9	120
reshaping a rectangular cloth	*P.Cairo* # 8	115
ridge pyramid, volume of	*TMS* 14	77, 245
right triangles, problems	*P.Genève 259*	220
segment area rule	*Jiu Zhang Suan Shu* 1:35-36	133
	Metrica I: 30	133, 199
	P.Cairo ## 36-37	129 ff
	P.Vindob. G 26740 # 5	133
semicircle (basket), area of	*P.Moscow* # 10	78
semicircle, constants	BM 85210	81
	TMS 3 (BR)	80
semicircle, problem for a	MLC 1354	80
semi-regular square sides	Plimpton 322, *UET 6/2* 222, *rev.*	156, fn. 27
series of exercises	YBC 4663	251
sharing problems	*P.Rhind* §§ 11-12, 27	29
'since he said'	VAT 8390	206
'since it was said to you'	MS 5112 §§ 1, 5	206
square in a circle	*P.Cairo* § 12 (# 37)	131
sum of 10 first square numbers	AO 6484 # 2	10
square in a circle	MS 3050	135
square side rule	*P.BM 10520* § 6	155
striped trapezoids	*P.Heidelberg 663*	174 ff
	VAT 7621 # 1	188
striped triangle	IM 43996	49, 102, 181
	P.Rhind # 53 a	45, 98
	Str. 364	48
symmetric trapezoid	VAT 7848 § 3	225
table of contents	*P.Akhmîm*	208
	P.Cairo J. E. 89127-30 +	106
	P.Moscow	69
	P.Rhind	27
	P.Vindob. G. 19996	234

———— and outline of the text	BM 85194	95
	BM 96954+102366+SÉ 93	249
	P.UC 32160.	98
	YBC 4698	61
tables of fractions	*P.Akhmîm*	210
table of powers	BM 22706, IM 73355, Ist. O 3826	6
————, descending	MS 2242, MS 3037	7
tables for reed measure	W 23291-x § 11	163
tabular array, combined market rate	MS 2268/19, MS 2832	207
	YBC 7353, YBC 11125	207
tabular array, combined work norm	MS 2221 *rev.*	70
tabular array, cost of excavation	UET 6/2 233, *rev.*	206
————, quadratic-linear equations	UET 6/2 274, *rev.*	84
————, systems of linear equations	*P.Mich. 620*	200 ff
————, sheep and lambs	YBC 7326	207
————, unequal shares	AO 8862	207
thousand-cubit-strip	*P.Rhind* § 17 (# 51)	40, 158
trailing part algorithm.	W 23021	8
trapezoid with vertical transversals	VAT 7531	182 ff
truncated cone	BM 96954+ § 4	264
	P.Vindob. G. 19996	234
truncated pyramid, square	BM 85194 § 16 (# 28)	243
	P.Moscow # 14, *P.Vindob. G. 19996*	74, 234
————, triangular	*P.Vindob. G. 19996*	234
————, with ridge	BM 96954 § 1 f	252
unequal sharing, loaves	*P.Rhind* § 12	36 ff
————, gold, silver, lead	*P.Rhind* # 62	67
work norm, volume of a ring wall	BM 85194 § 3 (# 4)	186, fn. 31

Bibliography

Preface

Baillet, J. (1892) *Le papyrus mathématique d'Akhmim* (Mémoires publiés par les membres de la Mission Archéologique Française au Caire, 9). Paris: Ernest Leroux.

Cavéing, M. (1994) *La constitution du type mathématique de l'idéalité dans la pensée grecque, 1. Essai sur le savoir mathématique dans la Mésopotamie et l'Égypte anciennes.* Lille: Presses Universitaires de Lille.

Chace, A. B., Bull, L., and Manning, H. (1927) *The Rhind Mathematical Papyrus, I.* Oberlin, Ohio: Mathematical Association of America.

Chambon, G. (2002) Trois documents pédagogiques de Mari. *Florilegium Marianum,* 6, 497-503.

Clagett, M. (1999) *Ancient Egyptian Science, a Source book. III: Ancient Egyptian Mathematics.* Philadelphia, PA: American Philosophical Society.

Couchoud, S. (1993) *Mathématiques égyptiennes. Recherches sur les connaissances mathématiques de l'Égypte pharaonique.* Paris: Le Léopard d'Or.

Damerow, P. (2001) Kannten die Babylonier den Satz des Pythagoras? Epistemologische Anmerkungen zur Natur der babylonischen Mathematik. In (Høyrup, J. and Damerow, P., Eds.) *Changing Views on Ancient Near Eastern Mathematics* (Berliner Beiträge zum Vorderen Orient, 19). Berlin: Dietrich Reimer Verlag, 219-310.

Englund, R. K. (12001) Grain accounting practices in archaic Mesopotamia. In (Høyrup, J. and Damerow, P., Eds.) *Changing Views on Ancient Near Eastern Mathematics* (Berliner Beiträge zum Vorderen Orient, 19). Berlin: Dietrich Reimer Verlag, 1-36.

Foster, B. R. and Robson, E. 2004. A new look at the Sargonic mathematical corpus. *Z. für Assyriologie* 94, 1-15.

Fowler, D. H. (1987; 2nd, revised ed. 1999) *The Mathematics of Plato's Academy. A New Reconstruction.* Oxford: Clarendon Press.

—— and Robson, E. (1998) Square root approximations in OB mathematics: YBC 7289 in context. *Historia Mathematica* 25, 366-378.

Friberg, J. (1985) Babylonian Mathematics. In (J. W. Dauben, Ed.) *The History of Mathematics from Antiquity to the Present. A Selective Bibliography.* New York, NY: American Mathematical Society, 37-51.

—— (1999) A Late Babylonian factorization algorithm for the computation of reciprocals of many-place sexagesimal numbers. *Baghdader Mitteilungen* 30, 139-161, 2 pl.

—— (1999/2000) Review of E. Robson, Mesopotamian Mathematics 2100-1600 BC. *Archiv für Orientforschung* 46/47, 309-317.

—— (2000) Korrigendum zum Friberg, *BaM* 30, 1999. *Baghdader Mitteilungen* 31, pl. 3.

—— (2000) Mesopotamian mathematics. In (J. W. Dauben, Ed.) *The history of mathematics from antiquity to the present: A selective bibliography.* (A. C. Lewis, Ed.) *Revised edition on CD-ROM.* American Mathematical Society.

—— (2001) Bricks and mud in metro-mathematical cuneiform texts. In (Høyrup, J. and Damerow, P., Eds.) *Changing Views on Ancient Near Eastern Mathematics* (Berliner Beiträge zum Vorderen Orient, 19). Berlin: Dietrich Reimer Verlag, 61-154.

—— (2005) *Mathematical Cuneiform Texts* (Pictographic and Cuneiform Tablets in the Schøyen Collection, 1). Oslo: Hermes Academic Publishing.

—— (2005) On the alleged counting with sexagesimal place value numbers in mathematical cuneiform texts from the third millennium BCE. *CDLJ* 2005/2. http://cdli.ucla.edu/Pubs/CDLJ/2005/002.html

Gerstinger, H. and Vogel, K. (1932) *Griechische literarische Papyri I.* Vienna: Österreichische Staatsdruckerei.

Glanville, S. R. K. (1927)The mathematical leather roll in the Britsh Museum. *J. of Egyptian Archaeology* 13, 232-238.

Griffith, F. L. (1898) *The Petrie Papyri: Hieratic Papyri from Kahun and Gurob.* London: Bernard Quaritch.

Høyrup, J. (1996) Changing trends in the historiography of Mesopotamian mathematics: An insider's view. *History of Science* 34, 1-32.

—— (1997) Review of Couchoud, Mathématiques égyptiennes. *Mathematical Reviews* 97c: 01005.

—— (2002) A note on OB computational techniques. *Historia Mathematica* 29, 193-198.

—— (2002) Reflections on the absence of a culture of mathematical problems in Ur III. In (Steele, J. M. and Imhausen, A., Eds.) *Under One Sky. Astronomy and Mathematics in the Ancient Near East.* Münster: Ugarit-Verlag, 121-145.

Imhausen, A. (2002) The algorithmic structure of the Egyptian mathematical problem texts. In (Steele, J. M. and Imhausen, A., Eds.) *Under One Sky. Astronomy and Mathematics in the Ancient Near East.* Münster: Ugarit-Verlag, 147-166.

—— (2003) Calculating the daily bread: rations in theory and practice. *Historia Mathematica* 30, 3-16.

—— (2003) *Ägyptische Algorithmen. Eine untersuchung zu den mittelägyptischen mathematischen Aufgabentexten* (Ägyptologische Abhandlungen 65). Wiesbaden: Harrassowitz Verlag.

—— and Ritter, J. (2004) Mathematical fragments. In (Collier, M. and Quirke, S., Eds.) *The UCL Lahun Papyri: Religious, Literary, Legal, Mathematical and Medical.* Oxford: Basingstoke Press.

Jursa, M. and K. Radner (1995/96) Keilschrifttexte aus Jerusalem. *Archiv für Orientfors-*

chung 42/43: 89-108.

Kaplony-Heckel, U. (1981) "Spätägyptische Mathematik." *Orientalistische Literatur-Zeitung* 76, 117-124.

Knorr, W. (1982) Techniques of fractions in ancient Egypt and Greece. *Historia Mathematica* 9, 133-171.

Melville, D. (2002) Ration computations at Fara: Multiplication or repeated addition? In (Steele, J. M. and Imhausen, A., Eds.) *Under One Sky. Astronomy and Mathematics in the Ancient Near East.* Münster: Ugarit-Verlag, 237-252.

―― (2004) Poles and walls in Mesopotamia and Egypt. *Historia Mathematica* 31, 148-162.

Muroi, K. (2000) Quadratic equations in the Susa mathematical text no. 21. *SCIAMVS* 1, 3-10.

―― (2001) Inheritance problems in the Susa mathematical text no. 26. *Historia Scientiarum* 10, 226-234.

―― (2001) Reexamination of the Susa mathematical text no. 12: A system of quartic equations. *SCIAMVS* 2, 3-8.

―― (2002) Expressions of multiplication in Babylonian mathematics. *Historia Scientiarum* 12, 115-120.

―― (2003) Excavation problems in Babylonian mathematics. Susa mathematical tablet no. 23 and others." *SCIAMVS* 4, 3-21.

―― (2003). Mathematical term takiltum and completing the square in Babylonian mathematics. *Historia Scientiarum* 12, 254-263.

―― (2003) Expressions of squaring in Babylonian mathematics."*Historia Scientiarum* 13, 111-119.

Nemet-Nejat, K. R. (2002) Square tablets in the Yale Babylonian Collection. In (Steele, J. M. and Imhausen, A., Eds.) *Under One Sky. Astronomy and Mathematics in the Ancient Near East.* Münster: Ugarit-Verlag, 253-281.

Oelsner, J. (2001) HS 201 – Eine Reziprokentabelle der Ur III-Zeit. In (Høyrup, J. and Damerow, P., Eds.) *Changing Views on Ancient Near Eastern Mathematics* (Berliner Beiträge zum Vorderen Orient, 19). Berlin: Dietrich Reimer Verlag, 53-59

Parker, R. A. (1959) A demotic mathematical papyrus fragment. *J. of Near Eastern Studies* 18, 275-279.

―― (1969) Some demotic mathematical papyri. *Centaurus* 14, 136-141.

―― (1972) *Demotic Mathematical Papyri.* Providence, RI & London: Brown University Press.

―― (1975) A mathematical exercise – pDem. Heidelberg 663. *J. of Egyptian Archaeology* 61, 189-196.

Peet, T. E. 1923 (reprinted 1978) *The Rhind Mathematical Papyrus. British Museum 10057 and 10058.* London.

Proust, C. (2000) La multiplication babylonienne: La part non écrite du calcul. *Revue d'histoire des mathématiques* 6, 293-303.

―― (2002) Numération centésimale de position à Mari. *Florilegium Marianum* 6, 513-

516.

——— (2004) *Tablettes mathématiques de Nippur (Mésopotamie, début du deuxième millénaire avant notre ère). 1. Reconstitution du cursus scolaire. 2. Édition des tablettes d'Istanbul.* Dissertation, Université Paris 7: Histoire des sciences.

Quillien, J. (2003) Deux cadastres de l'époque d'Ur III. *Revue d'Histoire des Mathématiques* 9, 9-31.

Ritter, J. (1989) Chacun sa vérité: les mathématiques en Égypte et en Mésopotamie. In (Serres, M., Ed.) *Éléments d'histoire des sciences.* Paris: Bordas, 39-61.

——— (1995) Measure for measure: Mathematics in Egypt and Mesopotamia. In (Serres, M., Ed.) *A History of Scientific Thought. Elements of a History of Science.* Oxford: Blackwell, 44-72.

——— (2000) Egyptian mathematics. Mathematics Across Cultures. In (Selin, H., Ed.) *The History of Non-Western Mathematics.* Dordrecht/Boston/London: Kluwer, 115-136.

——— (2002) Closing the eye of the Horus. In (Steele, J. M. and Imhausen, A., Eds.) *Under One Sky. Astronomy and Mathematics in the Ancient Near East.* Münster: Ugarit-Verlag, 297-323.

Robins, G. and Shute, C. C. D. (1987) *The Rhind Mathematical Papyrus: An Ancient Egyptian Text.* London: British Museum Publications.

Robson, E. (2000) Mathematical cuneiform texts in Philadelphia, Part I: Problems and calculations. *SCIAMVS* 1, 11-48.

——— (2002) More than metrology: Mathematics education in an OB scribal school. In (Steele, J. M. and Imhausen, A., Eds.) *Under One Sky. Astronomy and Mathematics in the Ancient Near East.* Münster: Ugarit-Verlag, 325-365.

——— (2003) Tables and tabular formatting in Sumer, Baylonia, and Assyria, 2500 BCE – 50 BCE. In (Campbell-Kelly, M., Croarken, M., Flood, R. G., and Robson, E., Eds.) *The History of Mathematical Tables from Sumer to Spreadsheets.* Oxford: University Press, 18-47.

——— (2004) Mathematical cuneiform texts in the Ashmolean Museum, Oxford. *SCIAMVS* 5, 3-65.

Schack-Schackenburg, H. (1900) Der Berliner Papyrus 6619. *Z. für Ägyptische Sprache* 38, 135-140.

Struve, W. W. (1930) *Mathematischer Papyrus des Staatlichen Museums der Schönen Künste in Moskau* (Quellen und Studien zur Geschichte der Mathematik, Abt. A, 1). Berlin: Springer Verlag.

Zauzich, K.-T. (1975) Review of Parker, Demotic Mathematical Papyri (1972). *Bibliotheca Orientalis* 32, 28-30.

Ch. 1. Two Curious Mathematical Cuneiform Texts from OB Mari

Arnaud, D. (1994)*Texte aus Larsa. Die epigraphischen Funde der 1. Kampagne In Senkereh-Larsa. 1933* (Berliner Beiträge zum Vorderen Orient. Texte, 3). Berlin: Dietrich

Reimer Verlag.
Baillet, J. (1892) *Le papyrus mathématique d'Akhmim* (Mémoires publiés par les membres de la Mission Archéologique Française au Caire, 9). Paris: Ernest Leroux.
Boyaval, B. (1971) Le pIFAO 88: Problèmes de conversion monétaire. *Z. für Papyrologie und Epigraphik*, 7, 165-168, pl. VI.
—— (1974) Nouvelles remarques sur le P.IFAO 88. *Z. für Papyrologie und Epigraphik*, 14, 61-66
Chambon, G. (2002) Trois documents pédagogiques de Mari. *Florilegium Marianum*, 6, 497-503.
Charpin, D. (1993) Données nouvelles sur la poliorcétique à l'époque paléo-babylonienne. *MARI, Annales de Recherches Interdisciplinaires*, 7, 193-197.
Chace, A. B., Bull, L., and Manning, H. (1927) *The Rhind Mathematical Papyrus, I*. Oberlin, Ohio: Mathematical Association of America.
Crawford, D. S. (1953) A mathematical tablet. *Aegyptus* 33, 222-240.
Fowler, D. H. (1987; 2nd ed. 1999) *The Mathematics of Plato's Academy. A New Reconstruction*. Oxford: Clarendon Press.
Friberg, J. (1986) The early roots of Babylonian mathematics 3: Three remarkable texts from ancient Ebla. *Vicino Oriente* 6, 3–25.
—— (1990) "Mathematik" (in English). In *Reallexikon der Assyriologie und Vorderasiatischen Archäologie*, 7. Berlin/New York: de Gruyter, 531-585.
—— (1999) A Late Babylonian factorization algorithm for the computation of reciprocals of many-place sexagesimal numbers. *Baghdader Mitteilungen* 30, 139-161, 2 pl.
—— (1999) Proto-literate counting and accounting in the Middle East. Examples from two new volumes of proto-cuneiform texts. *J. of Cuneiform Studies* 51, 107-137.
—— (2000) Korrigendum zum Friberg, *BaM* 30, 1999. *Baghdader Mitteilungen* 31, pl. 3.
—— (2005) *Mathematical Cuneiform Texts* (Pictographic and Cuneiform Tablets in the Schøyen Collection, 1). Oslo: Hermes Academic Publishing.
—— (2005) On the alleged counting with sexagesimal place value numbers in mathematical cuneiform texts from the third millennium BCE. *CDLJ* 2005/2. http://cdli.ucla.edu/Pubs/CDLJ/2005/002.html
Guichard, M. (1997) Quelques examples de l'attention portée à la nature: Comptes de fourmis (comment dénombrer la multitude). *MARI, Annales de Recherches Interdisciplinaires*, 8, 314-321.
Neugebauer, O. (1935-37; reprinted 1973) *Mathematische Keilschrift-Texte, I-III*. Berlin: Springer Verlag.
—— and A. Sachs (1945) *Mathematical Cuneiform Texts* (American Oriental Series, 29). New Haven, CT: American Oriental Society and the American Schools of Oriental Research.
Nissen, H. J., Damerow, P., and Englund, R. K. (1993) *Archaic Bookkeeping, Early Writing and Techniques of Economic Administration in the Ancient Near East*. Chicago, London: The University of Chicago Press.

Nougayrol, J. (1968) *Ugaritica 5* (Mission de Ras Shamra, 16). Paris.
Proust, C. (2002) Numération centésimale de position à Mari. *Florilegium Marianum*, 6, 513-516.
Rea, J. R. (1971) A reconsideration of P.IFAO 88. *Z. für Papyrologie und Epigraphik*, 8, 235-238.
Sachs, A. J. (1947) Reciprocals of regular sexagesimal numbers. *J. of Cuneiform Studies* 1, 219-240.
Soubeyran, D. (1984) Textes mathématiques de Mari. *Revue d'Assyriologie* 78, 19-48.

Ch. 2. Hieratic Mathematical Papyri

Baqir, T. (1951) Some more mathematical texts. *Sumer* 7: 28-45 & plates.
Blome, F. (1928) *du*/GAB in Angaben über die Grösse der Brote. *Orientalia* 34/35, 129-135.
Bruins, E. M. and Rutten, M. (1961) *Textes mathématiques de Suse* (Mémoires de la Mission Archéologique en Iran, 34). Paris: Paul Geuthner.
Bruins, E. M. 1964. *Codex Constantinopolitanus Palatii Veteris No. 1. Part 3: Translation and Commentary.*. Leiden: Brill
Cantor, M. (1898) Die mathematischen Papyrusfragmente von Kahun. *Orientalistische Literatur-Zeitung* 1, 306-308.
Chace, A. B., Bull, L., and Manning, H. (1927) *The Rhind Mathematical Papyrus, part 1*. Oberlin, Ohio: Mathematical Association of America.
——— (1929) *The Rhind Mathematical Papyrus, part 2. Photographs, Transcription, Transliteration, Literal Translation*. Oberlin, OH: Math. Association of America. (Abridged reprint 1979. Reston, VA: National Teachers of Mathematics.)
Clagett, M. (1999) *Ancient Egyptian Science, a Source book. III: Ancient Egyptian Mathematics*. Philadelphia, PA: American Philosophical Society.
Couchoud, S. (1993) *Mathématiques égyptiennes. Recherches sur les connaissances mathématiques de l'Égypte pharaonique*. Paris: Le Léopard d'Or.
Englund, R. K. (1996) *MSVO 4, Proto-Cuneiform Texts from Diverse Collections*. Berlin: Gebr. Mann Verlag.
Friberg, J. (1981) Plimpton 322, Pythagorean triples, and the Babylonian triangle parameter equations. *Historia Mathematica* 8, 277-318.
——— (1990) "Mathematik" (in English). In *Reallexikon der Assyriologie und Vorderasiatischen Archäologie*, 7. Berlin/New York: de Gruyter, 531-585.
——— (1996) Pyramids and cones in ancient mathematical texts. New hints of a common tradition. *Proceedings of the Cultural History of Mathematics* 6, 80-95.
——— (1997) 'Seed and Reeds Continued'. Another metro-mathematical topic text from Late Babylonian Uruk. *Baghdader Mitteilungen* 28, 251-365, pl. 45-46.
——— (1997/98) Round and almost round numbers in proto-literate metro-mathematical field texts. *Archiv für Orientforschung* 44/45, 1-58.
——— (1999) Proto-literate counting and accounting in the Middle East. Examples from

two new volumes of proto-cuneiform texts. *J. of Cuneiform Studies* 51, 107-137.

—— (2000) Mathematics at Ur in the OB period. *Revue d'assyriologie* 94, 97-188.

—— (2001) Bricks and mud in metro-mathematical cuneiform texts. In (Høyrup, J. and Damerow, P., Eds.) *Changing Views on Ancient Near Eastern Mathematics* (Berliner Beiträge zum Vorderen Orient, 19). Berlin: Dietrich Reimer Verlag, 61-154.

—— (2005) *Mathematical Cuneiform Texts* (Pictographic and Cuneiform Tablets in the Schøyen Collection, 1). Oslo: Hermes Publishing.

Gardiner, S. A. (1927; third, revised ed. 1957) *Egyptian Grammar, Being an Introduction to the Study of Hieroglyphics*. Oxford: David Stanford.

Gillings, R. J. (1982) *Mathematics in the Time of the Pharaohs*. Cambridge, MA: MIT Press. (Page references are to the 1972 reedition. New York, N. Y.: Dover Publications.)

Griffith, F. L. (1898) *The Petrie Papyri: Hieratic Papyri from Kahun and Gurob*. London: Bernard Quaritch.

Hoffmann, F. (1996) Die Aufgabe 10 des Moskauer mathematischen Papyrus. *Z. für Ägyptische Sprache* 123, 19-26.

Høyrup, J. (2002) *Lengths, Widths, Surfaces. A portrait of OB algebra and its kin*. New York, Berlin, *etc.*: Springer Verlag.

Imhausen, A. (2003) *Ägyptische Algorithmen. Eine untersuchung zu den mittelägyptischen mathematischen Aufgabentexten* (Ägyptologische Abhandlungen 65). Wiesbaden: Harrassowitz Verlag.

—— and Ritter, J. (2004) Mathematical fragments. In (Collier, M. and Quirke, S., Eds.) *The UCL Lahun Papyri: Religious, Literary, Legal, Mathematical and Medical*. Oxford: Basingstoke Press.

Kaplony-Heckel, U. (1981) Spätägyptische Mathematik. *Orientalistische Literatur-Zeitung* 76, 117-124.

Melville, D. (2002) Weighing stones in ancient Mesopotamia. *Historia Mathematica* 29, 1-12.

Muroi, K. (1988) Inheritance problems of Babylonian mathematics. *Historia Scientiarum* 34, 11-19.

—— (1994) The area of a semicircle in Babylonian mathematics: New interpretations of MLC 1354 and BM 85210 no. 8. *Sugakushi Kenkyu* 143, 50-60.

Neugebauer, O. (1931) *Die Geometrie der ägyptischen mathematischen Texte* (Quellen und Studien zur Geschichte der Mathematik, Abt. B, 1), 413-451.

—— (1935-37; reprinted 1973) *Mathematische Keilschrift-Texte, I-III*. Berlin: Springer Verlag.

—— and A. Sachs (1945) *Mathematical Cuneiform Texts* (American Oriental Series, 29). New Haven, CT: American Oriental Society and the American Schools of Oriental Research.

Nissen, H. J., Damerow, P., and Englund, R. K. (1993) *Archaic Bookkeeping, Early Writing and Techniques of Economic Administration in the Ancient Near East*. Chicago, London: The University of Chicago Press.

Parker, R. A. (1972) *Demotic Mathematical Papyri*. Providence, RI & London: Brown

University Press.
Peet, T. E. 1923 (reprinted 1978) *The Rhind Mathematical Papyrus. British Museum 10057 and 10058*. London.
—— (1931) Mathematics in ancient Egypt. *Bulletin of the John Rylands Library* 15, 409-441.
—— (1931) A problem in Egyptian geometry. *J. of Egyptian Archaeology* 17, 100-106.
Proust, C. (2004) *Tablettes mathématiques de Nippur (Mésopotamie, début du deuxième millénaire avant notre ère). 1. Reconstitution du cursus scolaire. 2. Édition des tablettes d'Istanbul*. Dissertation, Université Paris 7: Histoire des sciences.
Ritter, J. (1998) Reading Strasbourg 368: A thrice-told tale. *Preprint 103, Max-Planck Institut für Wissenschaftsgeschichte, Berlin*.
—— (2002) Closing the eye of the Horus. In (Steele, J. M. and Imhausen, A., Eds.) *Under One Sky. Astronomy and Mathematics in the Ancient Near East*. Münster: Ugarit-Verlag, 297-323.
Robins, G. and Shute, C. C. D. (1987) *The Rhind Mathematical Papyrus: An Ancient Egyptian Text*. London: British Museum Publications.
Robson, E. (1999) *Mesopotamian mathematics 2100-1600 BC: Technical Constants in Bureaucracy and Education*. Oxford: Clarendon Press.
Sachs, A. J. (1955) *Late Babylonian Astronomical and Related Texts. Copied by T. G. Pinches and J. N. Strassmaier*. Providence, R.I.: Brown University Press.
Schack-Schackenburg, H. (1900) Der Berliner Papyrus 6619. *Z. für Ägyptische Sprache* 38, 135-140.
Shen, K., Crossley, J. N., and Lun, A W.-C. (1999) *The Nine Chapters on the Mathematical Art – Companion and Commentary*. Oxford: Oxford University Press & Beijing: Science Press.
Spalinger, A. (1987) The grain system of Dynasty 18. *Studien zur altägyptischen Kultur* 14, 283-311.
—— (1990) The Rhind Mathematical Papyrus as a historical document. *Studien zur altägyptischen Kultur* 17, 295-335.
Struve, W. W. (1930) *Mathematischer Papyrus des Staatlichen Museums der Schönen Künste in Moskau* (Quellen und Studien zur Geschichte der Mathematik, Abt. A, 1). Berlin: Springer Verlag.
Thureau-Dangin, F. (1903) *Recueil de tablettes chaldéennes*. Paris: Ernest Leroux.
Weil, S. The first 28,915 odd primes. www.newdream.net/~sage/old/numbers/index.html

Ch. 3. Demotic Mathematical Papyri

Al-Rawi, F. N. H., & Roaf M. (1984) Ten OB mathematical problems from Tell Haddad, Himrin. *Sumer* 43, 195-218.
Baqir, T. (1950) Another important mathematical text from Tell Harmal. *Sumer* 6, 130-148.
Bruins, E. M., and Rutten, M. (1961) *Textes mathématiques de Suse* (Mémoires de la Mis-

sion Archéologique en Iran, 34). Paris: Paul Geuthner.

Fowler, D. H. (1987; 2nd ed. 1999) *The Mathematics of Plato's Academy. A New Reconstruction.* Oxford: Clarendon Press.

Friberg, J. (1990) "Mathematik" in *Reallexikon der Assyriologie und Vorderasiatischen Archäologie, Band* 7. Berlin/New York: de Gruyter, 531-585.

——, Hunger, H., & Al-Rawi, F. (1990) 'Seed and Reeds', a metro-mathematical topic text from Late Babylonian Uruk. *Baghdader Mitteilungen* 21, 483-557, pl. 46-48.

—— (1993) On the structure of cuneiform metrological table texts from the -1st millennium. In (H. D. Galter, Ed.) *Die Rolle der Astronomie in den Kulturen Mesopotamiens* (Grazer Morgenländische Studien, 3). Graz: Graz Kult, 383-405.

—— (1997) 'Seed and Reeds Continued'. Another metro-mathematical topic text from Late Babylonian Uruk. *Baghdader Mitteilungen* 28, 251-365, pl. 45-46.

—— (1997/98) Round and almost round numbers in proto-literate metro-mathematical field texts. *Archiv für Orientforschung* 44/45, 1-58.

—— (1999) A Late Babylonian factorization algorithm for the computation of reciprocals of many-place sexagesimal numbers. *Baghdader Mitteilungen* 30, 139-161, 2 pl.

—— (2000) Mathematics at Ur in the OB period. *Revue d'assyriologie* 94, 97-188.

—— (2005) *Mathematical Cuneiform Texts* (Pictographic and Cuneiform Tablets in the Schøyen Collection, 1). Oslo: Hermes Publishing.

Heath, S. T. (1921) *A History of Greek Mathematics, 1-2.* (All page references are to the reedition New York, N. Y.: Dover Publications, 1981.)

Høyrup, J. (2002) *Lengths, Widths, Surfaces. A portrait of OB algebra and its kin.* New York/Berlin/*etc.*: Springer.

Imhausen, A. (2003) *Ägyptische Algorithmen. Eine untersuchung zu den mittelägyptischen mathematischen Aufgabentexten* (Ägyptologische Abhandlungen 65). Wiesbaden: Harrassowitz Verlag.

Jones, A. (1993) Evidence for Babylonian arithmetical schemes in Greek astronomy. In (H. D. Galter, Ed.) *Die Rolle der Astronomie in den Kulturen Mesopotamiens* (Grazer Morgenländische Studien, 3). Graz: Graz Kult.

—— (2002) Babylonian Lunar Theory in Roman Egypt: Two new texts. In (Steele, J. M. and Imhausen, A., Eds.) *Under One Sky. Astronomy and Mathematics in the Ancient Near East.* Münster: Ugarit-Verlag, 167-174.

Kaplony-Heckel, U. (1981) "Spätägyptische Mathematik." *Orientalistische Literatur-Zeitung* 76, 117-124.

Knorr, W. (1982) Techniques of fractions in ancient Egypt and Greece. *Historia Mathematica* 9, 133-171.

Nemet-Nejat, K. R. (1982) *Late Babylonian Field Plans in the British Museum* (Studia Pohl, S. M. 11). Rome: Biblical Institute Press.

Neugebauer, O. (1930) Arithmetik und Rechentechnik der Ägypter. *Quellen und Studien zur Geschichte der Mathematik, Astronomie und Physik,* Abt. B 1, 301-380.

—— (1935-37) *Mathematische Keilschrift-Texte, 1-3.* Berlin: Springer Verlag. (All page references are to the reprinted edition, 1973.)

Nissen, H. J., Damerow, P., and Englund, R. K. (1993) *Archaic Bookkeeping, Early Writing and Techniques of Economic Administration in the Ancient Near East*. Chicago, London: The University of Chicago Press.

Parker, R. A. (1959) A demotic mathematical papyrus fragment. *J. of Near Eastern Studies* 18, 275-279.

——— (1969) Some demotic mathematical papyri. *Centaurus* 14, 136-141.

——— (1972) *Demotic Mathematical Papyri*. Providence, RI & London: Brown University Press.

——— (1975) A mathematical exercise – pDem. Heidelberg 663. *J. of Egyptian Archaeology* 61, 189-196.

Peet, T. E. (1923; reprinted 1978) *The Rhind Mathematical Papyrus. British Museum 10057 and 10058*. London.

Robson, E. (2000) Mathematical cuneiform texts in Philadelphia, Part I: Problems and calculations. *SCIAMVS* 1, 11-48.

——— (2004) Mathematical cuneiform texts in the Ashmolean Museum, Oxford. *SCIAMVS* 5: 3-65.

Sethe, K. (1916) *Von Zahlen und Zahlworten bei den alten Ägyptern und was für andere Völker und Sprachen daraus zu lernen ist. Ein Beitrag zur Geschichte von Rechenkunst und Sprache*. Straßburg: Karl J. Trübner.

Shen, K., Crossley, J. N., & Lun, A. W.-C. (1999) *The Nine Chapters on the Mathematical Art – Companion and Commentary*. Oxford: Oxford University Press, & Beijing: Science Press.

Thomas, I. (1939) *Selections Illustrating the History of Greek Mathematics. I Thales to Euklid* (Loeb Classical Library). Cambridge, MA: Harvard University Press, London: W. Heinemann.

Thompson, H. (1913) *Theban Ostraca. Part II. Demotic texts*. University of Toronto Library.

Thureau-Dangin, F. (1938) *Textes mathématiques babyloniens*. Leiden: E. J. Brill.

Vaiman, A. A. (1961) *Shumero-vavilonskaya matematika III - I tysyacheletiya do n. e.* Moscow: Izdatel'stvo vostochnoi literatury.

Zauzich, K.-T. (1975) Review of Parker, Demotic Mathematical Papyri (1972). *Bibliotheca Orientalis* 32, 28-30.

Ch. 4. Greek-Egyptian Mathematical Papyri

Baillet, J. (1892) *Le papyrus mathématique d'Akhmim* (Mémoires publiés par les membres de la Mission Archéologique Française au Caire, 9). Paris: E. Leroux.

Bell, H. I. (1917) *Greek Papyri in the British Museum, vol. 5*. London.

Bruins, E. M., and M. Rutten. (1961) *Textes mathématiques de Suse* (Mémoires de la Mission Archéologique en Iran, 34). Paris: Paul Geuthner.

Bruins, E. M., P.J. Sijpesteijn, and K. A. Worp (1974) A Greek mathematical papyrus.

Bibliography

Janus 61, 297-312.

Bülow-Jacobsen, A. and C. M. Taisbak (2003) P.Cornell inv. 69. Fragment from a handbook in geometry. In (A. Piltz, *et al.*, Eds.) *For Particular Reasons. Studies in Honour of Jerker Blomqvist*. Lund: Nordic Academy Press, 54-69.

Crawford, D. S. (1953) A mathematical tablet. *Aegyptus* 33, 222-240.

Fowler, D. H. (1987; 2nd ed. 1999) *The Mathematics of Plato's Academy. A New Reconstruction*. Oxford: Clarendon Press.

Friberg, J. (1981) An OB catalogue text with equations for squares and circles. *J. of Cuneiform Studies* 33, 57-64.

—— (1982) *A Survey of Publications on Sumero-Akkadian Mathematics, Metrology and Related Matters (1854–1982)*. Gothenburg: DepMath CTH–GU, 1982-17.

——, H. Hunger, and F. Al-Rawi (1990) 'Seed and Reeds', a metro-mathematical topic text from Late Babylonian Uruk. *Baghdader Mitteilungen* 21, 483-557, pl. 46-48.

—— (1990) "Mathematik" (in English) in *Reallexikon der Assyriologie und Vorderasiatischen Archäologie, Band 7*. Berlin/New York: de Gruyter, 531-585.

—— (1996) Pyramids and cones in ancient mathematical texts. New hints of a common tradition. *Proceedings of the Cultural History of Mathematics* 6, 80-95.

—— (1997) 'Seed and Reeds Continued'. Another metro-mathematical topic text from Late Babylonian Uruk. *Baghdader Mitteilungen* 28, 251-365, pl. 45-46.

—— (2000) Mathematics at Ur in the OB period. *Revue d'assyriologie* 94, 97-188.

—— (2005) *Mathematical Cuneiform Texts* (Pictographic and Cuneiform Tablets in the Schøyen Collection, 1). Oslo: Hermes Publishing.

Gerstinger, H. and K. Vogel (1932) *Griechische literarische Papyri I*. Vienna: Österreichische Staatsdruckerei.

Goodspeed, E. J. (1898) The Ayer Papyrus: A mathematical fragment. *American J. of Philology* 19, 25-39.

Greenfell, B. P.and A. S. Hunt (1903) *The Oxyrhynchus Papyri, Part III*. London: Egypt Exploration Fund.

Harrauer, H. and P.J. Sijpesteijn (1985) *Neue Texte aus dem antiken Unterricht*. Vienna: Brüder Hollinek.

Heath, T. L. (1921) *A History of Greek Mathematics, 1-2*. (All page references are to the reedition New York, N. Y.: Dover Publications, 1981.)

—— (1926) *Euclid. The Thirteen Books of The Elements, 1-3*. (All page references are to the reedition New York, N. Y.: Dover Publications, 1956.)

Høyrup, J. (1997) Sulla posizione della "formula di Erone" nei Metrica (con una nota platonica). *Bolletino di Storia delle Scienze Matematiche* 17, 3-11.

—— (1997) Hero, Ps-Hero, and Near Eastern practical geometry. An investigation of Metrica, Geometrica, and other treatises. *Antike Naturwissenschaft und ihre Rezeption* 7, 67-93.

—— (2002) *Lengths, Widths, Surfaces. A portrait of OB algebra and its kin*. New York/ Berlin/*etc*.: Springer.

Jursa, M. and K. Radner (1995/96) Keilschrifttexte aus Jerusalem. *Archiv für Orientfors-*

chung 42/43: 89-108.

Neugebauer, O. (1935-37) *Mathematische Keilschrift-Texte, 1-3*. Berlin: Springer Verlag. (All page references are to the reprinted edition, 1973.)

—— and A. Sachs (1945) *Mathematical Cuneiform Texts* (American Oriental Series, 29). New Haven, CT: American Oriental Society.

Robbins, F. E. (1929) P.Mich. 620: A series of arithmetical problems. *Classical Philology* 24, 321-329.

Robson, E. (1999) *Mesopotamian mathematics 2100-1600 BC: Technical Constants in Bureaucracy and Education*. Oxford: Clarendon Press.

Rudhart, J. (1978) Trois problèmes de géométrie, conservés par un papyrus genevois. *Museum Helveticum* 35, 233-241.

Sesiano, J. (1986) Sur un papyrus mathématique grec conservé à la Bibliothèque de Genève. *Museum Helveticum* 43, 74-79.

—— (1999) Sur le Papyrus graecus genevensis 259. *Museum Helveticum* 56, 26-32.

Shelton, J. (1981) Mathematical problems on a papyrus from the Gent collection (SB III 6951 verso). *Z. für Papyrologie und Epigraphik* 42, 91-94.

—— (1981) Land measures in VBP IV 92. *Z. für Papyrologie und Epigraphik* 42, 95-98.

—— (1981) Two notes on the artab. *Z. für Papyrologie und Epigraphik* 42, 99-106.

Thompson, H. (1913) *Theban Ostraca. Part II. Demotic texts*. University of Toronto Library.

Thureau-Dangin, F. (1938) *Textes mathématiques babyloniens*. Leiden: E. J. Brill.

Vogel, K. (1930) Die algebraischen Probleme des P.Mich. 620. *Classical Philology* 25, 373-375.

Weitzmann, K. (1959) *Ancient Book Illumination*. Cambridge, MA: Harvard University Press

Winter, J. G. (1936) *Papyri in the University of Michigan Collection. Miscellaneous papyri* (Michigan Papyri, 3). Ann Arbor: University of Michigan Press.